Analysis of ASME Boiler, Pressure Vessel, and Nuclear Components in the Creep Range

Second Edition

Wiley-ASME Press Series

Analysis of ASME Boiler, Pressure Vessel, and Nuclear Components in the Creep Range
Maan H. Jawad, Robert I. Jetter

Voltage-Enhanced Processing of Biomass and Biochar
Gerardo Diaz

Pressure Oscillation in Biomedical Diagnostics and Therapy
Ahmed Al-Jumaily, Lulu Wang

Robust Control: Youla Parameterization Method
Farhad Assadian, Kevin R. Mellon

Metrology and Instrumentation: Practical Applications for Engineering and Manufacturing
Samir Mekid

Fabrication of Process Equipment
Owen Greulich, Maan H. Jawad

Engineering Practice with Oilfield and Drilling Applications
Donald W. Dareing

Flow-Induced Vibration Handbook for Nuclear and Process Equipment Michel J. Pettigrew, Colette E. Taylor, Nigel J. Fisher

Vibrations of Linear Piezostructures
Andrew J. Kurdila, Pablo A. Tarazaga

Bearing Dynamic Coefficients in Rotordynamics: Computation Methods and Practical Applications
Lukasz Brenkacz

Advanced Multifunctional Lightweight Aerostructures: Design, Development, and Implementation
Kamran Behdinan, Rasool Moradi-Dastjerdi

Vibration Assisted Machining: Theory, Modelling and Applications
Li-Rong Zheng, Dr. Wanqun Chen, Dehong Huo

Two-Phase Heat Transfer
Mirza Mohammed Shah

Computer Vision for Structural Dynamics and Health Monitoring
Dongming Feng, Maria Q Feng

Theory of Solid-Propellant Nonsteady Combustion
Vasily B. Novozhilov, Boris V. Novozhilov

Introduction to Plastics Engineering
Vijay K. Stokes

Fundamentals of Heat Engines: Reciprocating and Gas Turbine Internal Combustion Engines
Jamil Ghojel

Offshore Compliant Platforms: Analysis, Design, and Experimental Studies
Srinivasan Chandrasekaran, R. Nagavinothini

Computer Aided Design and Manufacturing
Zhuming Bi, Xiaoqin Wang

Pumps and Compressors
Marc Borremans

Corrosion and Materials in Hydrocarbon Production: A Compendium of Operational and Engineering Aspects
Bijan Kermani, Don Harrop

Design and Analysis of Centrifugal Compressors
Rene Van den Braembussche

Case Studies in Fluid Mechanics with Sensitivities to Governing Variables
M. Kemal Atesmen

The Monte Carlo Ray-Trace Method in Radiation Heat Transfer and Applied Optics
J. Robert Mahan

Dynamics of Particles and Rigid Bodies: A Self-Learning Approach
Mohammed F. Daqaq

Primer on Engineering Standards, Expanded Textbook Edition
Maan H. Jawad, Owen R. Greulich

Engineering Optimization: Applications, Methods and Analysis
R. Russell Rhinehart

Compact Heat Exchangers: Analysis, Design and Optimization using FEM and CFD Approach
C. Ranganayakulu, Kankanhalli N. Seetharamu

Robust Adaptive Control for Fractional-Order Systems with Disturbance and Saturation
Mou Chen, Shuyi Shao, Peng Shi

Robot Manipulator Redundancy Resolution
Yunong Zhang, Long Jin

Stress in ASME Pressure Vessels, Boilers, and Nuclear Components
Maan H. Jawad

Combined Cooling, Heating, and Power Systems: Modeling, Optimization, and Operation
Yang Shi, Mingxi Liu, Fang Fang

Applications of Mathematical Heat Transfer and Fluid Flow Models in Engineering and Medicine
Abram S. Dorfman

Bioprocessing Piping and Equipment Design: A Companion Guide for the ASME BPE Standard
William M. (Bill) Huitt

Nonlinear Regression Modeling for Engineering Applications: Modeling, Model Validation, and Enabling Design of Experiments
R. Russell Rhinehart

Geothermal Heat Pump and Heat Engine Systems: Theory and Practice
Andrew D. Chiasson

Fundamentals of Mechanical Vibrations
Liang-Wu Cai

Introduction to Dynamics and Control in Mechanical Engineering Systems
Cho W.S. To

Analysis of ASME Boiler, Pressure Vessel, and Nuclear Components in the Creep Range

Maan H. Jawad
Bothell, Washington

Robert I. Jetter
Pleasanton, California

Second Edition

This work is a co-publication between ASME Press and John Wiley & Sons, Inc.

Copyright © 2022 by ASME
All rights reserved.

This work is a co-publication between ASME Press and John Wiley & Sons, Inc.
Published simultaneously in Canada.

No part of this publication may be reproduced, stored in a retrieval system, or transmitted in any form or by any means, electronic, mechanical, photocopying, recording, scanning, or otherwise, except as permitted under Section 107 or 108 of the 1976 United States Copyright Act, without either the prior written permission of the Publisher, or authorization through payment of the appropriate per-copy fee to the Copyright Clearance Center, Inc., 222 Rosewood Drive, Danvers, MA 01923, (978) 750-8400, fax (978) 750-4470, or on the web at www.copyright.com. Requests to the Publisher for permission should be addressed to the Permissions Department, John Wiley & Sons, Inc., 111 River Street, Hoboken, NJ 07030, (201) 748-6011, fax (201) 748-6008, or online at https://www.wiley.com/go/permission.

Limit of Liability/Disclaimer of Warranty: While the publisher and author have used their best efforts in preparing this book, they make no representations or warranties with respect to the accuracy or completeness of the contents of this book and specifically disclaim any implied warranties of merchantability or fitness for a particular purpose. No warranty may be created or extended by sales representatives or written sales materials. The advice and strategies contained herein may not be suitable for your situation. You should consult with a professional where appropriate. Neither the publisher nor author shall be liable for any loss of profit or any other commercial damages, including but not limited to special, incidental, consequential, or other damages. Further, readers should be aware that websites listed in this work may have changed or disappeared between when this work was written and when it is read. Neither the publisher nor authors shall be liable for any loss of profit or any other commercial damages, including but not limited to special, incidental, consequential, or other damages.

For general information on our other products and services or for technical support, please contact our Customer Care Department within the United States at (800) 762-2974, outside the United States at (317) 572-3993 or fax (317) 572-4002.

Wiley also publishes its books in a variety of electronic formats. Some content that appears in print may not be available in electronic formats. For more information about Wiley products, visit our web site at www.wiley.com.

A catalogue record for this book is available from the Library of Congress

Hardback ISBN: 9781119679462; ePub ISBN: 9781119679486; ePDF ISBN: 9781119679448; oBook ISBN: 9781119679493

Cover image: © Photo smile/Shutterstock
Cover design by Wiley

Set in 9.5/12.5pt STIXTwoText by Integra Software Services Pvt. Ltd, Pondicherry, India

SKY10035633_080522

In memory of
Betty Jetter
1940–2020

Contents

Preface

Many structures in chemical plants, refineries, and power generation plants operate at elevated temperatures where creep and rupture are a design consideration. At such elevated temperatures, the material tends to undergo gradual strain with time, which could eventually lead to failure. Thus, the design of such components must take into consideration the creep and rupture of the material. In this book a brief introduction to the general principles of design at elevated temperatures is given with extensive references cited for further in-depth understanding of the subject. A key feature of the book is the use of numerous examples to illustrate the practical application of the design and analysis methods presented.

This book is divided into nine chapters. The first chapter is an introduction to various creep topics such as allowable stresses, creep properties, elastic analog, and reference stress methods, as well as a few introductory topics needed in various subsequent chapters.

Structural members in the creep range are covered in Chapters 2 and 3. In Chapter 2, the subject of structural tension members is presented. Such members are encountered in pressure vessels as hangers, tray supports, braces, and other miscellaneous components. Chapter 3 covers beams and plates in bending. Components such as piping loops, tray support beams, internal piping, nozzle covers, and flat heads are included. A brief discussion of the requirements of ANSI B31.1 and B31.3 in the creep regime is given.

Chapters 4 and 5 discuss stress analysis of shells in the creep range. In Chapter 4, various stress categories are defined and an analysis of various components using "load-controlled limits," as defined in ASME VIII-2, is discussed. Comparisons are also given between the design criteria in ASME VIII-2 and ASME III-5 and the limitations encountered in ASME VIII-2 when designing in the creep range. Chapter 5 covers the analysis of pressure components using "strain and deformation-controlled limits." Discussion includes the requirements and limitation of the "A Test" and "B Test" outlined in ASME VIII-2.

Cyclic loading in the creep-fatigue regime, using the "Strain Method," is discussed in Chapter 6. Both repetitive and non-repetitive cycles are presented with some examples illustrating the applicability and intent of ASME VIII-2 in non-nuclear applications.

Chapter 7 gives a brief presentation of creep fatigue analysis using the "Remaining Life Method" outlined in the API 759/ASME FFS-1 code. A comparison of the results obtained from this method, versus the results obtained from the "Strain Method," is made.

Chapter 8 outlines the requirements for creep analysis in nuclear components given in ASME III-5. Some of the differences between these requirements and those of ASME VIII-2 Creep Rules are presented.

Compressive stress in components is discussed in Chapter 9. External pressure charts obtained from isochronous curves, as well as from the Remaining Life method, are presented. Cylindrical and spherical shells, as well as axial structural members, are discussed. Simplified methods are presented for design purposes. The assumptions and limitations required to derive the simplified methods are also given.

The book also includes six appendices. Appendix A lists ASME VIII-2 supplemental creep rules, as shown in ASME Code Case 2843. Appendix B lists the equations for constructing the Isochronous Stress-Strain Curves presently used in ASME. Appendix C shows some equations for the tangent modulus, Et. Appendix D outlines the derivation of the Bree diagram in ASME. Appendix E gives some constants used in the remaining life methods, and Appendix F gives some conversion factors.

The rules for creep analysis of pressure vessels in ASME VIII-2 are presently in ASME Code Case 2843. These rules are a simplification of the rules in the nuclear code, ASME III-5, which are more extensive since they cover broader applications such as piping and valves. The rules of Code case 2843 are intended to be placed in the body of ASME VIII-2. In doing so, the paragraph numbers, as well as the table and figure numbers, will change. In order to keep the discussion of the topics in this book consistent with Code Case 2843, a copy of the code case is shown in Appendix A of this book. In order to avoid confusion, reference in this book is made to ASME VIII-2 supplemental creep rules to indicate creep rules presently in Code Case 2843 that will eventually be incorporated in ASME VIII-2. The equations in Chapter 9 for external pressure are taken, in part, from ASME Code Case 2964, that will eventually be incorporated in ASME VIII as well.

Frequently referenced ASME standards in this book are abbreviated for simplicity. ASME Section I is abbreviated as ASME I, ASME Section VIII, Division 1 is abbreviated as ASME VIII-1 and ASME Section VIII, Division 2 is abbreviated as ASME VIII-2. Similarly, ASME Section III, Division 5 is abbreviated as ASME III-5.

The units expressed in this book are mainly in the customary English units such as oF, ksi, inches, and lbs. Equivalent SI units are also shown, such as oC, MPa, mm, and kgs. Example problems are solved in either customary or SI units.

Maan H. Jawad
Bothell, Washington

Robert I. Jetter
Pleasanton, California

Acknowledgement for the Original Edition

This book could not have been written without the help of numerous people and we give our thanks to all of them. Special thanks are given to Pete Molvie, Bob Schueller, and the late John Fischer for providing background information on Section I, and to George Antaki, Chuck Becht, and Don Broekelmann for supplying valuable information on piping codes B31.1 and B31.3.

Our thanks also extend to Don Griffin, Vern Severud, and Doug Marriott for providing insight into the background of various creep criteria and equations in III-NH, and for their guidance.

Special acknowledgement is also given to Craig Boyak for his generous help with various segments of the book, to Joe Kelchner for providing a substantial number of the figures, to Wayne Mueller and Jack Anderson for supplying information regarding the operation of power boilers and heat recovery steam generators, to Mike Bytnar, Don Chronister, and Ralph Killen for providing various photographs, to Basil Kattula for checking some of the column-buckling equations, and to Ms. Dianne Morgan of the Camas Public Library for magically producing references and other older publications obtained from faraway places.

A special thanks is also given to Mary Grace Stefanchick and Tara Smith of ASME for their valuable help and guidance in editing and assembling the book.

Acknowledgement for this Edition

Our thanks to the many people who helped us while writing this edition of the book. Mark Messner of Argonne National Laboratory helped with providing equations for various isochronous stress-strain curves and providing comparisons between rigorous and simplified elastic follow-up analysis. Kevin Jawad helped by providing the derivatives for various complicated equations, and determining the tangent modulus used in external pressure calculations, using a Symbolic Math program, while Chithranjan Nadarajah provided the finite element outputs used in various chapters.

We would also like to thank Donald Griffin for his valuable input to Chapter 9 regarding compressive stress and to Yanli Wang of Oak Ridge National Laboratory for her help with providing various resources.

A special acknowledgement is given to the following for providing various photographs, figures, and other support: Mike Bytnar and Chris Cimarolli of Nooter Construction, Wayne Mueller of Ameren Missouri, Peter Carter, Mike Cohen of Terra Power, Mike Arcaro of GE Power, Argonne National Laboratory, and Harlan Bowers of X-Energy.

Many thanks are also extended to the staff of Wiley and ASME for editing and producing this edition of the book.

Abbreviations for Organizations

AISC	American Institute of Steel Construction
ANSI	American National Standards Institute
API	American Petroleum Institute
ASM	American Society of Metals
ASME	American Society of Mechanical Engineers
ASTM	American Society for Testing and Materials
BS	British Standard
EN	European Standard
IBC	International Building Code
MPC	Materials Properties Council
WRC	Welding Research Council

1

Basic Concepts

OPERATING UNIT IN A REFINERY

Analysis of ASME Boiler, Pressure Vessel, and Nuclear Components in the Creep Range, Second Edition. Maan H. Jawad and Robert I. Jetter.
© 2022 John Wiley & Sons Ltd. Published 2022 by John Wiley & Sons Ltd.

1.1 Introduction

Many vessels and equipment components encounter elevated temperatures during their operation. Such exposure to elevated temperature could result in a slow continuous deformation and creep of the equipment material under sustained loads. Examples of such equipment include hydrocrackers at refineries, power boiler components at electric generating plants, turbine blades in engines, and components in nuclear plants. The temperature at which creep becomes significant is a function of material composition and load magnitude and duration.

Components under loading are usually stressed in tension, compression, bending, torsion, or a combination of such modes. Most design codes provide allowable stress values at room temperature or at temperatures well below the creep range; for example, the codes for civil structures such as the American Institute of Steel Construction and International Building Code. Pressure vessel codes such as the ASME Boiler and Pressure Vessel Code, British, and the European Standard BS EN 13445 contain sections that cover temperatures from the cryogenic range to much higher temperatures where effects of creep are the dominant failure mode. For temperatures and loading conditions in the creep regime, the designer must rely on either in-house criteria or use a pressure vessel code that covers the temperature range of interest. Table 1.1 gives a general perspective on when creep becomes a design consideration for various materials. It is broadly based on the temperature at which creep properties begin to govern allowable stress values in the ASME Boiler and Pressure Vessel Code. There may

Table 1.1 Approximate temperatures[1] at which creep becomes a design consideration in various materials.

	Temperature	
Material	**°F**	**°C**
Carbon and low alloy steel	700–900	370–480
Stainless steels	800–1000	425–535
Aluminum alloys	300	150
Copper alloys	300	150
Nickel alloys	900–1100	480–595
Titanium and zirconium alloys	600–650	315–345
Lead	Room temperature	

[1] These temperatures may vary significantly for the specific product chemistry and failure mode under consideration.

be other specific considerations for a particular design situation, e.g., a short duration load at a temperature above the threshold values shown in Table 1.1. These considerations will be discussed later in this chapter in more detail.

It will be assumed in this book that material properties are not degraded due to process conditions. Such degradation can have a significant effect on creep and rupture properties. Items such as exfoliation Thielsch (1977), hydrogen sulfide Dillon (2000), hydrogen embrittlement, nuclear radiation, and other environment impacts may have great influence on the creep rupture of an alloy; engineers have to rely on experience and field data to supplement theoretical analysis.

One of the concerns for design engineers is the recent increase in allowable stress values in both ASME VIII-1 and VIII-2 and their effect on equipment design, such as hydrotreaters. The recent increase in allowable stress reduces the temperature at which creep controls and upgrading older equipment based on the newer allowable stress requires the knowledge of creep design covered in this book.

1.2 Creep in Metals

1.2.1 Description and Measurement

Creep is the continuous, time-dependent deformation of a material at a given temperature and applied load. Although, conceptually, creep will occur at any stress level and temperature if the measurements are taken over very long periods, there are practical measures of when creep becomes significant for engineering considerations in metallic structures.

Metallurgically, creep is associated with the generation and movement of dislocations, cavities, grain boundary sliding, and mass transport by diffusion. There are many studies of these phenomena and there is extensive literature on the subject. Fortunately for the practicing engineer, a detailed mastery of the metallurgical aspects of creep is not required to design reliable structures and components at elevated temperature. What is required is a basic understanding of how creep is characterized and how creep behavior is translated into design rules for components operating at elevated temperatures.

A creep curve at a given temperature is experimentally obtained by loading a specimen at a given stress level and measuring the strain as a function of time until rupture. Figure 1.1 conceptually shows a standard creep testing machine. A constant force is applied to the specimen through a lever and deadweight load. Typically, the test specimen is surrounded by an electrically controlled furnace. Because creep is highly temperature-dependent, considerable care must be taken to ensure that the specimen temperature is maintained at a constant value, both spatially and temporally.

There are various methods for measuring strain. Figure 1.2 shows one such arrangement suitable for higher temperatures and longer times, which uses

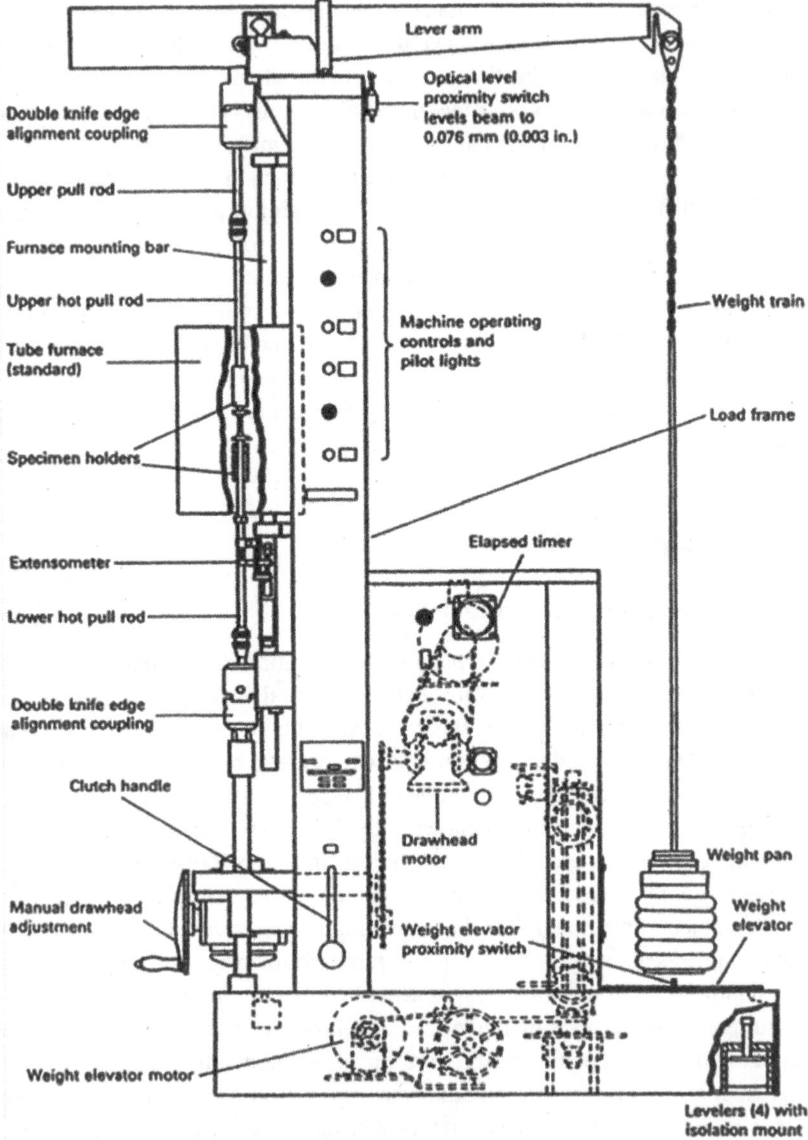

Figure 1.1 Standard creep testing machine [ASM 2000].

two or three extensometers arranged concentrically around the specimen. Penny and Marriott (1995) have summarized the effects of test variables on typical test results. They concluded that faulty measurement of mean stress and temperature are the largest sources of error and that these measurements should be accurate to better than 1% and 1.25%, respectively, to achieve creep

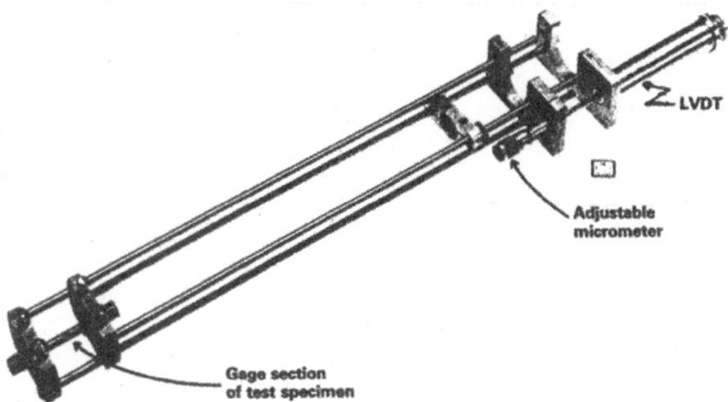

Figure 1.2 Extensometer for elevated temperature creep testing [ASM 2000].

strain measurement accuracy to within 10%. For example, it is recommended that, in order to minimize bending effects, tolerances to within 0.002 in. (0.05 mm) must be achieved in aligning a 1.25-in. (6.4 mm)-diameter specimen.

The above highlights an aspect of material behavior in the creep regime that influences design factors and approximations when establishing allowable design parameters. There can be considerable scatter in measured creep behavior; not only considering the measurement issues addressed above but also the role of alloy composition and the impact of fabrication processes. Basically, it is the consensus evolution of design methods and corresponding margins to account for material variability that leads to component configurations that will robustly withstand the applied loading conditions throughout the intended life of the component. Quoting from the Foreword in each Code Book, "*The objective of the rules is to afford reasonably certain protection of life and property, and to provide a margin for deterioration in service to give a reasonably long, safe period of usefulness. Advancements in design and materials and evidence of experience have been recognized.*"

1.2.2 Elevated Temperature Material Behavior

The distinguishing feature of elevated temperature material behavior is whether significant creep effects are present. Consider a uniaxial tensile specimen with a constant applied load at a given temperature. As shown in Figure 1.3, if the temperature is low enough that there is no significant creep then the stresses and strain achieve their maximum values at time t_0 and remain constant as long as the load is maintained. The stresses and strain are thus *time-independent*. However, as shown in Figure 1.4, if the test temperature is high enough for significant creep effects, the strain will increase with time and eventually, depending on time, temperature, and load, rupture will occur. In the latter case, the strain is time-dependent.

In the previous example, the load was held constant. Now, consider the case with the specimen stretched to a constant displacement. In this case, as shown

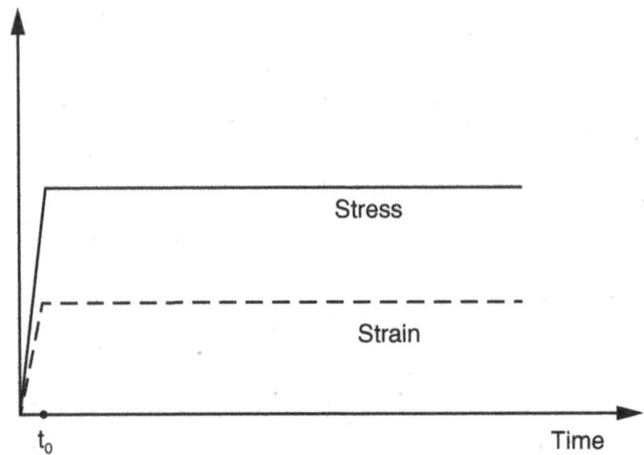

Figure 1.3 Load-controlled loading at low temperature.

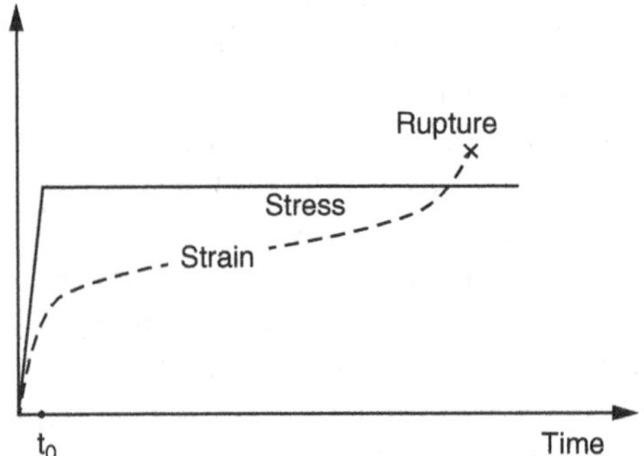

Figure 1.4 Load-controlled loading at elevated temperature.

in Figure 1.5, Line (a), if the temperature is low enough that there is no significant creep, then both the stresses and strain will be constant. However, if the temperature is high enough for significant creep, the stress will relax while the strain is constant (Line b). The behavior illustrated by Line (a) is *time-independent* and by Line (b) *time-dependent.*

Note also the difference in structural response between the constant applied load and the constant applied displacement. In the first case, referred to as *load-controlled*, the stress did not relax and, at elevated temperature, the strain increased until the specimen ruptured. The membrane stress in a pressurized cylinder is an example of load-controlled stress. In the second case, referred to as *deformation-controlled*, the strain was constant and the stress relaxed without causing rupture. Certain stresses resulting from the temperature distribution in a structure are an example of deformation-controlled stresses. Load-controlled

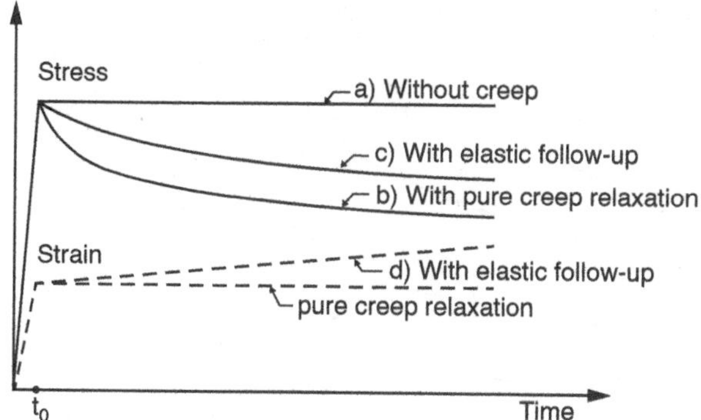

Figure 1.5 Strain-controlled loading at elevated temperature.

stresses can result in failure in one sustained application, whereas failure due to a deformation-controlled stress usually results from repeated load applications. However, due to stress and strain redistribution effects (discussed in more detail in subsequent chapters), the behavior of actual structures is more complex. For example, if there is elastic follow-up, then stress relaxation will slow and there will be an increase in strain, as shown in Figure 1.5, Lines (c) and (d), respectively. Thus, elastic follow-up, depending on the magnitude of the effect, can cause deformation-controlled stresses to approach the characteristics of load-controlled stresses. The distinction between load-controlled and displacement-controlled response and the role of elastic follow-up – or, more generally, time-dependent stress and strain redistribution – is central to the development and implementation of elevated temperature design criteria.

1.2.3 Creep Characteristics

A representative set of creep curves is shown in Figure 1.6 for carbon steel. As shown in Figure 1.7, the curve is usually divided into three zones. The first zone is called primary creep and is characterized by a relatively high initial creep rate that slows to a constant rate. This constant rate characterizes the second zone, called secondary creep. For many materials, the major portion of the test duration is spent in secondary creep. The third zone is called tertiary creep and is characterized by an increasing creep rate that culminates in creep rupture. Although for many materials most of the test is spent in secondary creep, for some materials – for example, certain nickel-based alloys at very high temperatures – primary and secondary creep are virtually negligible, Figure 1.8, and almost the entire test is in the third, or tertiary creep, stage.

As described more fully in Section 1.4.6, it is sometimes assumed that deformations and stresses in the primary creep regime do not significantly contribute

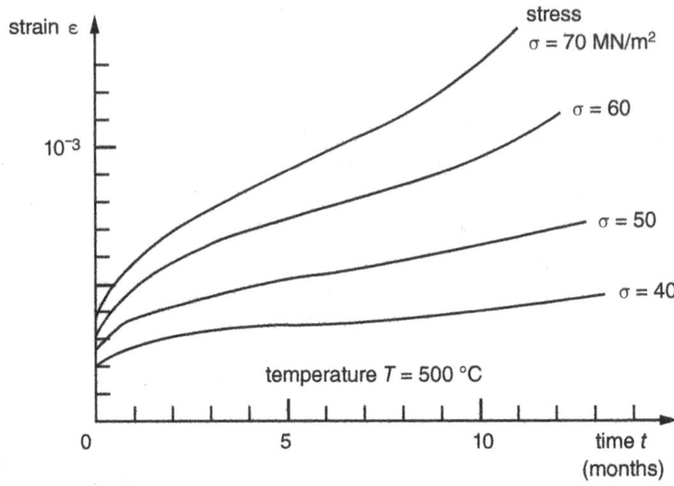

Figure 1.6 Creep curves for carbon steel Hult (1966).

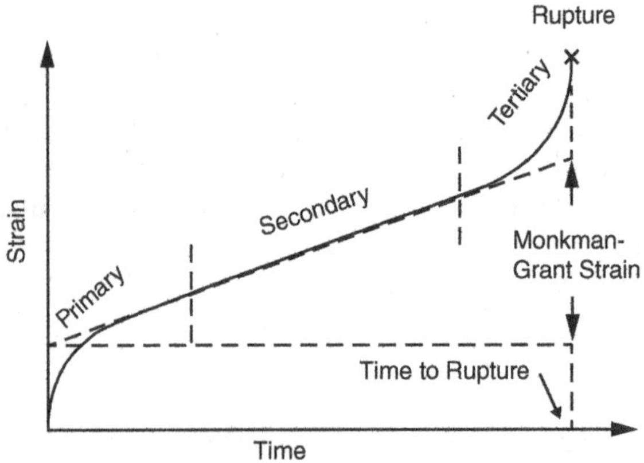

Figure 1.7 Creep regimes – strain vs. time at constant stress.

to accumulated creep rupture damage. An interesting application of the above assumption occurs in the assessment of the impact of heat treatment on structural integrity. For very large components, in particular, the complete time for the whole heat-treating cycle can be quite significant. Thus, if it were possible to ensure that the heat-treating cycle did not exceed the time duration of primary creep, then one could rationalize that the time spent in heat treatment would not significantly compromise the functional structural integrity of the component. Clearly, the key to this approach is to have an estimate of the time

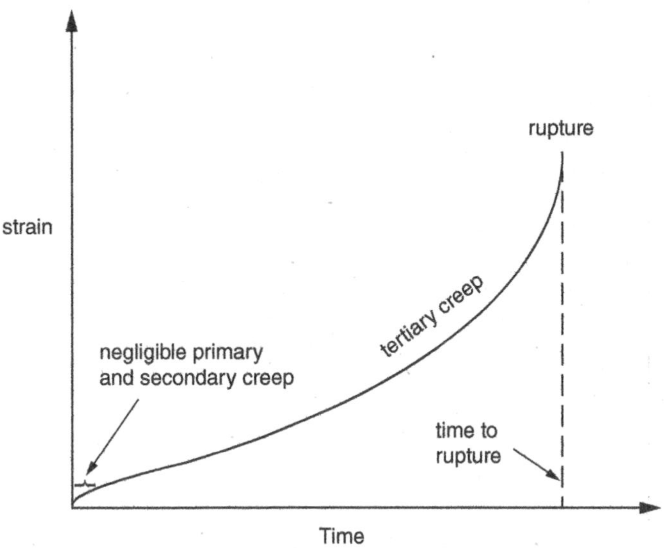

Figure 1.8 Material with negligible primary and secondary creep regimes.

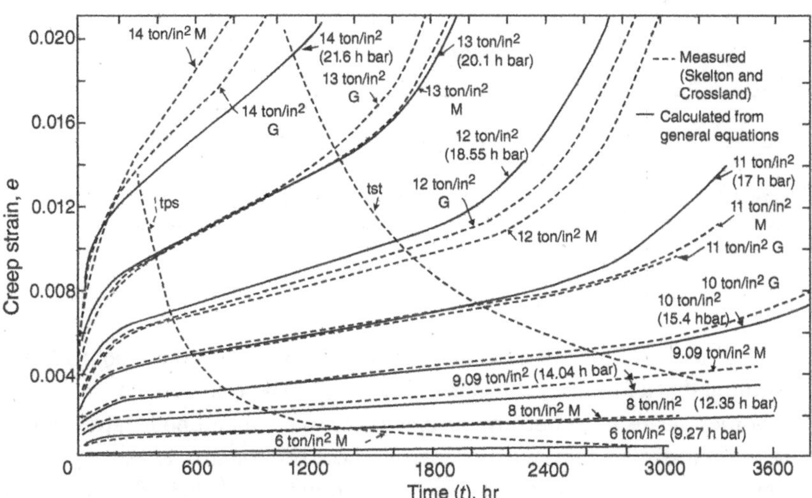

Figure 1.9 Measured and calculated tensile creep curves – primary creep duration Smith and Nicolson (1971).

to the point at which primary creep ends and secondary begins. To obtain a general idea of the relevant time duration of primary creep, there is an evaluation by Larke and Parker in a volume edited by Smith and Nicolson (1971) where they have plotted creep data and analytical correlations for a 0.19% carbon steel at 842°F (450°C). In Figure 1.9, it can be seen that the duration of

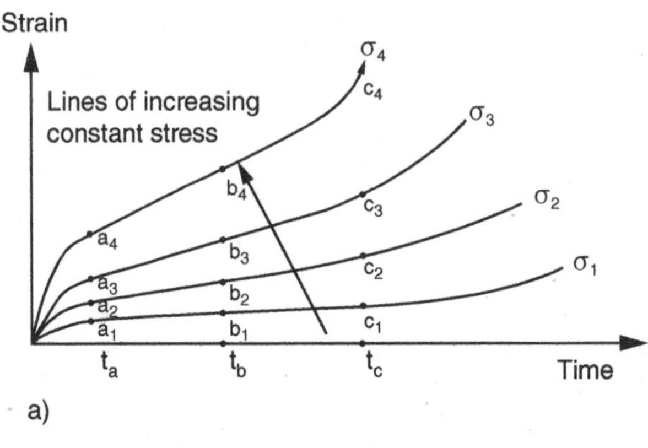

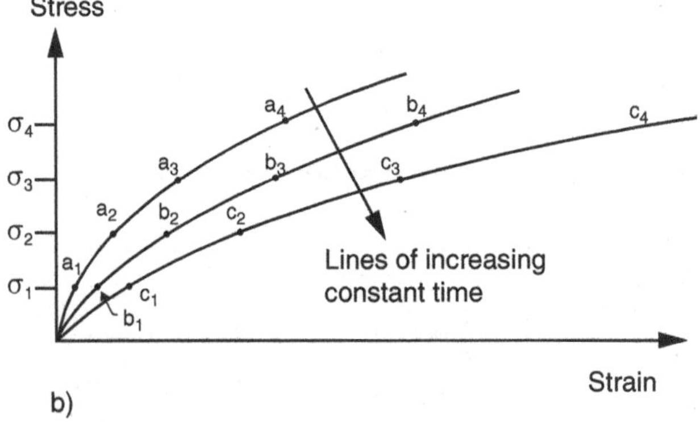

Figure 1.10 (a) A family of creep curves conventionally plotted as strain vs. time at constant stress. (b) The resultant stress-strain curves, plotted as stress vs. strain at constant time.

primary creep depends on the stress level, varying from 300 to about 1200 h, indicating that a total cycle time of 150–200 h should be acceptable.

Another means of characterizing creep is to plot "isochronous" stress-strain curves. Outwardly, these curves resemble conventional stress-strain curves except that the strain on the abscissa is the strain that would be developed in a given time by the stress given on the ordinate, as shown in Figure 1.10. These stress-strain values are usually plotted as a family of curves, each for a constant time, as shown in Figure 1.11 for 316 stainless steel at 1200°F (649°C). Although, conceptually, these curves could be directly plotted from data, the curves are usually generated from creep laws which are in turn derived from experimental data and correlate stress, strain, and time at a constant temperature. These curves can be very useful when designing elevated-temperature

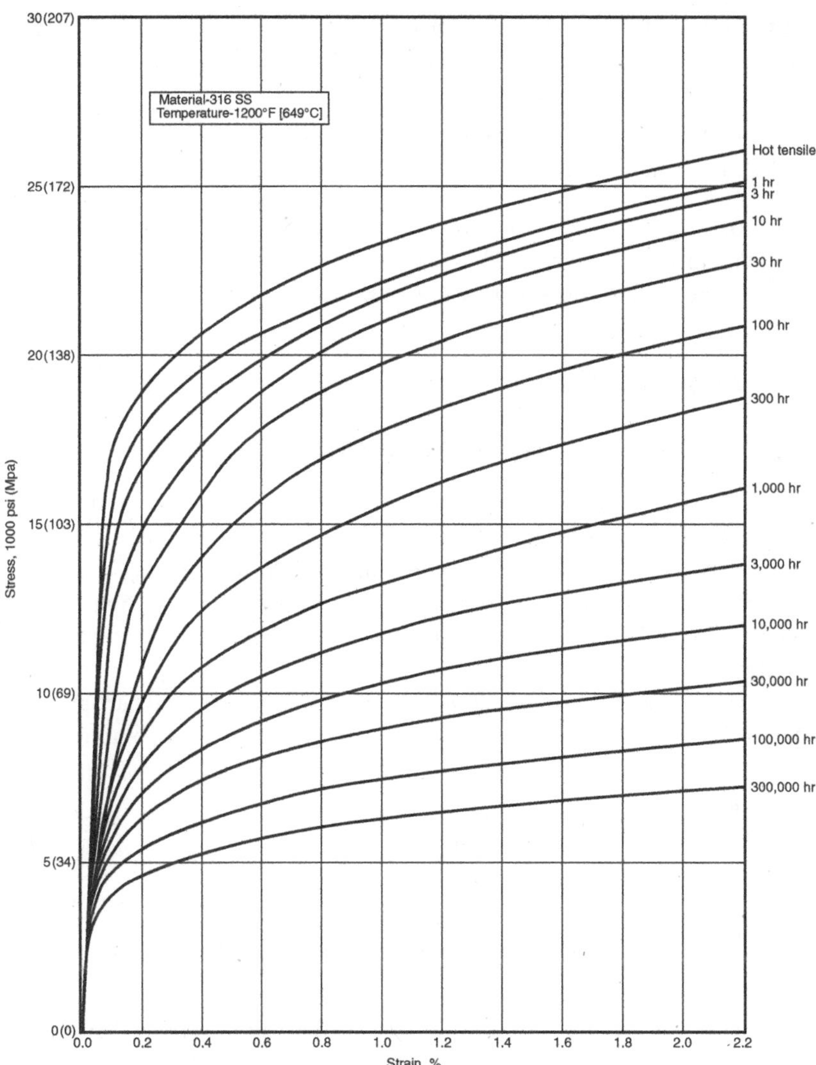

Figure 1.11 Isochronous stress-strain curves [ASME II-D].

structures where they can be used similarly to a conventional stress-strain curve in some situations; i.e., evaluating buckling and instability, and as a means of approximating accumulated strain. The equations for Figure 1.11's isochronous curves are shown in Appendix B.

Example 1.1 The effective stress in a pressure vessel component is 8000 psi. The material and temperature are shown in Figure 1.11. What is the expected design life of the component if:

(a) A strain limit of 0.5% is allowed?
(b) A strain limit of 1.0% is allowed?

Solution
(a) In Figure 1.11, the expected life is 15,000 h.
(b) In Figure 1.11, the expected life is 65,000 h.

1.3 Allowable Stress

1.3.1 ASME Boiler and Pressure Vessel Code

The ASME Boiler and Pressure Vessel Code lists numerous materials that meet the ASTM, as well as other European and Asian specifications. It provides allowable stresses for the various sections of the Code for temperatures below the creep range and at temperatures where creep is significant. For non-nuclear applications, by far the most common, these allowable stress levels are provided as a function of temperature in ASME II-D.

For ASME I and VIII-1 applications, the allowable stress criteria are given in Appendix 1 of ASME II-D. The allowable stress at elevated temperature is the

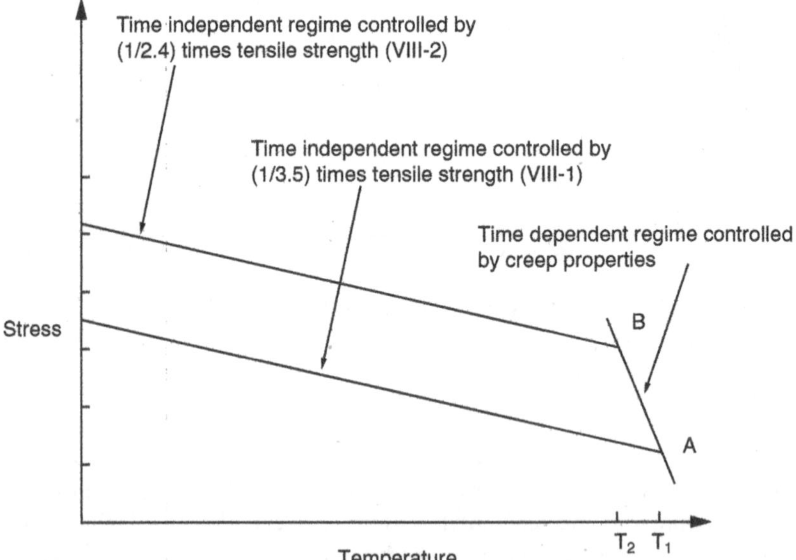

Figure 1.12 Shift of time-dependent stress with factor of safety.

lesser of: (1) the allowable stress given by the criteria based on yield and ultimate strength, (2) 67% of the average stress to cause rupture in 100,000 h, (3) 80% of the minimum stress to cause rupture in 100,000 h, and (4) 100% of the stress to cause a minimum creep rate of 0.01%/1000 h. Above 1500°F, however, the factor on average stress to rupture is adjusted to provide the same time margin on stress to rupture as existed at 1500°F (815°C). Although the allowable stress is a function of the creep rupture strength at 100,000 h, this is not intended to imply that there is a specified design life for these applications. There are additional criteria for welded pipes and tubes that are 85% of the above values. A very large number of materials are covered in these tables.

Unlike previous editions, the 2007 edition of ASME VIII-2, and subsequent editions, cover temperatures in the creep regime. The time-dependent allowable stress criteria for ASME VIII-2 are the same as for ASME VIII-1. However, because the time-independent criteria are less conservative (tensile strength divided by a factor of 2.4 vs. 3.5), the temperature at which the allowable stress is governed by time-dependent properties is lower in ASME VIII-2 than ASME VIII-1, as shown in Figure 1.12.

The allowable stress criteria for components of Class A nuclear systems, covered by ASME III-5, are different than for non-nuclear components. For nuclear components, the allowable stress at operating conditions for a particular material is a function of the load duration and is the lesser of: (1) the allowable stress for Class A nuclear systems based on yield and ultimate strength; (2) 67% of the minimum stress to rupture over time T; (3) 80% of the minimum stress to cause initiation of third-stage creep over time T; and (4) 100% of the average stress to cause a total (elastic, plastic, and creep) strain of 1% over time T. Note that these allowable stress criteria are more conservative than for non-nuclear systems for the same 100,000h reference time. However, these allowable stresses apply to operating loads and temperatures (Service Conditions in ASME III terminology) that are, in general, not defined as conservatively as the Design Conditions for non-nuclear applications. There are additional criteria for allowable stresses at welds and their heat-affected zones. All these allowable stresses are given in ASME III-5 for a quite limited number of materials.

The allowable stresses for Class B elevated-temperature nuclear systems are in general similar to those for non-nuclear systems, and are provided in ASME III-5.

ASME III-NB covers Class 1 nuclear components in the temperature range where creep effects do not need to be considered. Specifically, ASME III-NB is limited to temperatures for which applicable allowable stress values are provided in ASME II-D. These temperature limits are 700°F (370°C) for ferritic steels and 800°F (425°C) for austenitic steels and nickel-based alloys.

Unlike ASME I and VIII-1 components, the design procedures for nuclear components, particularly Class A are significantly different at elevated temperatures compared to the requirements for nuclear components below the creep regime. This is due in part to the time-dependence of allowable stresses, but more significantly to the influence of creep on cyclic life. As compared to ASME I and VIII-1 components, ASME III-5 explicitly considers cyclic failure modes at elevated temperature, whereas the former do not. ASME VIII-3 does address cyclic failure modes below the creep range. The provisions of ASME III-5, particularly with respect to VIII-2 are discussed in greater detail in Chapter 8.

ASME VIII-2 addresses cyclic failure modes and, as previously noted, currently covers temperatures in the creep regime above the previous limits of 700°F (370°C) and 800°F (425°C) for ferritic and austenitic materials, respectively. ASME VIII-2 also stipulates either meeting the requirements for exemption from fatigue analysis, or, if that requirement is not satisfied, meeting the requirements for fatigue analysis. However, above the 700/800°F (370/425°C) limit, the only available option is to satisfy the exemption from fatigue analysis requirements because the fatigue curves required for a full fatigue analysis are limited to 700°F and 800°F (370°C and 425°C).

1.3.2 European Standard EN 13445

EN 13445 applies to unfired pressure vessels. It is analogous to ASME VIII-1 and -2 in that it covers both Design by Formula (DBF), similarly to ASME VIII-1, and Design by Analysis (DBA), similarly to ASME VIII-2. It is unlike ASME VIII in several important respects. First, the EN 13445 allowable stresses are time-dependent, as in ASME III-5. They are also a function of whether there is in-service monitoring of compliance with design conditions. Provisions are also made for weld strength reduction factors, as in ASME III-5. Unlike the DBF rules in ASME VIII-1, those in the EN code are only applicable when the number of full pressure cycles is limited to 500.

The basic allowable stress parameters in EN 13445 in the creep range are: the mean creep rupture strength in time, t, and the mean stress to cause a creep strain of 1% in time, t. For DBF rules, the safety factor applied to the mean creep rupture stress is 1/1.5 if there is no in-service monitoring, and 1/1.25 if there is. There is no safety factor on the 1% strain criteria. If there is in-service monitoring then the strain limit does not apply, but strain monitoring is required. Thus, for a design life of 100,000 h in the EN code, without in-service monitoring, the base metal design allowable stress will be the same as in ASME VIII-1, which is to say, the allowable stresses are governed by creep rupture strength (remembering that ASME VIII-1 allowable stresses are based on 100,000h properties, even though there is no specified design life in ASME VIII-1).

There are two DBA methodologies defined in the EN 13445, the "Direct Route" and the "Method based on stress categories." Conceptually, the stress category methodology is similar to the methodology defined in ASME VIII-2 and III-NB for temperatures below the creep range and in ASME III-5 for elevated temperatures; however, there are many differences in the details of their application. The basic allowable stresses for the stress category DBA methodology are the same as for the DBF rules, and are dependent on whether there is in-service monitoring. The "Direct Route" is based on limit analysis and reference stress concepts. It is quite complex. Indeed, there is a warning in the introduction cautioning that, "Due to the advanced methods applied, until sufficient in-house experience can be demonstrated, the involvement of an independent body, appropriately qualified in the field of DBA, in the assessment of the design (calculations)." On that basis, a detailed discussion of the "Direct Route" DBA rules in EN 13445 will be considered beyond the scope of this presentation; however, there is a further discussion of the reference stress concept in Section 1.5.3.2.

Example 1.2 Figure 1.13 shows a representative plot of creep rupture data with extrapolation to 100,000 h. Figure 1.14 shows a plot of creep strength

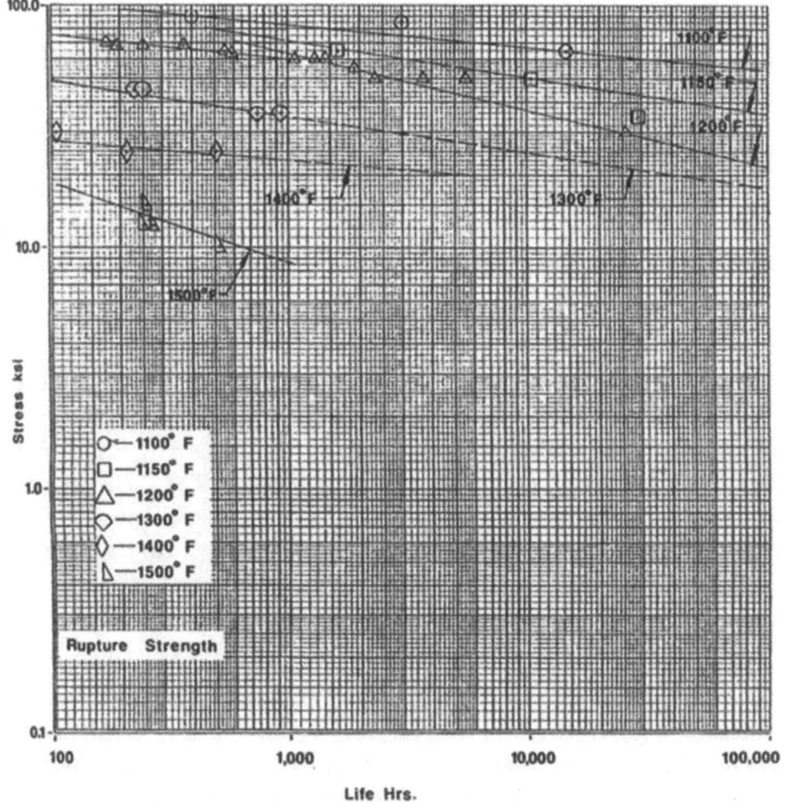

Figure 1.13 Rupture strength Jawad and Farr (2019).

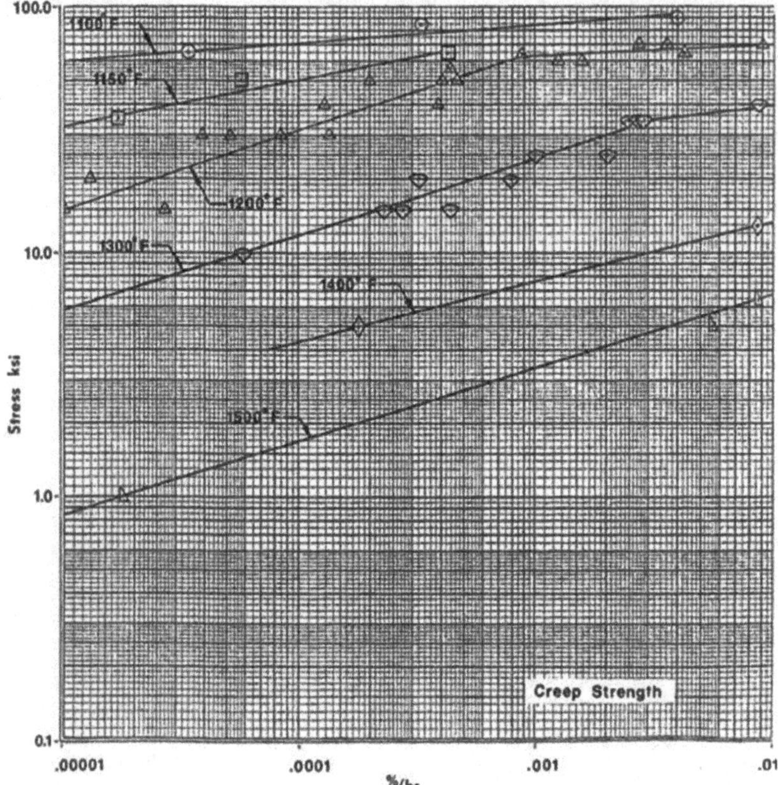

Figure 1.14 Creep strength Jawad and Farr (2019).

(minimum creep rate) for the same material. Curve fitting procedures are usually used for the extrapolation. Based on the creep properties shown in Figures 1.13 and 1.14, calculate the allowable stress at 1200°F for an ASME VIII-1 application. (Note: This is for a non-ASME Code application. The Code has published values for Code applications.) Compare the results to the allowable stress for an EN 13445 application with 100,000 h design life and with in-service monitoring. (Note: Under EN 13445, allowable stress values are established by the user based on published properties as described therein.)

Solution

In Figure 1.13, the average stress to rupture in 100,000 h is 22 ksi and the allowable stress for ASME VIII-1, based on average creep rupture, is 22 × (0.67) = 14.7 ksi assuming a minimum value of creep rupture based on a 20% scatter band gives a minimum creep rupture strength of 17.6 ksi, and an allowable stress based on minimum creep rupture of 17.6 × (0.8) = 14.1 ksi. In Figure 1.14, the stress for a minimum creep rate of 0.01% in 1000 h is 15 ksi, which gives an allowable stress of 15 ksi. Therefore, the applicable stress for an ASME VIII-1 application is 14.1 ksi governed by the minimum creep rupture strength.

For EN 13445 applications with in-service monitoring, the safety factor is 1.25 on mean creep rupture strength (assumed equal to average strength plotted in Figure 1.13), so the allowable stress for a design life of 100,000 h is 22/(1.25) = 17.6 ksi.

1.4 Creep Properties

1.4.1 ASME Code Methodology

One of the issues facing the designer of elevated temperature components is how to extrapolate limited time duration test data to the service lives representative of most design applications or, in the case of developing allowable stresses for ASME Code applications, 100,000 h. The method for the development of ASME Code elevated temperature allowable stresses for non-nuclear applications is described in Chapter 3 ("Basis for Tensile and Yield Strength Values") of *the Companion Guide to the ASME Boiler & Pressure Vessel Code* (Jetter 2018). Quoting from that source:

> At the elevated temperature range in which the tensile properties become time-dependent, the data is analyzed to determine the stress to cause a secondary creep rate of 0.01% in 1000 h and the stress needed to produce rupture in 100,000 h. This data must be from material that is representative of the product specification, requirements for melting practice, chemical composition, heat treatment, and product form. The data is plotted on log-log coordinates at various temperatures. The 0.01%/1000 h creep stress and the 100,000 h rupture stress are determined from such curves by extrapolation at the various temperatures of interest. The values are then plotted on semi log coordinates to show the variation with temperature. The minimum trend curve defines the lower bound for 95% of the data.

As part of the ASME Code methodology, data for development of allowable stress values is required for long times, usually at least 10,000 h for some data, and at temperatures above the range of interest, usually 100°F (56°C) higher. Considerable judgment is exercised in the development of ASME Code allowable stress values and the use of these values is required for ASME Code-stamped construction. As a corollary, if the material of interest is not listed in the Code for the applicable type of construction, or at the desired temperature, then it is not possible to qualify the component for a Code stamp. The designer may, however, use this method for non-Code applications.

A somewhat different approach is taken in EN 13445. There, the mean creep rupture strength and mean stress for a 1% strain limit are listed in referenced standards for approved materials for various times and temperatures. The requirements for extrapolation or interpolation to other conditions are defined and the safety factors to be applied are defined as a function of the application, as described above.

1.4.2 Larson-Miller Parameter

As might be expected, there are numerous methods Conway (1969) for extrapolation of creep data; the ASME procedure described above is the one used for establishment of allowable stresses shown in ASME II-D. Generally, the use of other extrapolation techniques would only be required for non-coded construction, or for evaluation of failure modes beyond the scope of the applicable code. Penny and Marriott (1995) provide an extensive assessment of various extrapolation techniques, including the widely used Larson-Miller parameter, which they characterize as simple and convenient, but not particularly accurate.

The starting point for development of the Larson-Miller Grant and Mullendore (1965) parameter is to assume that creep is a rate process governed by the Arrhenius equation

$$d\varepsilon_c / dt = Ae^{(-Q/RT)} \tag{1.1}$$

where

A = constant
Q = activation energy for the creep process, assumed a function of stress only
R = universal gas constant
T = absolute temperature (460 +°F)
t = time.

Noting that from the Monkman-Grant relationship the time to rupture, t_r, multiplied by the minimum creep rate, can be assumed to be constant, Eq. (1.1) can be rewritten as

$$AT_r e^{(-Q/R_T)} = \text{constant}.$$

Taking the logarithms for each side, the Larson-Miller parameter, P_{LM}, can be expressed as

$$P_{LM} = T(C + \log_{10} t) \tag{1.2}$$

where P_{LM} is a function of time and temperature and independent of stress. It was also assumed that C is independent of both stress and temperature and is a function of material only. Experimental data shows that the range of C for various materials is between 15 and 27. Most steels have a C value of 20. Hence, Eq. (1.2) can be expressed as

$$P_{LM} = (460 + {}^{\circ}F)(20 + \log_{10} t). \tag{1.3}$$

The Larson-Miller parameter is also used to correlate creep data using specific values of C that are material and temperature range-dependent, thus minimizing some of the uncertainties. Further details of the Larson-Miller parameter are given in Chapters 2 and 6.

Another important use of the Larson-Miller parameter is determination of equivalent time at temperature, as shown by the following examples.

Example 1.3 A pressure vessel component was designed at 1200°F with a life expectancy of 100,000 h. What is the expected life if the design were lowered to 1175°F?

Solution

The Larson-Miller parameter for the original design condition is obtained from Eq. (1.3) as

$$\begin{aligned} P_{LM} &= (460 + 1200)(20 + \log_{10} 100{,}000). \\ &= 41{,}500 \end{aligned}$$

Using this value for the new design condition yields

$$41{,}500 = (460 + 1175)(20 + \log_{10} t)$$

or

$$\begin{aligned} \log_{10} t &= 5.382 \\ t &= 241{,}100 \text{ h} \end{aligned}$$

which indicates a 2.4-fold increase in the life of the component when the temperature drops 25°F.

Example 1.4 A pressure vessel shell is constructed of 2.25Cr-1Mo steel. The thickness is 4 in. and requires post-weld heat treating at 1300°F for 4 h. The fabricator requires two separate post-weld heat treatments (8 h) and the user needs three more (12 h) for future repair. Hence, a total of 20 h is needed. The material supplier furnishes the steel plates with material properties guaranteed for a minimum of 20 h of post-weld heat treating. During the manufacturer's second post-weld heat treatment, the temperature spiked to 1325°F for 2 h. How many hours are left for the user?

Solution

Calculate P_{LM} from Eq. (1.3) for 2 h at 1325°F.

$$\begin{aligned} P_{LM} &= (460+1325)(20+\log_{10} 2) \\ &= 36{,}237 \end{aligned}.$$

Substitute back into Eq. (1.3) to calculate the equivalent time for 1300°F.

$$\begin{aligned} 36{,}237 &= (460+1300)(20+\log_{10} t) \\ t &= 3.9 \text{ h} \end{aligned}.$$

Thus, the fabricator used a total of 4.0 + (2.0 + 3.9) = 9.9 h.

Available hours for the user = 20 − 9.9 = 10.1 h. This corresponds to two full and one partial post-weld heat treatment.

1.4.3 Omega Method

The Omega Method Prager (2000) is based on a different model for creep behavior than that described above for the Larson-Miller parameter. Originally developed to address the issue of determining the accumulated damage, and thus the remaining life of service-exposed equipment, the Omega Method is based on the observation that, at design stress levels, both the primary creep and secondary creep phases are of relatively short duration, with small strain accumulation, and that most of the component life is spent in the third stage, where the strain rate increases with time and accumulated strain. In the Omega Method, the creep strain rate is accelerated in accordance with the following relationship:

$$\ln(d\varepsilon / dt) = \ln(d\varepsilon_o / dt) + \Omega_p \varepsilon \tag{1.4}$$

where

$(d\varepsilon/dt)$ and $(d\varepsilon_0/dt)$ = current and initial strain rates, respectively

Ω_p = Omega parameter

ε = current strain level.

From this relationship, various parameters relating to accumulated damage and remaining life may be developed. The Omega Method has been incorporated into ASME FFS-1 for remaining life assessments. Chapter 7 discusses this method in more detail.

1.4.4 Negligible Creep Criteria

Another issue of interest is the temperature at which creep becomes significant. To answer this quantitatively, the key point is – significant compared to what?

There are no single, rigorous criteria for assessing when creep effects are negligible. However, in each of the design codes of interest, the criteria for negligible creep applicable to that particular design code are defined.

For ASME I and VIII-1, the comparison is between the results provided by the allowable stress criteria based on short-time tensile tests without creep as compared to long-term tests with creep. When the allowable stress as a function of temperature is governed by creep properties, the stress value is italicized in ASME II-D, Table 1. However, in this case, even though the allowable stress is governed by creep properties, the design evaluation procedures do not change.

The situation is different with ASME III-5. In ASME III-5, there are two sets of allowable stress for primary (load-controlled) stresses to be used in the evaluation of Service Conditions. One set, S_m, is time-independent and a function of short-time tensile tests. The other set, S_t, is time-dependent and a function of creep. As will be discussed in more detail later, the design rules for time-independent and time-dependent allowable stress levels are different. However, the rules for displacement-controlled stress, such as thermally induced stress, state that the criteria for negligible creep are the most restrictive.

The ASME III-5 criteria for negligible creep for displacement-controlled stresses are based on the idea that, under maximum stress conditions, creep effects should not compromise the design rules for strain limits or creep-fatigue damage. The key consideration from that perspective is that actual stress in a localized area can be much greater due to discontinuities, stress concentrations, and thermal stress than the wall-averaged primary stresses in equilibrium with external loads. Basically, the magnitude of the localized stress will be limited by the material's actual yield stress because it is at this stress level that the material will deform to accommodate higher stresses due to structural discontinuities or thermal gradients. Thus, the objective of the negligible creep criteria for localized stresses is to ensure that the damage due to the effects of creep at the material's yield strength will not significantly impact the design rules for the failure mode of concern. For example, there are two resulting criteria, one based on negligible creep damage and the other on negligible strain. Negligible creep damage can be approximated by:

$$\sum(t_i / t_{id}) \leq 0.1 \tag{1.5}$$

where

t_i = the time duration at high temperature

t_{id} = time duration at a stress level as defined in III-5 and VIII-2 but nominally 1.5 times the yield stress, S_y, except for 9Cr-1Mo-V steel.

For negligible strain, the criteria are given by

$$\sum\varepsilon_i \leq 0.2\% \tag{1.6}$$

where

ε_l = the creep strain at a stress of 1.25 times yield strength, S_y.

In ASME III-5 Part HCB, which provides elevated temperature design rules for Class B nuclear components, Appendix HCB-III contains a figure that shows time temperature limits below which creep effects need not be considered in evaluating deformation-controlled limits. These curves are lower, smoothed versions of the ASME III-5 criteria for negligible creep for a limited number of materials: cast and wrought 304 and 316 stainless steel, nickel-based Alloy 800H, low alloy steel, and carbon steel. The advantage of these curves is that no computations are required.

The French code for elevated temperature nuclear components, RCC-MR, also provides criteria for negligible creep, which is somewhat different than that in ASME III-5. The procedures are more involved than those in ASME III-5, but the resulting values for long-term service are similar to the temperature limits of ASME III-NB – 700°F (371°C) for ferritic and 800°F (427°C) for austenitic and nickel-based alloys. For 316L(N) stainless steel, whose creep properties are fairly close to 316 stainless steel, the time–temperature limit curve is generally in agreement with the curve shown in ASME III-5 Part HCB for 316 stainless steel.

1.4.5 Environmental Effects

As stated in its Foreword, the ASME Boiler and Pressure Vessel Code does not specifically address environmental effects. However, non-mandatory general guidance is provided in several sections. ASME II-D, Appendix A provides guidance on metallurgical effects, including a number of references on corrosion and stress-corrosion cracking. ASME VIII-1, Appendix E suggests good practice for determining corrosion allowances, which are the responsibility of the user to specify based on the equipment's intended service. It is noted that the corrosion allowance is in addition to the minimum required thickness. ASME III, Appendix W, has a comprehensive discussion of environmental effects. Included for each phenomenon is a discussion of the mechanism, materials, design, mitigating actions, and references.

In the context of elevated temperature applications, the designer should be particularly aware of environments that can reduce a material's creep rupture life and/or ductility. For example, it has been shown that short-term exposure to oxygen at temperatures exceeding 1650°F (900°C) could lead to embrittlement at intermediate temperatures of 1300°F–1500°F (705°C–815°C), which was attributed to intergranular diffusion of oxygen. Hydrogen, chlorine, and sulfur may also cause embrittlement due to penetration. Sulfur is of particular concern because it diffuses more rapidly and embrittles more severely than oxygen.

1.4.6 Monkman-Grant Strain

Another parameter of interest is the strain computed by multiplying the time to rupture by the secondary creep rate. This strain parameter, shown diagrammatically in Figure 1.7, is sometimes known as the Monkman-Grant strain. As discussed by Penny and Marriott (1995), this computed strain has been shown to be useful in correlating rupture under variable loading conditions. A corollary of this approach is that it implies that the primary creep strain may be disregarded in assessing damage accumulation.

It has also been suggested that a relevant measure of creep ductility for the application of reference stress methods (Section 1.5.3.2) in the presence of local stress discontinuities is for the material of interest to show a ratio of total strain at failure to the Monkman-Grant strain of at least 5:1.

1.5 Required Pressure-Retaining Wall Thickness

There are basically two approaches in general use in design for determining the wall thickness required to resist internal pressure and applied external loads. The first is usually referred to as Design by Rule, or Design by Formula (DBF in the European Standard terminology), and the second is Design by Analysis (DBA). As an alternative to DBA, there are other approaches based on experimental methods; however, those methods are generally not applicable in the creep regime. In addition to the above approaches, there are many pressure-retaining components that have standardized allowable pressure ratings as a function of design temperature. Typically, these include flanges, piping components, and valve bodies. In general, these pressure/temperature ratings do not include the effects of loadings other than internal pressure. The following discussion will provide an overview of these methodologies; the specific requirements for their implementation will be discussed in later chapters.

1.5.1 Design by Rule

In this approach, formulas are provided for the required thickness as a function of the design pressure, allowable stress, and applicable parameters defining the geometry of interest. Numerous diagrams are provided to define the requirements for specific configurations; for example, reinforcement of openings, head-to-cylinder joints, and weldments. This is the approach used, for example, in ASME VIII-1 "Unfired Pressure Vessels," and ASME I "Power Boilers."

1.5.2 Design by Analysis

In the DBA approach, stress levels are determined at various critical locations in the structure and compared to allowable stress levels, which are a function of the applied loading conditions and failure mode under consideration. The most commonly used methodology, particularly at elevated temperatures, is based on elastically calculated stresses, which are sequentially categorized based on the relevant failure mode. Primary stresses (those that normally determine wall thickness) are first determined by separating the structure into simpler segments (free bodies) in equilibrium with external loads. Next, secondary and peak stresses (which in combination with primary stresses normally determine cyclic life) are determined from stresses at structural discontinuities and induced thermal stresses. Different allowable stresses are assigned to the different stress categories based on the failure mode of concern.

1.5.3 Approximate Methods

There is another category of Design-by-Analysis methodologies that are approximate in the sense that they approximate the "true" time-dependent stress and strain history in a component. In fact, considering the variations in creep behavior and difficulties encountered in defining comprehensive models of material behavior, they can be quite useful under appropriate circumstances. Two main approaches will be described. The first is the elastic analog or stationary creep solution and the second is the reference stress approach, which is somewhat analogous to limit analysis.

1.5.3.1 Stationary Creep – Elastic Analog

Subject to certain restrictions on representation of creep behavior, a structure subjected to a constant load will reach a condition where the stress distribution does not change with time, thus the term "stationary creep." The fundamental restriction on material representation is that the creep strain is the product of independent functions of stress and time. Conceptually, stationary creep is valid when the strains and strain rates due to creep are large compared to elastic strains and strain rates.

If the structure is statically determinate throughout, then the initial stress distribution will not change with time; subject to the applicability of small displacement theory, which applies to the large majority of practical design problems. Examples would be a single bar with a constant tension load and the stresses in the wall of a thin-walled cylinder, remote from discontinuities, subjected to a constant internal pressure.

However, it is with indeterminate structures that the stationary creep concept is of most value. It has been shown that, in a structure with redundant load paths or subject to local redistribution (i.e., a beam in bending), the stress redistribution will take place relatively quickly: on the order of the time it takes for the creep strain to equal twice the initial elastic strain. For a set of variables representative of pressure vessels in current use, Penny and Marriott (1995) calculated an effective redistribution time of about 100 h. Although this would be a long time if the vessel were subject to significant daily cycles, it is short compared to the long times of extended operation.

A number of investigators have shown that, because the stress distribution in stationary creep does not vary with time (and thus corresponding creep rates are constant), the stationary creep stress distribution is analogous to non-linear elastic stress distribution, so solutions to the creep problem can be obtained from solutions to the non-linear elastic stress distribution problem. This is usually referred to as the "elastic analog." Although the elastic analog has been shown as valid in more general terms, a more convenient representation is analogous to a simple power law representation of steady, secondary creep in which primary creep is considered negligible

$$d\varepsilon / dt = k'\sigma^n, \tag{1.7}$$

which results in the following expression for accumulated creep strain:

$$\varepsilon = K'\Delta t\sigma^n. \tag{1.8}$$

This is analogous to the equation for non-linear elasticity

$$\varepsilon = K'\sigma^n. \tag{1.9}$$

An example of stationary creep solutions for various values of the power law exponent, n, is shown in Figure 1.15. This is the non-dimensional stationary creep solution for a beam in bending with a constant applied moment. Note that for $n = 1$ the stress distribution is elastic, and for $n \to \infty$ the distribution corresponds to that for the assumption of ideal plasticity. All the distributions pass through a point partway through the wall, which is referred to as the "skeletal point." The reduction in steady creep stress, as compared to the initial elastic distribution, is the basis for the reduction of the elastically calculated bending stress by a section factor when comparing the calculated stresses to allowable stress levels in ASME III-5. Further description of the elastic analog is given in Chapters 2 and 4.

1.5.3.2 Reference Stress

The initial idea of a reference stress is that the creep behavior of a structure could be evaluated against a single creep test at its reference stress. Initially applied to problems of creep deformation, there were a number of analytical solutions

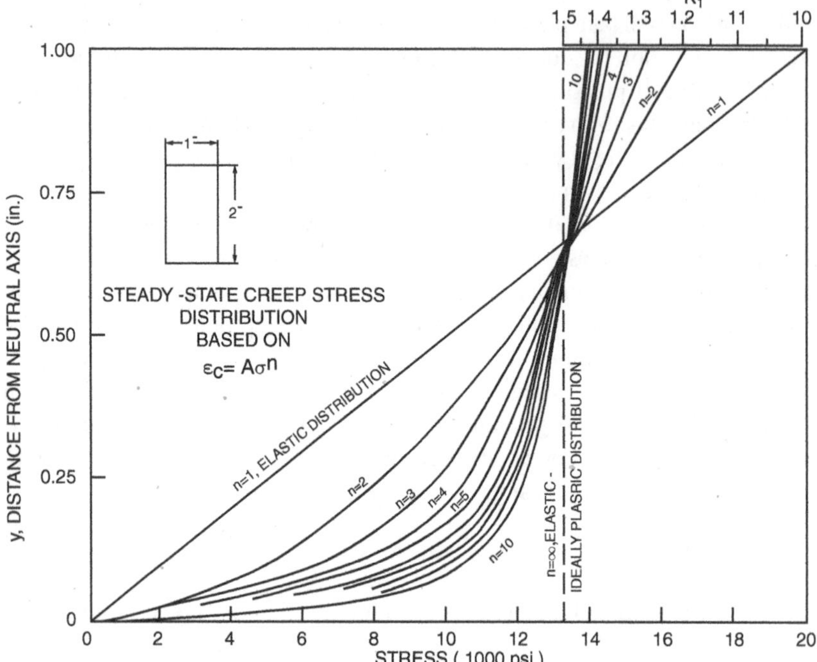

Figure 1.15 Steady-state creep stress distribution across a rectangular beam in pure bending and having a steady-state creep law of the form $\varepsilon_c = A\sigma^n$ [Oak Ridge National Laboratory].

developed for specific geometries. However, Sim (1968), noting that reference stress is independent of the creep exponent, and also that the solution for an infinite creep exponent is analogous to a limit solution which corresponds to ideal plasticity, proposed that the reference stress could be conservatively obtained from

$$\sigma_{R=}(P / P_L)\sigma_y \tag{1.10}$$

where

$P =$ load on the structure
$P_L =$ limit value of the load
$\sigma_R =$ reference stress
$\sigma_y =$ the yield stress.

There have been numerous comparisons between the results of this approach and experimental and rigorous analyses of the same component or test article. In general, the results are quite favorable. Although the reference stress approach has not been incorporated into the ASME Boiler and Pressure Vessel Code, it has been used in the British elevated temperature design code for nuclear systems, R5, and in the recent European Standard EN 13445. However, the British standard recommends an adjusted reference stress for design given by a factor of 1.2 times the reference stress, as in Sim's relationship above. Also, as previously noted, the EN standard cautions against the use of the reference stress method by those not familiar with its application.

Part of the reason for this concern is inherent in the basis for both limit loads and reference stress determination. Both are based on structural instability considerations and not local damage. As such, there is an inherent requirement that the material under consideration be sufficiently ductile. This is easier to achieve at temperatures below the creep range. Within the creep range, ductility decreases, particularly at the lower stress levels associated with design conditions. There have been some studies to more specifically identify creep ductility requirements, but current thinking would put it in the range of 5%–10% for balanced structures which don't have extreme strain-concentrating mechanisms.

The following example highlights the differences between an elastically calculated stress distribution, a steady stationary creep stress distribution, and the reference stress distribution.

Example 1.5 Consider the two-bar model shown in Figure 1.16. As explained in Chapter 2, this is actually representative of the way in which cyclone separators are sometimes hung from vessels. For this example, the two, parallel, uniaxial bars are of equal area, A, unequal lengths L_1 and L_2, and attached to a rigid boss constrained to move in the vertical direction only. The assembly is loaded with a constant force F. Compare the (1) initial elastic stress, (2) stationary stress, and (3) reference stress in each bar. Assume that creep is modeled with a power law with exponent $n = 3$. Consider two cases. In the first case, $L_1 = L_2/8$ and in the second case, $L_1 = L_2/2$.

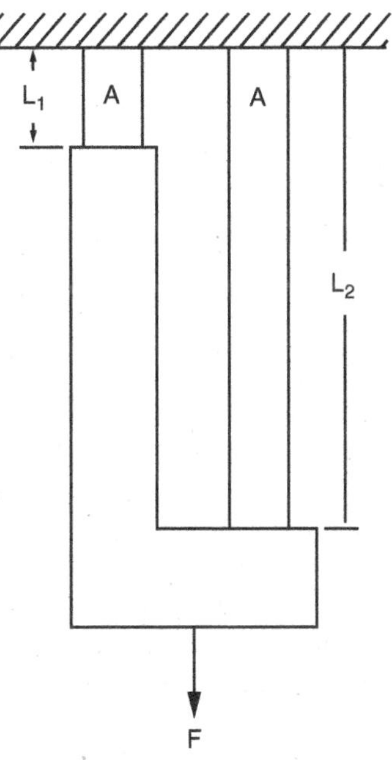

Figure 1.16 Two-bar model with constant load.

Solution

(1) Elastic Analysis

The initial elastic distribution can be expressed as

$$\varepsilon_1(0) = \sigma_1(0)/E \text{ and } \varepsilon_2(0) = \sigma_2(0)/E \quad \text{(a)}$$

where

$\varepsilon_1(0)$, $\sigma_1(0)$ and $\varepsilon_2(0)$, $\sigma_2(0)$ = initial strain and stress in bars 1 and 2, respectively
E = modulus of elasticity.

From equilibrium

$$F = A(\sigma_1 + \sigma_2). \tag{b}$$

From displacement compatibility

$$L_1\varepsilon_1(0) = L_2\varepsilon_2(0). \tag{c}$$

Substituting Eq. (a) into Eq. (c) gives

$$L_1\sigma_1 = L_2\sigma_2 \text{ or } \sigma_2 = (L_1 / L_2)\sigma_1. \tag{d}$$

Substituting Eq. (d) into Eq. (b) gives

$$F = A[\sigma_1 + (L_1 / L_2)\sigma_1]. \tag{e}$$

From which, solving for the initial elastic stress distribution gives

$$\sigma_1(0) = [L_2 / (L_1 + L_2)]F / A \tag{f}$$

$$\sigma_2(0) = [L_1 / (L_1 + L_2)]F / A. \tag{g}$$

(2) Stationary Stress Analysis
The solution for the stationary stress distribution in each bar proceeds in a similar fashion: the equilibrium equation (a) remains the same; the strain rate at any time can be expressed as the sum of elastic and creep strain rates

$$\mathrm{d}\varepsilon_1 / \mathrm{d}t = k'\sigma_1^n + (\mathrm{d}\sigma_1 / \mathrm{d}t) / E \quad \text{and} \quad \mathrm{d}\varepsilon_2 / \mathrm{d}t = k'\sigma_2^n + (\mathrm{d}\sigma_2 / \mathrm{d}t) / E. \tag{h}$$

From displacement rate compatibility

$$L_1\mathrm{d}\varepsilon_1 / \mathrm{d}t = L_2\mathrm{d}\varepsilon_2 / \mathrm{d}t. \tag{i}$$

Using the above relationships, an equation can be developed involving functions of σ_1; but, as noted by Kraus (1980), it cannot be solved in closed form. However, we are interested in the stationary creep solution, where $\mathrm{d}\sigma/\mathrm{d}t \to 0$ and $(\mathrm{d}\sigma/\mathrm{d}t)/E < K\sigma^n$. Thus, Eq. (h) in the stationary creep regime becomes

$$\mathrm{d}\varepsilon_1(\infty) / \mathrm{d}t = k'\sigma_1(\infty)^n \quad \text{and} \quad \mathrm{d}\varepsilon_2(\infty) / \mathrm{d}t = k'\sigma_2(\infty)^n. \tag{j}$$

Substituting Eq. (j) into Eq. (i) provides the following relationship for $\sigma_2(\infty)$

$$\sigma_2(\infty) = (L_1^{1/n} / L_2^{1/n})\sigma_1(\infty). \tag{k}$$

From the above, and the equilibrium equation (b), the following expressions for stationary stress distribution may be obtained

$$\sigma_1(\infty) = \left[\left(L_2^{1/n}\right) / \left(L_1^{1/n} + L_2^{1/n}\right)\right] F / A \tag{l}$$

$$\sigma_2(\infty) = \left[\left(L_1^{1/n}\right) / \left(L_1^{1/n} + L_2^{1/n}\right)\right] F / A. \tag{m}$$

(Note that, from the elastic analog, the initial elastic stress distribution corresponds to the steady creep solution with $n = 1$.)

(3) Reference Stress Analysis

The reference stress is obtained from Eq. (1.10), noting that the limit load in each bar is equal to $A\sigma_y$

$$\sigma_1(R) = \sigma_2(R) = F / (2A). \tag{n}$$

(Note that a similar result is obtained by letting $n \to \infty$ in the stationary creep stress solution, as predicted by the Sim hypothesis.)

The following results are obtained for Case #1, where $L_1 = L_2/8$:

Initial elastic stress: $\sigma_1(0) = (8/9)F/A$, $\sigma_2(0) = (1/9)F/A$
Stationary creep stress: $\sigma_1(\infty) = (2/3)F/A$, $\sigma_2(\infty) = (1/3)F/A$
Reference stress: $\sigma_1(R) = (1/2)F/A$, $\sigma_2(R) = (1/2)F/A$.

In a similar fashion, for Case #2, where $L_1 = L_2/2$:

Initial elastic stress: $\sigma_1(0) = (2/3)F/A$, $\sigma_2(0) = (1/3)F/A$
Stationary creep stress: $\sigma_1(\infty) = (0.56)F/A$, $\sigma_2(\infty) = (0.44)F/A$
Reference stress: $\sigma_1(R) = (1/2)F/A$, $\sigma_2(R) = (1/2)F/A$.

Discussion

Case #1 is representative of a highly unbalanced system with an extreme stress concentration. The initial elastically calculated stresses differ by a factor of eight. The stationary creep stresses differ by a factor of two and the reference stress is equal in both bars. This phenomenon is sometimes referred to as "load shedding." However, for this highly unbalanced system, the strain in the shorter, stiffer bar is also a factor of eight higher than the lower stressed bar. As a result, the question arises as to whether there is sufficient creep ductility in the shorter bar to eventually realize the lower reference stress level. If the shorter bar fails prematurely, the entire load will shift to the longer, more lightly loaded bar, causing the stress in that bar to increase to twice the reference stress level. Depending on the creep ductility and how far into the loading cycle the shorter bar fails, the result could be a premature failure of the two-bar system.

Case #2 represents a more balanced system without an extreme stress concentration. The stationary creep solution is within 6% of the reference stresses and the strain ratio is only a factor of two. In this case, one would not expect a premature failure.

Although it is difficult to develop quantified guidance from these two cases, the clear lesson is that considerable caution should be taken in applying reference stress methods to highly unbalanced systems, particularly if the creep ductility of the construction material is suspect.

Further descriptions of this method are given in Chapters 2, 3, and 4.

1.6 Effects of Structural Discontinuities and Cyclic Loading

This is a general discussion to acquaint the designer with the structural phenomena which are significant when evaluating elevated temperature failure modes associated with structural discontinuities and cyclic loading. Specific rules and procedures will be presented in subsequent chapters.

1.6.1 Elastic Follow-Up

Elastic follow-up can cause larger strains in a structure with applied displacement-controlled loading than would be calculated using elastic analysis. These strain concentrations may result when structural parts of different flexibility are connected in series while loaded with an applied displacement, and the flexible portions are highly stressed. In order for follow-up to occur, in a two-bar model (as shown in Figure 1.17) a less stressed, flexible element can generate inelastic deformation in a more highly stressed adjacent element. The other requirement is that the lower stressed remainder of the structure should be capable of transmitting a significant deformation to the more highly stressed portion of the structure undergoing inelastic deformation.

In the two-bar example, a displacement applied to the end of the smaller diameter bar, B, will initially cause an elastic deformation in both A and B, with B being the more highly stressed bar. Although under creep conditions the stress in A and B will both relax, the higher stress in B will cause further creep deformation in B and some of the initial elastic deformation in A will be absorbed in B – hence the term "elastic follow-up." This process is shown

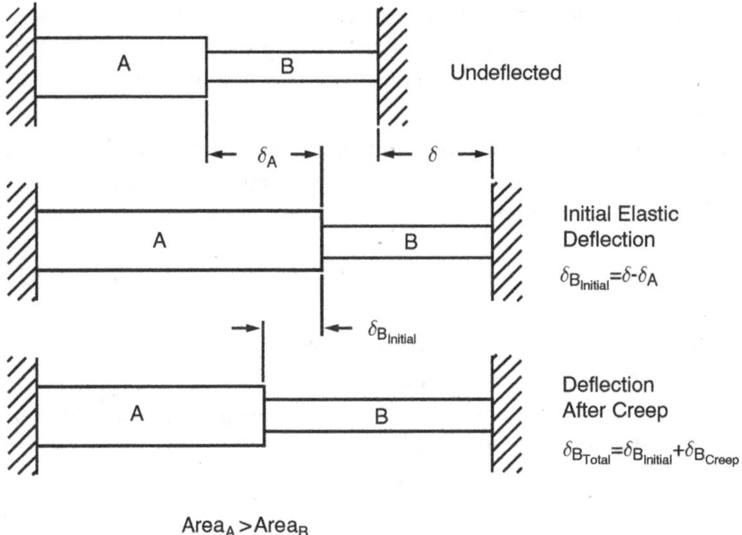

Figure 1.17 Two-bar model with elastic follow-up.

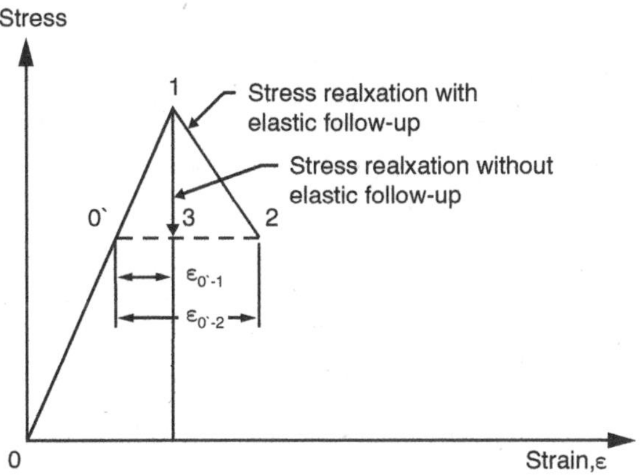

Figure 1.18 Definition of elastic follow-up.

schematically in Figure 1.18. The path 0–1 is the initial elastic loading and the path 1–2 shows the stress relaxation in time, *t*. If there were no elastic follow-up – i.e., the stresses in A and B were equal – there would be "pure" relaxation without follow-up along Path 1–3.

Also, if A is very much stiffer than B, such that the force in B will cause a negligible deformation in A, then there will also be pure relaxation in B. This is why the stress due to a local hot spot in a vessel wall is classified as a peak thermal stress.

One method for defining elastic follow-up is to compute the ratio of 0′-2, the creep strain in B at time, t, to the creep strain that would have occurred under pure relaxation, 0′-1. Thus, the elastic follow-up, q, is given by

$$q = \varepsilon_{0'-2} / \varepsilon_{0'-1}. \tag{1.11}$$

Note that, for $q = 1$ there is no follow-up, just pure relaxation, and if $q \rightarrow \infty$ the stress in B behaves as though loaded by a sustained load – i.e., load-controlled rather than displacement-controlled. For most geometries loaded in the displacement-controlled mode, $q = 2$ or less with a reasonable upper bound of $q \leq 3$. A more representative case of elastic follow-up is illustrated by Figure 1.19, a tube sheet connected to a shell at a different temperature. Elastic follow-up effects can increase the strain and stress levels at the tube-sheet-to-shell junction as shown in Figure 1.20, which shows a representative hysteresis loop with and without elastic follow-up.

The main consequence of elastic follow-up is to reduce the predicted cyclic life as compared to the life that would be predicted from an elastic analysis

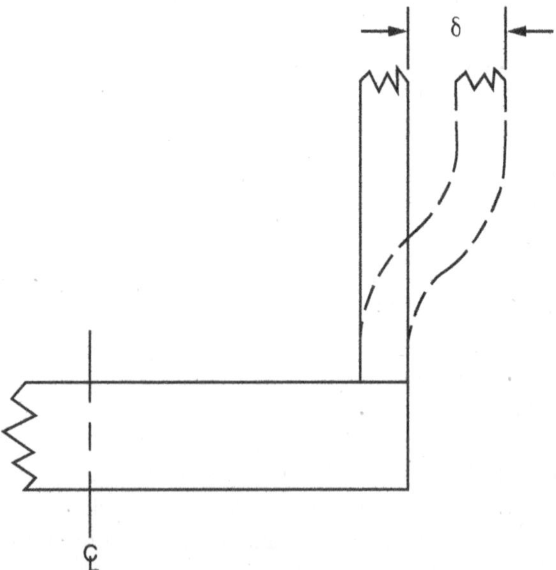

Figure 1.19 Tube-sheet-to-shell junction with relative deflection, δ, due to temperature.

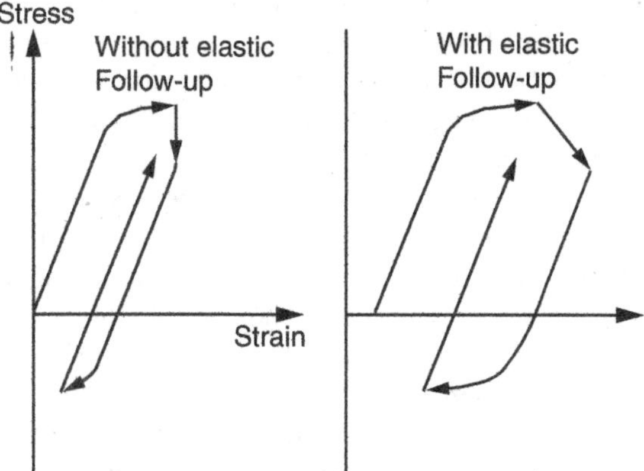

Figure 1.20 Hysteresis loop with and without elastic follow-up.

without consideration of elastic follow-up. This is due to two effects as shown in Figure 1.20. The actual strain range will be greater than that predicted by an unadjusted elastic analysis and the stress level will be higher due to the slowed rate of stress relaxation.

In general, there are two approaches to dealing with this problem. The first and most rigorous approach is to do a full inelastic analysis, which predicts the stresses and strain at critical points in the structure as a function of time. The disadvantage of this approach is that it requires complex models of material behavior, which may only have been established for a quite limited number of materials. These models require substantial judgment in their selection and use, and the actual computation times and effort involved in interpreting the results can be significant. However, in addition to general guidance on inelastic analysis, ASME III-5, Appendix Z, provides explicit guidance for development of constitutive equations and specific equations for 316H stainless steel, 9Cr-1Mo-V steel, and Alloy 617.

The second approach is based on elastic analysis without directly considering the effects of inelastic behavior. In this approach, adjustments are made to the elastic analysis results to compensate for the effects of inelastic behavior. The disadvantage of this approach is that the simpler methods tend to be overly conservative and the more complex methods can, themselves, be difficult to interpret and implement.

1.6.2 Pressure-Induced Discontinuity Stresses

In Section 1.5.2, the first step was to separate the structure into "free bodies" and compute the primary stresses in equilibrium with primary loads. The second step is to establish structural continuity by applying self-equilibrating loads to the boundaries of the "free body" segments. The stresses resulting from these self-equilibrating loads are called discontinuity stresses. This procedure for calculating discontinuity stresses is described in detail in Article A-6000 of the ASME III-1 Appendices.

At elevated temperature where creep is significant, it has been shown by analysis and experiment that the discontinuity stresses resulting from applied pressure do not relax as might be expected from self-equilibrating loads. Figure 1.21 Becht et al. (1989) is a comparison of the analytically predicted stress history for several structural configurations and loading conditions. A key comparison is Case #5 for a built-in cylinder (radial and rotational constraints at the edge) versus Case #1 for pure strain-controlled stress relaxation. After an initial redistribution of stress across the thickness, the discontinuity stress at the built-in edge is essentially constant, analogous to a primary stress. The explanation for this is that under creep conditions the "free body" segments of the structure undergo continuous deformation due to creep, with the resultant relative displacement continually increasing at the interfaces. This increasing relative displacement prevents relaxation of the interface loads and the consequent discontinuity stresses. Although this phenomenon does not exactly fit the elastic follow-up model, the resulting non-relaxing discontinuity

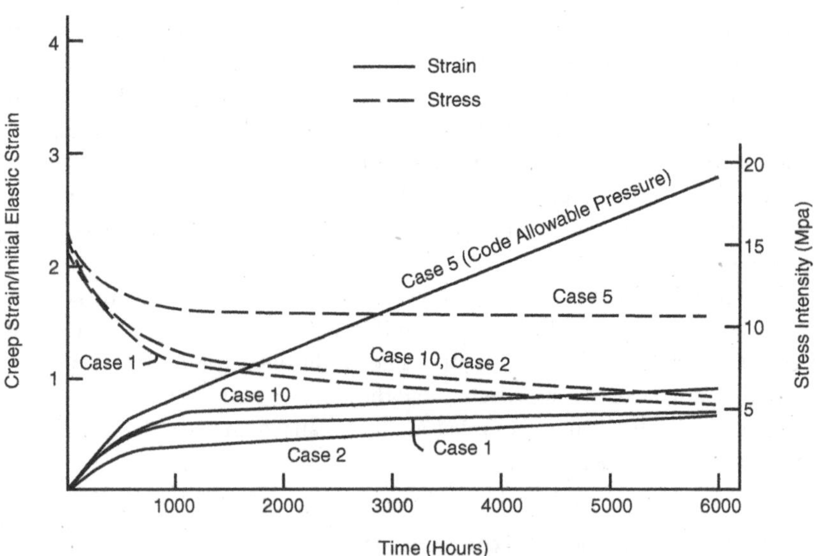

Figure 1.21 Stress relaxation and strain accumulation for pressure discontinuity (Case 5), thermal discontinuity (Case 10), and secondary stress (Cases 1 and 2). Rao (2018).

stress is analogous to the case where $q \to \infty$ and indicative of a sustained, non-relaxing load – e.g., a primary stress.

Pressure (and mechanical load)-induced discontinuity stresses do not affect the required wall thickness, but they can affect the strain and accumulated creep damage at structural discontinuities. Thus, in the rules in ASME III-5, and most other elevated temperature nuclear code criteria, these stresses are classified as primary when evaluating strain limits and creep-fatigue damage using the results of elastic analyses. If strain limits and creep-fatigue damage are evaluated using inelastic analysis while accounting for creep, this effect is automatically included.

For criteria based on Design-by-Rule, there are some restrictions that qualitatively address non-relaxing pressure-induced discontinuity stresses. For example, in the design rules of ASME VIII-1, this issue is addressed for a number of configurations by the requirement for a 3:1 taper when joining plates of unequal thickness (UW-9(c)) and head to shell joints (UW-13).

1.6.3 Shakedown and Ratcheting

Ratcheting can be described as progressive incremental deformation, and shakedown as the absence of ratcheting. A similar definition of shakedown is used in the criteria for ASME III, Class 1 and ASME VIII-2. In this case, shakedown is considered to occur when there is negligible plasticity after a few loading cycles. This later approach is illustrated by Figures 1.22a and b for elastic-plastic materials with no strain hardening or creep. Consider a tensile

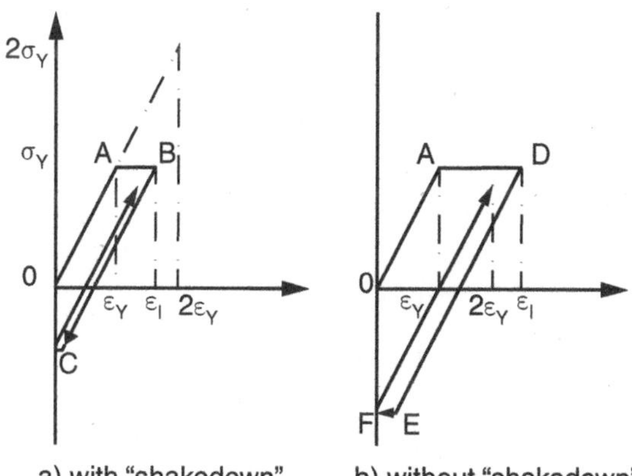

Figure 1.22 Stress and strain for cyclic strain-controlled loading without creep.

specimen that is strained in tension to a value ε_1, as shown in Figure 1.22a, which is somewhat greater than the strain at yield, ε_y, and less than twice ε_y. The initial loading will follow path OAB, initially yielding at A and continuing to plastically deform until the maximum tensile strain is reached at B. The unloading portion of the cycle consists of reversing the applied strain to the original staring point, O, following path BC without yielding. Subsequent loading for the same strain range, $0 \rightarrow \varepsilon_t \rightarrow 0$, will cycle along the path BC without yielding, hence the term "shakedown." If, on the other hand, the applied strain range is greater than twice the strain at yield, as shown in Figure 1.22b, the loading will follow the path OADEF and there will be yielding from E to F on the unloading cycle. Subsequent cycles of the same strain range will trace out a hysteresis loop with plasticity at each end of the cycle. What enables shakedown when the strain range, ε_t, is less than twice the strain at the yield strength is the establishment of a residual stress extending the strain range that can be achieved without yielding in cyclic strain-controlled loading. However, because the residual stress is limited to the yield strength, if the applied strain range exceeds twice the strain at yield there will be straining in tension and compression at either end of the cycle and shakedown will not occur.

In the creep regime, the residual stresses will relax and the strain range that can be achieved without yielding on each cycle will be reduced. There is, however, a quite useful elevated temperature analogy to the low-temperature shakedown concept. This is illustrated in Figures 1.23 and 1.25, which are plots of stress versus time, and Figures 1.24 and 1.26, which are the corresponding plots of stress versus strain for strain-controlled loading.

As in the preceding example without creep, in the case shown in Figures 1.23 and 1.24 the initial loading follows the path OAB but now the strain is held constant for a certain period and the stress relaxes to B′. The strain is then reversed to point 0, the initial starting point, reaching a compressive stress at

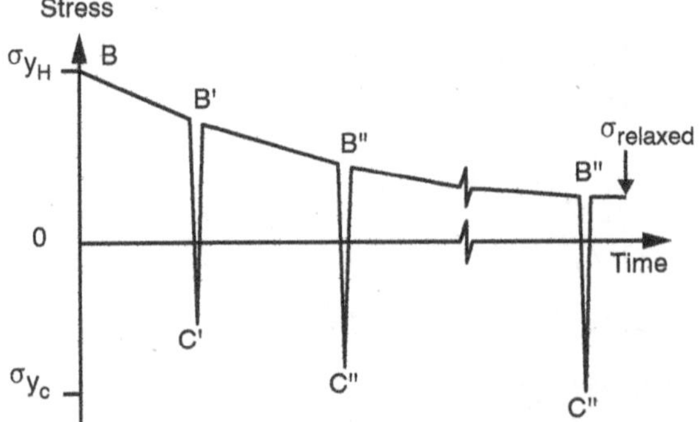

Figure 1.23 Stress history for cyclic strain control with creep and shakedown.

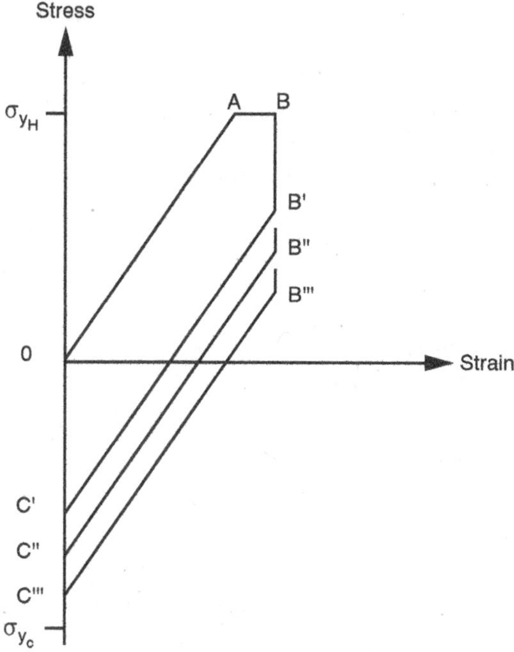

Figure 1.24 Stress and strain cyclic strain control with creep and shakedown.

C′. If the initial strain beyond yield is sufficiently limited, there will be no yielding when the applied strain is reversed and, assuming that there is no creep at the reversed end of the cycle (either the temperature is below the creep range or the duration is short), there will be no subsequent yielding upon reloading to point B′. During the next tensile hold portion of the cycle, the stress will relax to B″ and will subsequently reach C″ when the strain cycle is reversed. Under these conditions, the cyclic history will be as shown in Figures 1.23 and 1.24, and the criteria for shakedown will be that the stress range associated with the applied strain range does not exceed the yield stress at the cold (or short duration) end of the cycle plus the stress remaining after full relaxation of the yield stress over the life of the component. Thus, a criterion for "shakedown" in the creep regime becomes

$$\varepsilon_t \leq \sigma_{yc} / E_{\text{cold}} + \sigma_r / E_{\text{hot}} \tag{1.12}$$

where

E_{cold} and E_{hot} = the modulus of elasticity at the cold and hot ends of the cycle, respectively

ε_t = applied strain range

σ_{yc} = yield strength at the cold (or short time) end of the cycle, approximately equal to 1.5 S_{cold}

σ_r = relaxation strength, the stress remaining after the strain at yield strength is held constant for the total life under consideration. It can be conservatively approximated by 0.5 S_{hot}.

Equation (1.12) can also be expressed in terms of calculated stress as

$$(P_L + P_b + Q) \leq \sigma_{yc} + \sigma_r \tag{1.13}$$

where

P_b = primary bending stress intensity
P_L = primary local membrane stress intensity
Q = secondary stress.

and P_L, P_b, and Q are computed using the values of E which correspond to the operating condition being evaluated. Note that, generally, the use of E_{cold} will be conservative.

The above is a very useful concept for the design of elevated temperature components, as will be discussed in more detail in subsequent chapters, particularly Chapters 4 and 5.

Figures 1.25 and 1.26 illustrate the case where the strain range is large and there is yielding at the end of the strain unloading cycle. The load path goes from 0 to A with subsequent yielding until D. During the hold time at D the stress relaxes to E. This is followed by the unloading, or reversed strain portion, of the cycle that results in yielding at F until the cycle is completed at G. For the cycle shown there will be subsequent yielding on the tensile portion of the cycle and the stress at the start of the hold period, D, will equal the hot yield

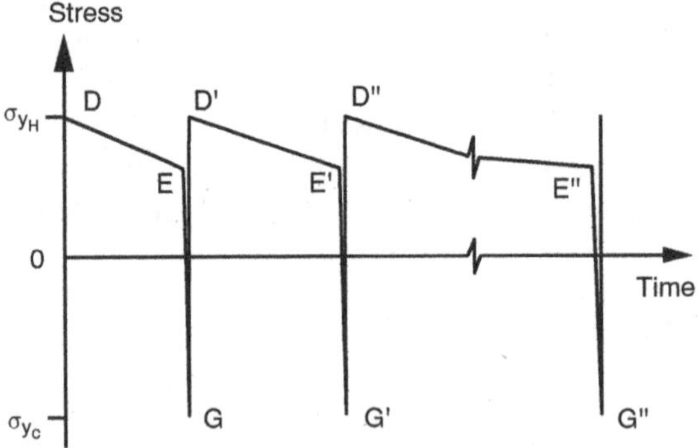

Figure 1.25 Stress history for cyclic strain control with creep and without shakedown.

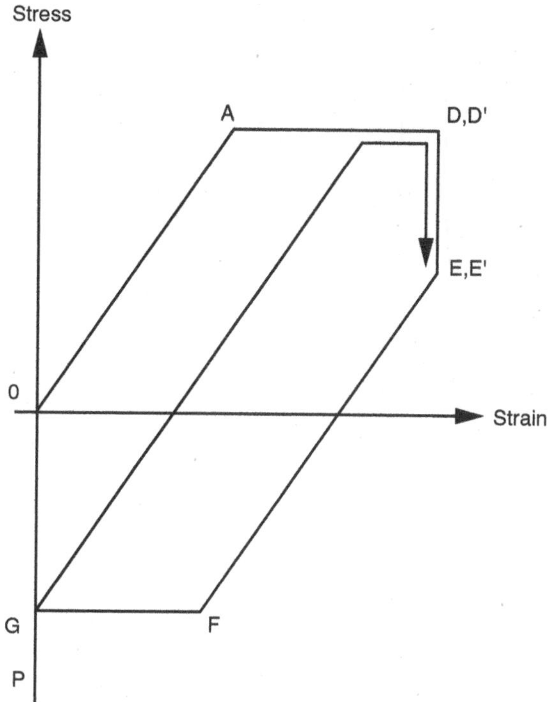

Figure 1.26 Cyclic strain control with creep and without shakedown.

strength. In effect, the creep damage for each cyclic hold time will be reinitiated at the yield strength, thus greatly increasing the creep damage, as compared to the previous case in which the stress at the start of the cycle is the same as the relaxed stress at the end of the previous hold time. In the case of the higher strain range, a hysteresis loop is established with creep strain and yielding in each cycle. One of the significant differences between the two cases is that, for shakedown, the creep damage is only that associated with monotonic stress relaxation throughout the life of the component. In the alternate case, where shakedown is not achieved, the creep damage is accumulated at a significantly higher stress level.

As noted above, another use for the term "shakedown" is to denote freedom from ratcheting – i.e., progressive incremental deformation. In the preceding example, the loading was considered to be a fully reversed strain-controlled cycle. However, in normal design practice there are both primary loads and secondary loads which, in combination, can cause ratcheting. (Purely displacement-controlled thermal stresses can also result in ratcheting, but these cases are usually associated with complete through-the-wall yielding.)

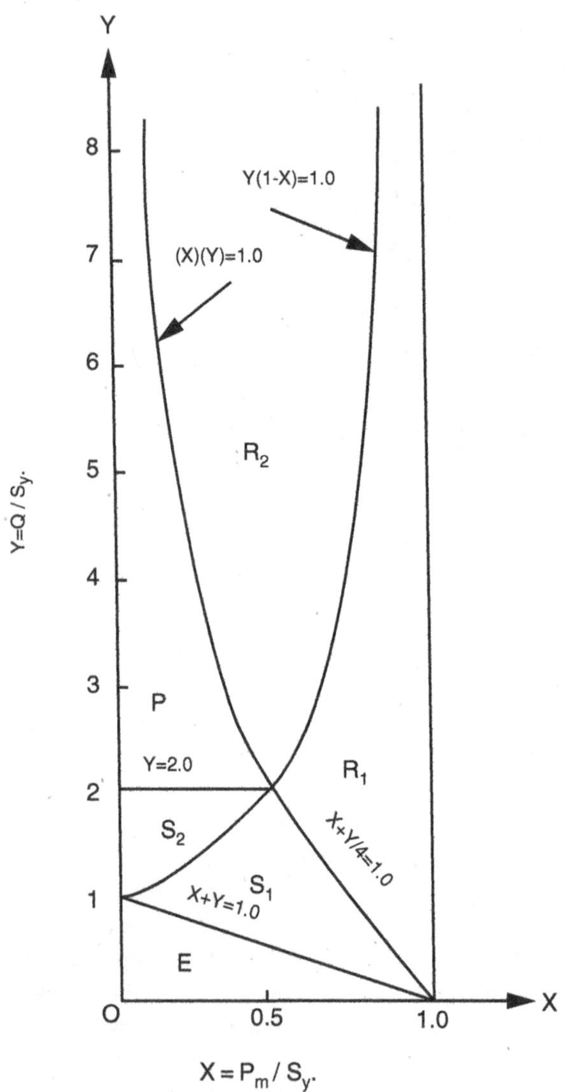

Figure 1.27 Bree diagram Bree (1967).

Bree (1968) evaluated potential ratcheting in cylinders under constant pressure and with a cyclic linear radial thermal gradient. He developed the relationships shown in Figure 1.27, which identify various regimes of behavior as a function of the relative magnitude of the elastically calculated thermal stress and pressure stress divided by the yield stress. Six areas of behavior were identified. Regions R_1 and R_2 resulted in ratcheting or incremental growth, even without creep. Region P resulted in plastic cycling in the absence of creep, and the regions S_1 and S_2 shook down to elastic action after one or two cycles – again in the absence of creep. When creep is considered only one region, E, resulted in structural behavior that could be considered as not being subject to incremental growth. Although useful, the simple approach of aiming to remain in the elastic region, E, can be too restrictive.

Based on a Bree-type model, O'Donnell and Porowski (1974) developed a less conservative approach to assess the strains accumulated under pressure (primary stresses) and cyclic thermal gradients (secondary stresses). This technique is a methodology for putting an upper bound on the strains that can accumulate due to ratcheting. The key feature of this technique is the identification of an elastic core in a component which is subjected to both primary and cyclic secondary loads. Once the magnitude of this elastic core has been established, the deformation of the component can be bounded by noting that the elastic core stress governs the net deformation of the section. Deformation in the ratcheting, R, regions of the Bree diagram can also be estimated by considering individual cyclic deformation. The resulting modifications to the basic Bree diagram are shown in Figure 1.28.

A much more comprehensive development of the Bree and O'Donnell/Porowski assessment of ratcheting is presented in Appendix D.

1.6.4 Fatigue and Creep-Fatigue

A very important consideration in elevated temperature design is the reduction in cyclic life due to the effects of creep. This is illustrated by Figure 1.29 showing the effects of hold time on the cyclic life of 304 stainless steel at 1100°F (595°C). The loading cycle constantly increases tensile strain, followed by a hold time at a fixed strain, and then a constantly decreasing strain back to the original starting point. This is referred to as a strain-controlled test, as compared to a load-controlled test in which the load is increased to a fixed level and then reversed. During the hold period at a fixed strain the specimen undergoes pure relaxation with no elastic follow-up. As can be seen from the figure, as the hold time increases the cycles to failure reduce. For a hold time of one hour, the reduction in life in this test is around a factor of 10, and it is not clear from the data if longer hold times will result in a further reduction in life. Fortunately, for most materials, as the hold time increases the stress relaxes and the rate damage accumulation slows until the effect essentially saturates. Such a saturation effect is shown in Figure 1.30, which is based on 304 stainless steel data at 1200°F (650°C). At this temperature, relaxation is fairly rapid and saturation occurs in the range of roughly one to 10 h, depending on the strain range.

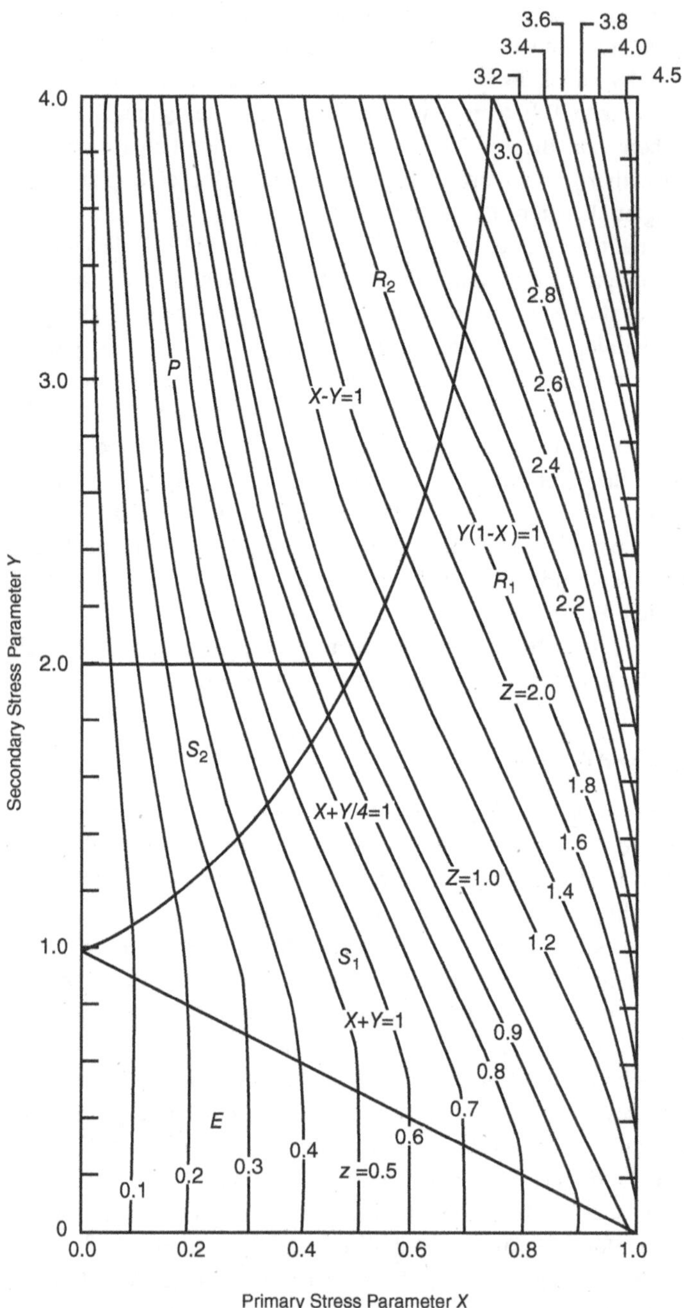

Figure 1.28 O'Donnell and Porowski's modified Bree diagram [ASME VIII-2].

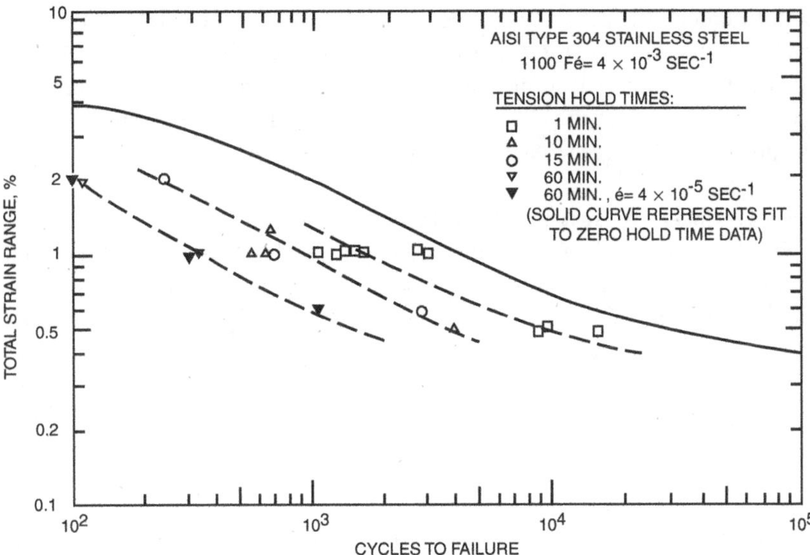

Figure 1.29 Effect of tension hold-time on the fatigue life of AISI Type 304 stainless steel at 1000°F (538°C) in air Weeks (1973).

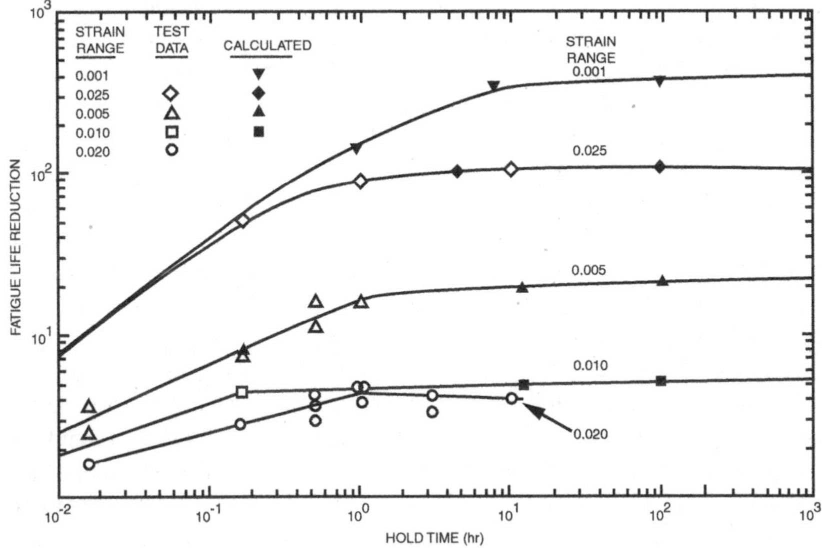

Figure 1.30 Hold-time effects on fatigue life reduction for 304 and 316 stainless steel ASME, December (1976).

Unfortunately, the actual loads encountered in design are rarely strain-controlled alone, because there are usually follow-up effects and non-relaxing stresses from primary loading conditions. The development of design methodologies to account for the effect of these varying loading mechanisms is one of the greatest challenges in elevated temperature design.

1.6.4.1 Linear Life Fraction – Time Fraction

A number of methods have been explored to correlate creep-fatigue test data and to evaluate cyclic life in design. The method chosen for III-5 is linear damage summation based on linear life fraction for creep damage and Miner's rule for fatigue damage

$$\sum(\Delta t / t_r) + \sum(n / N_f) \leq D \tag{1.14}$$

where

Δt = time at a given stress level
t_r = allowable time to rupture at that stress level
n = number of cycles of a given strain range
N_f = allowable number of cycles at that strain range
D = factor to account for the interaction of creep and fatigue damage.

In practice, safety factors are applied to the given stress level to determine a conservative time to rupture and to the allowable number of cycles.

For a design evaluation using the above relationship, it is necessary to determine the stresses and strain as a function of time at critical points in the structure. This is conceptually straightforward, using an inelastic analysis that models stress and strain behavior as a function of time. However, as previously noted, the disadvantage of this approach is that it requires complex models of material behavior, which may only have been established for a quite limited number of materials. These models require substantial judgment in their selection and use, and the actual computation times and effort involved in interpreting the results can be significant.

Alternatively, one can use elastic analysis results in mechanistic models to bound the stress-strain history, without directly considering the effects of inelastic behavior. In this approach, adjustments are made to the elastic analysis results to compensate for the effects of inelastic behavior. The disadvantage of this approach is that the simpler methods tend to be overly conservative and the more complex methods can, themselves, be difficult to interpret and implement.

1.6.4.2 Ductility Exhaustion

Another frequently used approach to the assessment of creep damage during cyclic loading is ductility exhaustion. In its simplest form, the combined effect of creep and fatigue damage may be expressed as

$$\sum(\Delta\varepsilon_c / d_r) + \sum\left(n / N_f\right) \leq D \quad (1.15)$$

where

$\Delta\varepsilon_c$ = strain increment
d_c = creep ductility
n = number of cycles of a given strain range
N_f = allowable number of cycles at that strain range
D_{cf} = factor to account for the interaction of creep and fatigue damage.

Most of the above comments regarding the application of the time fraction approach also apply to the ductility exhaustion approach. Conceptually, it should be easier to calculate creep damage via ductility exhaustion, as compared to time fractions, because the time to rupture is quite sensitive to calculated stress; by contrast, creep ductility can be relatively constant, depending on the material. However, in practice there have been numerous modifications to the ductility exhaustion approach to take into account the variations of creep ductility and the significance of when strain accumulation occurs during the loading cycle. Experimentally, some studies show better correlation for some materials using ductility exhaustion, depending on selected modifications, but there has not been a clear indication of universal applicability. The ductility exhaustion approach tends to see greater use as a damage assessment tool in failure analyses than as a design tool.

1.7 Buckling and Instability

There are two types of buckling that need to be considered: elastic or elastic-plastic, that may occur instantaneously at any time in life; and, creep buckling, which may be caused by enhancement of initial imperfections with time resulting in geometric instability. The essential difference between elastic and elastic-plastic buckling and creep buckling is that the former occurs with increasing load independent of time, whereas creep buckling is time-dependent and may occur even when loads are constant. Elastic and elastic-plastic buckling depends only on the geometric configuration and short-time material response at the time of application. Creep buckling occurs at loads below the elastic and elastic-plastic buckling loads as a result of creep strain accumulation over time.

The sensitivity of creep buckling to initial imperfections is illustrated by the deformation-time relationships shown in Figure 1.31. Although typical of the behavior of axially compressed columns and externally pressurized cylinders, these curves are representative of most structures. In general, a structural component will deviate initially from a perfect geometrical structure by some small amount. Under a system of loads, below those that would cause elastic or

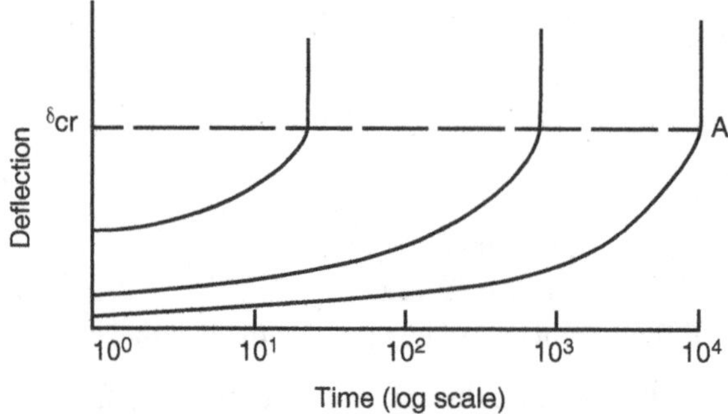

Figure 1.31 Creep buckling deflection-time characteristics of tubes with different ovalities *ASME, May (1976).*

inelastic instability, the initial deflection is magnified over time due to creep. The deflection increases until the geometrical configuration becomes unstable, as shown by point A in Figure 1.31, and buckling occurs.

ASME III-5 provides figures that define temperature and time limits within which creep effects need not be considered when evaluating buckling and instability.

Chapter 9 gives a more comprehensive coverage of creep buckling.

Problems

1.1 The maximum effective strain in the longitudinal weld of a steam drum is limited to 0.5%. What is the effective operating stress in the cylinder, from Figure 1.11, that will result in an expected life of

a. 100,000 h?

b. 300,000 h?

1.2 What is the allowable stress of the material shown in Figures 1.13 and 1.14 at 1100°F, based on creep and rupture criteria?

1.3 A pressure vessel component operating at 850°F has an expected life of 300,000 h. What is the expected life if the temperature is inadvertently raised to 900°F while maintaining the same stress level?

2

Axially Loaded Structural Members

Lifting lugs on channel sections [Nooter Construction, Inc].

Analysis of ASME Boiler, Pressure Vessel, and Nuclear Components in the Creep Range, Second Edition. Maan H. Jawad and Robert I. Jetter.
© 2022 John Wiley & Sons Ltd. Published 2022 by John Wiley & Sons Ltd.

2.1 Introduction

Axially loaded members subjected to elevated temperatures are encountered in many structures, such as internal cyclone hangers in process equipment, structural supports for internal trays, and pressure vessel supports. It is assumed that buckling, which is discussed in Chapter 9, is not a consideration in this chapter. Theoretically, analysis of uniaxially loaded members operating in the creep range follows Norton's relationship, which correlates the stress and strain rates in the creep regime, given by

$$d\varepsilon / dt = k'\sigma^n \tag{2.1}$$

where

k' = constant

n = creep exponent, which is a function of material property and temperature

$d\varepsilon/dt$ = strain rate

σ = stress.

This equation, however, is impractical for most problems encountered by the engineer. Its complexity arises from the nonlinear relationship between stress and strain rates. In addition, the equation has to be integrated to obtain strain, and thus deflections. A simpler method is normally used to solve uniaxially loaded members. This method, referred to as the "stationary stress" or "elastic analog" method, consists of using a viscoelastic stress-strain equation to evaluate stress due to creep rather than the more complicated creep equation, which relates the strain rate to stress. The viscoelastic equation is given by

$$\varepsilon = K'\sigma^n \tag{2.2}$$

where K' is a constant. The results obtained by the stationary stress, Eq. (2.2), are approximate but adequate for most pressure vessel engineering calculations. This method was rigorously discussed by Hoff (1958) and mentioned in numerous articles such as those by Hult (1966), Penny and Marriott (1995), and Finnie and Heller (1959). Hoff proved that Norton's Eq. (2.1) can be replaced with the classical viscoelastic stress-strain relationship of Eq. (2.2) for a wide range of structures encountered by the engineer when the following conditions and assumptions are satisfied:

- Creep strain can be interchanged with total strain.
- Primary strain is ignored and only secondary strain is considered.
- Strain obtained from Eq. (2.2) is numerically equal to the strain rate in Eq. (2.1).
- The material property is in accordance with Eq. (2.1).
- The stress field in the solid must remain constant with time. This is usually achieved after initial stress redistribution in an indeterminate structure due to load or temperature application.

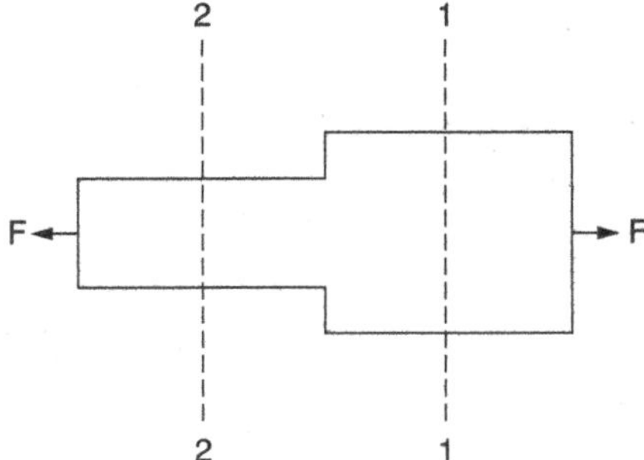

Figure 2.1 Axial tension member.

- Both the viscoelastic structure and the creep structure are loaded in the same manner and have the same boundary conditions.

The justification for using Eq. (2.2) in lieu of Eq. (2.1), with the stated assumptions, can be illustrated for the simple case shown in Figure 2.1. The stress at Sections 1-1 and 2-2 is given by

$$\sigma_1 = F / A_1 \quad \text{and} \quad \sigma_2 = F / A_2$$

where

A_1, A_2 = cross-sectional areas at locations 1-1 and 2-2
F = force
σ_1, σ_2 = stress at locations 1-1 and 2-2.

Integration of Eq. (2.1) with respect to time, at locations 1-1 and 2-2 and using the above expressions, gives

$$\varepsilon_1 = k'(F / A_1)^n \Delta t \quad \text{and} \quad \varepsilon_2 = k'(F / A_2)^n \Delta t \tag{2.3}$$

where

Δt = time increment.

The ratio of these two expressions is

$$\varepsilon_1 / \varepsilon_2 = (F / A_1)^n / (F / A_2)^n. \tag{2.4}$$

The same expression, given by Eq. (2.4), can also be obtained from Eq. (2.2) at these two locations. Thus, the relationship between stress and strain in a structure, subject to the assumptions made above, is the same whether calculated from Eq. (2.1) or Eq. (2.2). Accordingly, Eq. (2.2) is used because it is more practical to solve. The designer must realize that the strains obtained from Eq. (2.2) actually correspond to the strain rates obtained from Eq. (2.1).

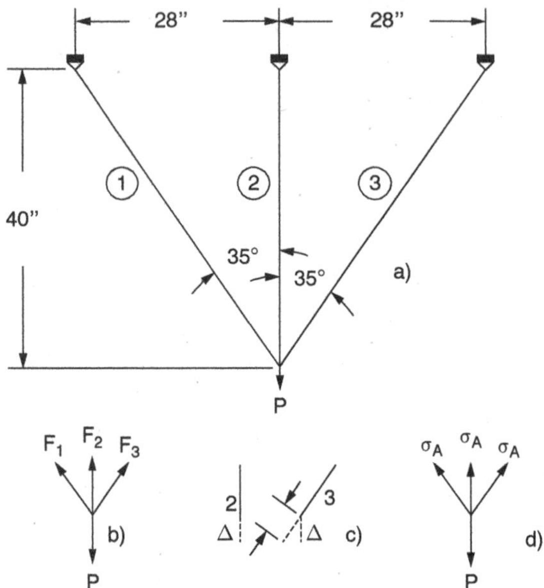

Figure 2.2 Structural frame.

The application of Eq. (2.2) to regular engineering problems, although simpler than Eq. (2.1), is still very complicated due to the nonlinear relationship between stress and strain. This is illustrated in the following example.

Example 2.1 What are the forces in the cyclone support rods shown in Figure 2.2a using (1) elastic analysis, (2) plastic analysis, and (3) creep analysis? Assume all members have the same cross-sectional area, A.

Solution

(a) Elastic analysis

Referring to Figure 2.2b, summation of the forces in the horizontal direction gives

$$F_1 = F_3. \tag{1}$$

Summation of forces in the vertical direction gives

$$F_2 = P - 1.638F_3. \tag{2}$$

From Figure 2.2c, the deflection, Δ, for Member 2 is expressed as

$$\Delta = F_2(40)/AE. \tag{3}$$

Table 2.1 F/P values for Example 2.1.

Analysis	n	F_1/P	F_2/P	F_3/P
Elastic		0.320	0.476	0.320
Creep	1	0.320	0.476	0.320
	2	0.350	0.427	0.350
	3	0.360	0.411	0.360
	4	0.365	0.403	0.365
	5	0.368	0.398	0.368
	10	0.373	0.389	0.373
	100	0.379	0.380	0.379
Plastic		0.379	0.379	0.379

From Figure 2.2c, the deflection, Δ, for Member 3 is expressed as

$$\Delta(\cos 35) = F_3(40/\cos 35)/AE$$

or

$$\Delta = 59.612F_3/AE. \tag{4}$$

Equating Eqs. (3) and (4) gives

$$F_2 = 1.490\ F_3. \tag{5}$$

And from Eqs. (1), (2), and (5)

$$F_1 = 0.320\ P,\quad F_2 = 0.476\ P,\quad F_3 = 0.320\ P.$$

These values are shown in the first row of Table 2.1.

(b) Plastic analysis

In a perfectly plastic analysis, it is assumed that all members have reached their yield stress value. The history of the structural plastic theory may be found in various books, such as Baker and Heyman (1969); Beedle (1966). Summation of the forces (Figure 2.2d) in the vertical direction gives

$$2\sigma A(\cos 35) + \sigma_A A = P$$

or

$$F_1 = F_2 = F_3 = 0.379P.$$

A comparison between the maximum force in parts (a) and (b) shows a 20% ((0.476–0.379)/0.426) reduction. The results of the plastic analysis are shown in the last row of Table 2.1.

(c) Creep analysis

Equations (1) and (2) using summation of forces are still applicable

$$F_1 = F_3$$

or

$$\sigma_1 = \sigma_3 \tag{6}$$

and

$$F_2 = P - 1.638F_3$$

or

$$\sigma_2 = P/A - 1.638\sigma_3. \tag{7}$$

Equations (3) and (4) cannot be used in creep analysis because the relationship $\Delta = FL/AE$ applies only in the elastic range. The strain in Members 3 and 2 are expressed as

$$\varepsilon_3 = \Delta(\cos 35)/L_3 = \Delta(\cos 35)/48.831 \tag{8}$$

and

$$\varepsilon_2 = \Delta/40.0. \tag{9}$$

The stationary stress Eq. (2.2) will have to be used to correlate stress and strain. Substituting Eqs. (7), (8), and (9) into Eq. (2.2) results in the following two expressions

$$\Delta(\cos 35)/48.831 = K'\sigma_3^n \tag{10}$$

and

$$\Delta/40 = K'(P/A - 1.638\sigma_3)^n. \tag{11}$$

Deleting Δ from these two equations gives

$$\sigma_3[1.638 + (1.490)^{1/n}] = P/A$$

and the forces are equal to

$$F_1 = [1.638 + (1.490)^{1/n}]^{-1}P$$
$$F_2 = \left\{1 - 1.638[1.638 + (1.490)^{1/n}]^{-1}\right\}P$$
$$F_3 = [1.638 + (1.490)^{1/n}]^{-1}P$$

These expressions are substantially more complex than those obtained from the elastic or plastic analysis. The complexity increases exponentially as the number of members and degrees of freedom increase to the point where it is impractical to use this manual approach to solve large problems.

The magnitudes of F_1 and F_2 for various n values are shown in Table 2.1. For $n = 1$, the values of F_1 and F_2 are the same as those obtained from an elastic analysis. Also, as n approaches infinity, the values of F_1 and F_2 approach those obtained from a plastic analysis. Most materials used for pressure vessel construction operating at elevated temperatures have an n value in the neighborhood of 2.5–10. Table 2.1 shows that the difference between the maximum member force having $n = 5$ and that obtained from plastic design is only about 9%. This difference reduces to about 5% for $n = 10$. Accordingly, many engineers in the boiler and pressure vessel area use plastic analysis to evaluate structures in the creep range, taking into consideration the accuracy of the analysis. Also, plastic analysis is much simpler to use compared to creep analysis. The designer can use Eq. (2.2), or a nonlinear finite element analysis, when a more accurate result is needed.

Table 2.1 also shows that high member forces obtained from an elastic analysis tend to reduce in magnitude due to creep, whereas low member forces tend to increase in magnitude due to creep. Accordingly, designing members with high forces obtained from an elastic analysis at elevated temperature results in a conservative design. Conversely, designing members with low forces obtained from an elastic analysis at elevated temperature results in an unconservative design.

It must also be kept in mind that both the creep and plastic analyses result in some member forces that change from tension to compression as the applied loads are reduced or eliminated during a cycle. Accordingly, special consideration must be given to bracing.

2.2 Stress Analysis

Table 2.1 shows the magnitude of the forces in a truss obtained by creep analysis converges to that obtained from plastic analysis as the creep exponent, n, increases. Accordingly, plastic analysis is often utilized for the solution of trusses in the creep regime since it is much easier to formulate plastic equations than creep equations. The finite element analysis is a convenient tool for obtaining the forces, thus stresses, in a truss operating in the creep regime. Elastic finite element analysis is commonly used to obtain a plastic solution by incrementing the applied loads as described below. The formulation of an elastic finite element analysis for a truss is shown next.

Figure 2.3 shows a typical truss member. Positive directions of forces, applied loads, joint deflections, and angles are shown in Figure 2.3. Member 1, shown in the figure, starts at Node 1 and ends at Node 2. At Node 1, forces P_1 and P_2, or deflections X_1 and X_2 in the x and y directions, may be applied as shown in the figure. Similarly, forces P_3 and P_4, or deflections X_3 and X_4, can also be applied at Node 2. The local member rotation angle, α, with respect to the global coordinate

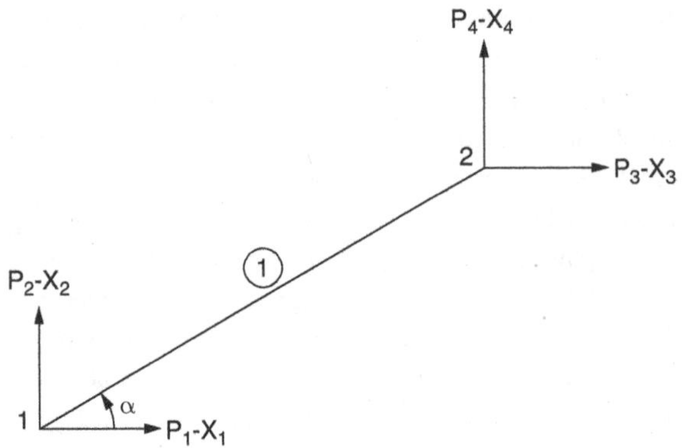

Figure 2.3 Sign convention for axial members.

system of the truss, is measured positive counterclockwise from the positive x direction at the beginning of the member. Tensile force is as shown in the figure.

The general relationship between applied joint loads and forces in a truss is given by

$$[P] = [K][X] \tag{2.5}$$

where

$[P]$ = applied loads at truss joints
$[K]$ = global stiffness matrix of truss
$[X]$ = deformation of truss joints.

The global stiffness matrix of the truss Weaver and Gere (1990) is assembled from the stiffness of each individual truss $[K]$. The stiffness matrix, $[K]$, of any member in a truss is expressed as

$$[k] = \begin{bmatrix} k_1 & k_2 & -k_1 & -k_2 \\ k_2 & k_3 & -k_2 & -k_3 \\ -k_1 & -k_2 & k_1 & k_2 \\ -k_2 & -k_3 & k_2 & k_3 \end{bmatrix} \tag{2.6}$$

where

$k_1 = AE\,(\cos^2\alpha)/L$
$k_2 = AE\,(\sin\alpha)(\cos\alpha)/L$
$k_3 = AE\,(\sin^2\alpha)/L.$

Equation (2.6) is a 4×4 matrix because there are a total of four degrees of freedom at the ends of the member. They are X_1, X_2, X_3, and X_4, or P_1, P_2, P_3, and P_4. The stiffness matrix of the truss is assembled by combining the stiffness matrices of all the individual members into one matrix, called the global matrix. Once the global stiffness matrix [K] is developed, Eq. (2.5) can be solved for the unknown joint deflections [X]. This assumes that the load matrix [P] at the nodal points is known.

$$[X]=[K]^{(-1)}[P]. \tag{2.7}$$

The forces in the individual members of the truss are then calculated from the equation

$$[F]=[D][X_{\mathrm{m}}] \tag{2.8}$$

where

[F] = force in member
[D] = force-deflection matrix of a member
[X_{m}] = matrix for deflection at ends of individual members.

The matrix [D] is defined as

$$[D]=[-D_1 \; -D_2 \; D_1 \; D_2] \tag{2.9}$$

where

$$D_1 = AE(\cos\alpha)/L$$
$$D_2 = AE(\sin\alpha)/L$$

Analysis of a structure in the creep regime consists of performing a finite element analysis through a number of cycles. In Cycle 1 a unit load, P_1, is applied to the truss and the forces in the members using Eqs. (2.5) through (2.9) are calculated. The expression for the force in the member with the largest resultant force is

$$F_1 = K_{11}P_1 \tag{2.10}$$

where

F_1 = force in Member 1 having the largest force magnitude
K_{11} = force coefficient in Member 1 obtained from finite element analysis in Cycle 1 with applied unit load P_1
P_1 = applied unit load in Cycle 1.

The force in the member with the second largest resultant force is

$$F_2 = K_{21}P_1 \tag{2.11}$$

where

F_2 = force in Member 2 having the second largest force magnitude
K_{21} = force coefficient in Member 2 obtained from finite element analysis in Cycle 1 with applied unit load P_1.

The force in the member with the third largest resultant force is

$$F_3 = K_{13}P_1, \text{ and so forth} \tag{2.12}$$

where

F_3 = force in Member 3 having the third largest force magnitude
K_{31} = force coefficient in Member 3 obtained from finite element analysis in Cycle 1 with applied unit load P_1.

The stress in member F_1 is

$$S = F_1 / A \tag{2.13}$$

where

A = cross-sectional area of member
S = limiting stress.

Combining Eqs. (2.10) and (2.13) gives

$$P_1 = (A)(S) / K_{11}. \tag{2.14}$$

In Eq. (2.14) it is assumed that Member 1 has reached its maximum load capacity and cannot sustain additional loading. Member 2 can still take an additional force of magnitude $P_1(K_{11}–K_{21})$. Member 3 still has a reserve force of $P_1(K_{11}–K_{31})$, and so forth.

In the second cycle, Member 1 is eliminated from the truss since it cannot carry any further force. This is accomplished by setting the stiffness matrix for Member 1 to zero in Eq. (2.6). A unit load P_2 is then applied to the truss and the forces in the remaining members are calculated. Their magnitude is $F_2 = K_{22}P_2$, $F_3 = K_{32}P_2$, etc.,

where

F_2 = force in Member 2
F_3 = force in Member 3
K_{22} = force coefficient in Member 2 obtained from finite element analysis in Cycle 2 with applied unit load P_2
K_{32} = force coefficient in Member 3 obtained from finite element analysis in Cycle 2 with applied unit load P_2
P_2 = applied unit load in Cycle 1.

The additional stress that can be sustained by member F_2 is

$$\left[(K_{22})(P_2)/(A)\right] = S\left[1-(K_{21}/K_{11})\right]$$

or

$$P_2 = (A)(S)\left[1-(K_{21}/K_{11})\right]/K_{22}. \tag{2.15}$$

In the third cycle Members 1 and 2 are eliminated from the truss since they cannot carry any further forces. This is accomplished by setting their stiffness matrix to zero in Eq. (2.6). A unit load P_3 is then applied to the truss and the forces in the remaining members are calculated. This process is repeated until the truss becomes unstable and no further cycles are possible. The total number of applied unit loads are added and equated to the total load in the structure

$$P_1 + P_2 + P_3 + \ldots. = P.$$

Substituting for the values of P_i Eqs. (2.14), (2.15), etc.

$$(A)(S)/K_{11} + (A)(S)\left[1-(K_{21}/K_{11})\right]/K_{22} + \ldots. = P. \quad (2.16)$$

The only unknown in Eq. (2.16) is the area A which can be determined directly.

The following example illustrates this procedure.

(It should be noted that this methodology works for a support system where the rods are of relatively similar stiffness and adequate ductility. If one rod is substantially stiffer than the others, it will carry most of the initial load and can rupture before it sheds load to the other rods. This is similar to the discussion of reference stress concepts in Chapter 1.)

Example 2.2 An internal cyclone is supported by three braces as shown in Figure 2.4a. All braces have the same cross-sectional area. The total weight of the cyclone and its contents is 10,000 lb. Determine the forces in the braces and their size. The material of the braces is 304 stainless steel. The design temperature is 1450°F and the modulus of elasticity is 18,700 ksi. The allowable stress in accordance with ASME VIII-1 is 1800 psi at 1450°F and 20,000 psi at room temperature.

Solution

Cycle 1

Because all members have the same cross section, they will be assumed to have a cross-sectional area of 1.0 in.^{2}. The actual area will be determined when the member forces in the truss are obtained from the analysis. To simplify the discussion in this chapter, and to compare the results with subsequent examples, the braces are assumed to be fabricated from round rods. In actual practice, pipes, channels, or angles are used.

The load on the truss and the angle of orientation of each truss member are shown in Figure 2.4b. The global stiffness is obtained by combining all local member stiffnesses.

- Member I (Nodal Points 1, 2, 3, and 4)
  $\alpha = 60°$, $L = 69.28$ in. From Eq. (2.6), the stiffness matrix for Member 1 is

a)

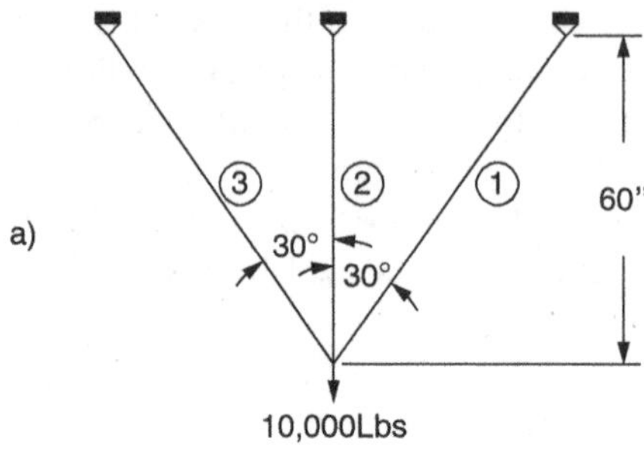

b)

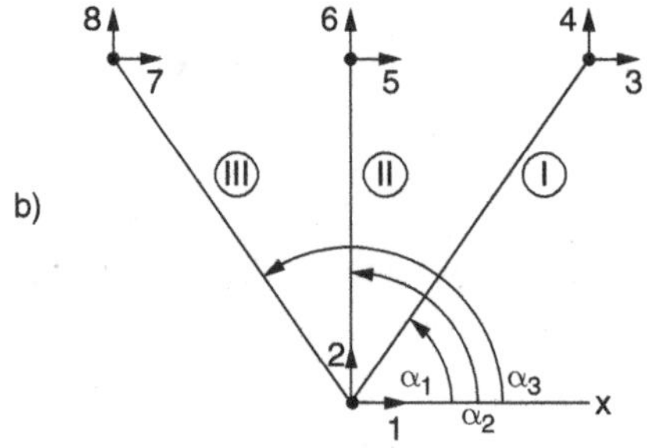

Figure 2.4 Cyclone support frame.

$$[K]_I = \begin{bmatrix} 67,480 & 116,879 & -67,480 & -116,879 \\ & 202,439 & -116,879 & -202,439 \\ & & 67,480 & 116,879 \\ \text{symmetric} & & & 202,439 \end{bmatrix}$$

- Member II (Nodal Points 1, 2, 5, and 6)
  $\alpha = 90°$, $L = 60.00$ in. From Eq. (2.6), the stiffness matrix for Member 2 is

$$[K]_{II} = \begin{bmatrix} 0 & 0 & 0 & 0 \\ & 311,667 & 0 & -311,667 \\ & & 0 & 0 \\ \text{symmetric} & & & 311,667 \end{bmatrix}$$

- Member III (Nodal Points 1, 2, 7, and 8)
  $\alpha = 120°$, $L = 69.28$ in. From Eq. (2.6), the stiffness matrix for Member 3 is

$$[K]_{III} = \begin{bmatrix} 67,480 & -116,879 & -67,480 & 116,879 \\ & 202,439 & 116,879 & -202,439 \\ & & 67,480 & -116,879 \\ \text{symmetric} & & & 202,439 \end{bmatrix}.$$

Combining the above three matrices, and deleting the terms that pertain to Nodal Points 3 through 8 in Figure 2.4b because they are attached to the head, and hence fixed against movement, gives

$$[K]=\begin{bmatrix}134{,}960 & 0\\ 0 & 716{,}545\end{bmatrix}.$$

The load matrix based on an applied unit load is

$$[P]=\begin{bmatrix}0.0\\ -1.0\end{bmatrix}.$$

The deflections at locations 1 and 2 are obtained from Eq. (2.7) as

$$\begin{bmatrix}X_1\\ X_2\end{bmatrix}=\begin{bmatrix}0.0\\ -1.39x10^{-6}\end{bmatrix}.$$

Then, from Eq. (2.8), the member forces are

$$\begin{bmatrix}F_1\\ F_2\\ F_3\end{bmatrix}=\begin{bmatrix}0.326\text{ lb}\\ 0.435\text{ lb}\\ 0.326\text{ lb}\end{bmatrix}.$$

The largest force is in Member 2. From Eq.(2.14),

$$\begin{aligned}P_1&=(A)(1800)/0.435\\ &=4137.9A\end{aligned}.$$

Cycle 2

The stiffness matrices for Members I and III are the same as those in Cycle 1 while the stiffness matrix for Member II is set to zero since the member cannot carry any further load. Hence the total matrix for Cycle 2 is

$$[K]=\begin{bmatrix}134{,}960 & 0\\ 0 & 404{,}878\end{bmatrix}.$$

The applied load matrix is

$$[P]=\begin{bmatrix}0.0\\ -1.0\end{bmatrix}.$$

The deflections at locations 1 and 2 are obtained from Eq. (2.7) as

$$\begin{bmatrix}X_1\\ X_2\end{bmatrix}=\begin{bmatrix}0.0\\ -2.470x10^{-6}\end{bmatrix}.$$

Then from Eq. (2.12) the member forces are

$$\begin{bmatrix} F_1 \\ F_2 \\ F_3 \end{bmatrix} = \begin{bmatrix} 0.577 \text{ lb} \\ 0 \\ 0.577 \text{ lb} \end{bmatrix}.$$

From Eq. (2.15)

$$P_2 = (A)(1800)\left[1-(0.326/0.435)\right]/0.577.$$
$$= 781.7A$$

From Eq. (2.16)

$$P_1 + P_2 = P$$
$$4137.9A + 781.7A = 10{,}000$$

or

$$A = 2.03 \text{ in.}^2.$$

Hence, the required diameter of the rods operating in the creep regime is

$$d = [(2.03)(4)/\pi]^{0.5} = 1.61 \text{ in.}$$

It is of interest to note that the required area using elastic analysis is obtained from Cycle 1 as

$$A = (0.435)(10{,}000)/1800 = 2.42 \text{ in.}^2.$$

Hence, performing a plastic analysis to simulate creep analysis results in about a 16% reduction in the required area ((2.42 – 2.03)/2.42).

2.3 Design of Structural Components Using ASME I and VIII-1 as a Guide

The design of axially loaded structural components at elevated temperatures in ASME I and VIII-1 construction is straightforward. These two codes require an elastic analysis with the allowable stress obtained from ASME II-D. At elevated temperatures, the allowable stress in ASME II-D is based on the creep and rupture criteria discussed in Chapter 1. Hence, at elevated temperatures, ASME I and VIII-1 permit elastic analysis with the allowable stresses obtained from creep and rupture data. This approach, although commonly used, has many drawbacks, which include a lack of limits on strain and deformation due to creep and ignoring thermal stresses that may lead to excessive strain in the creep regime due to cyclic conditions.

As described in Section 1.6.3, there is a quite useful concept in the creep regime that is analogous to the shakedown concept below the creep regime. The resultant "shakedown" concept in the creep regime is given by Eq. (1.12) as

$$\varepsilon_t \leq \sigma_{yc} / E_{cold} + \sigma_r / E_{hot} \tag{2.17}$$

where

E_{cold} and E_{hot} = modulus of elasticity at the cold and hot ends of the cycle, respectively
ε_t = applied strain range
σ_{yc} = yield strength at the cold (or short time) end of the cycle; it is approximately equal to 1.5 S_{cold}
σ_r = relaxation strength, the stress remaining after the strain at yield strength is held constant for the total life under consideration; it can be conservatively approximated by $0.5S_{hot}$.

Equation (2.17) can also be expressed in terms of calculated stress as

$$(P_L + P_b + Q) \leq \sigma_{yc} + \sigma_r \tag{2.18}$$

where

P_b = primary bending equivalent stress
P_L = primary local membrane equivalent stress
Q = secondary equivalent stress.

And P_L, P_b, and Q are computed using the values of E corresponding to the operating condition being evaluated. Note that, in general, the use of E_{cold} will be conservative. For axial members, P_b is zero and Eq. (2.18) becomes

$$(P_L + Q) \leq (1.5S_C + S_H / 2). \tag{2.19}$$

Where, S_c and S_H are the allowable stress at the cold and hot ends of the cycle. It is of interest to note that the axial equation for stresses below the creep range is given in ASME VIII-2 creep rules as

$$\left(P_L + Q\right) \leq \left(1.5S_C + 1.5S_H\right). \tag{2.20}$$

Equation (2.19) was developed by ASME with the intention of assuring shakedown after a few cycles. Comparison of the creep (2.19) and ASME VIII-2 Eq. (2.20) shows them to be similar, with the exception of the last term.

2.4 Temperature Effect

The effect of temperature variation on the member forces in a truss can easily be incorporated into the matrix equations as well. The procedure consists of calculating the following equivalent joint load for members that have a change in temperature

$$F = (\alpha)(\Delta T)(A)(E) \tag{2.21}$$

where

A = cross-sectional area of member
E = modulus of elasticity
F = equivalent member force
α = coefficient of thermal expansion
(ΔT) = change of temperature in member.

The equivalent F force is then entered as a nodal force and the truss nodal deflections are calculated from Eq. (2.7). The final forces are those calculated from Eq. (2.8) minus the quantity from Eq. (2.21) for those members that have temperature changes.

For axial members, it is assumed that the mechanical stresses are primary in nature and thermal stresses are secondary in accordance with the general criteria of ASME VIII-2 Part 5. The analysis procedure is illustrated in the following example.

Example 2.3 Member 2 in Figure 2.4 is subjected to a 25°F increase in temperature excursion. What are the forces in all members due to this temperature increase? Let $\alpha = 10.8 \times 10^{-6}$ in./in.°F, $E = 18{,}700$ ksi. Assume the area for all members to be $A = 2.42$ in.^{2}. The yield stress is 30,000 psi at room temperature and 10,600 psi at 1450°F. The allowable stress is 20,000 psi at room temperature and 1800 psi at 1450°F.

Solution
The global stiffness is obtained from Cycle 1 of Example 2.2 by multiplying the stiffness $[K]$, which is based on an area of members equal to 1.0 in.^{2} by 2.42, which is the area of the members in this problem. The result is

$$[K] = \begin{bmatrix} 326{,}600 & 0 \\ 0 & 1{,}734{,}000 \end{bmatrix}.$$

From Eq. (2.21)

$$\begin{aligned} F &= \left(10.8 \times 10^{-6}\right)(25)(2.42)(18{,}700{,}000) \\ &= 12{,}220 \text{ lb} \end{aligned}$$

and the applied load matrix is

$$[P] = \begin{bmatrix} 0 \\ -12{,}220 \end{bmatrix}.$$

The deflections at locations 1 and 2 are obtained from Eq. (2.7) as

$$\begin{bmatrix} X_1 \\ X_2 \end{bmatrix} = \begin{bmatrix} 0 \\ -0.705 \times 10^{-2} \end{bmatrix}.$$

Then, from Eq. (2.8), the member forces are

$$\begin{bmatrix} F_1 \\ F_2 \\ F_3 \end{bmatrix} = \begin{bmatrix} 3987 \\ -6905 \\ 3987 \end{bmatrix} \text{lb}$$

and the corresponding thermal stress, based on a cross-sectional area of 2.42 in.^{2}, is

$$\sigma = \begin{bmatrix} 1650 \\ -2850 \\ 1650 \end{bmatrix} \text{psi}.$$

It is of interest to note that a small temperature increase in one of the members results in member stresses that are approximately of the same magnitude as those obtained from a large applied mechanical load. This condition illustrates the importance of maintaining a constant temperature, whenever possible, in all members during operation.

The total sum of mechanical stresses, P_L, from Cycle 1 of Example 2.2 and thermal stress, Q, from above is

$$(P_L + Q) = \begin{bmatrix} 3000 \\ -1050 \\ 3000 \end{bmatrix} \text{psi}.$$

The ASME VIII-2 criteria for acceptable stress levels require that

1) For temperatures below the creep range, the mechanical plus thermal stress are less than three times the allowable stress, or two times the average yield stress, whichever is greater (ASME VIII-2 Part 5). The allowable and yield stress are taken as the average value at the high and low temperature extremes of the cycle. This criterion, given by Eq. (2.20), is to ensure stress shakedown.
2) For temperatures above the creep range, the mechanical plus thermal stress is less than three times the allowable stress (ASME VIII-2 creep rules). The allowable stress is taken as the average value at the high and low temperature extremes of the cycle. This criterion, given by Eq. (2.19), is to ensure stress creep.

In this example the temperature at one extreme of the cycle is in the creep range and Eq. (2.19) is applicable. The right-hand side of this equation is

$$1.5(20{,}000) + (1800)/2 = 30{,}900 \text{ psi.}$$

All $(P_L + Q)$ values are well below the allowable stress of 30,900 psi.

2.5 Design of Structural Components Using ASME I, III-5, and VIII as a Guide – Creep Life and Deformation Limits

Presently, ASME III-5 has no rules for design of structural members at elevated temperature. Structural members and structural supports in ASME III are presently designed in the noncreep range. Axially loaded members in ASME I and VIII-1 are occasionally required to be analyzed for fatigue and creep at elevated temperatures. The analysis is usually based on commonly accepted practice because there are no published standards that cover this area. The procedure for assessing creep rupture life of a structural member is:

1) Calculate the strain corresponding to the maximum stress in the member due to mechanical and thermal stresses in accordance with the equation

$$\Delta\varepsilon = 2S_{alt}/E \tag{2.22}$$

where

E = modulus of elasticity at the maximum metal temperature experienced during the cycle

$2S_{alt}$ = maximum stress range during the cycle due to mechanical and thermal loads

$\Delta\varepsilon$ = maximum equivalent strain range.

2) Calculate the fatigue ratio

$$\Sigma(n / N_{\mathrm{f}})_{\mathrm{j}} \tag{2.23}$$

where

$(n)_{\mathrm{j}}$ = number of applied repetitions of cycle type j
$(N_{\mathrm{f}})_{\mathrm{j}}$ = number of design allowable cycles for cycle type j, obtained from a design fatigue table such as that shown in Table 2.2.

3) Calculate the mechanical stress P_{L}.
4) Enter an isochronous chart similar to the one shown in Figure 2.5 with stress P_{L} and determine the strain corresponding to the number of expected life hours. The obtained strain should not exceed 1%.
5) Obtain the number of hours, t_{m}, corresponding to mechanical stress, P_{L}, from a stress-to-rupture table similar to the one shown in Table 2.3.

Table 2.2 Design fatigue strain range, ε_{t}, for 304 stainless steel [ASME II-D].

Number of cycles, N_d [Note (1)]	(ε_t in./in.) Strain range at temperature US customary units						
	100°F	800°F	900°F	1000°F	1100°F	1200°F	1300°F
10	0.051	0.050	0.0465	0.0425	0.0382	0.0335	0.0297
20	0.036	0.0345	0.0315	0.0284	0.025	0.0217	0.0186
40	0.0263	0.0246	0.0222	0.0197	0.017	0.0146	0.0123
10^2	0.018	0.0164	0.0146	0.0128	0.011	0.0093	0.0077
2×10^2	0.0142	0.0125	0.011	0.0096	0.0082	0.0069	0.0057
4×10^2	0.0113	0.00965	0.00845	0.00735	0.0063	0.00525	0.00443
10^3	0.00845	0.00725	0.0063	0.0055	0.0047	0.00385	0.00333
2×10^3	0.0067	0.0059	0.0051	0.0045	0.0038	0.00315	0.00276
4×10^3	0.00545	0.00485	0.0042	0.00373	0.0032	0.00263	0.0023
10^4	0.0043	0.00385	0.00335	0.00298	0.0026	0.00215	0.00185
2×10^4	0.0037	0.0033	0.0029	0.00256	0.00226	0.00187	0.00158
4×10^4	0.0032	0.00287	0.00254	0.00224	0.00197	0.00162	0.00138
10^5	0.00272	0.00242	0.00213	0.00188	0.00164	0.00140	0.00117
2×10^5	0.0024	0.00215	0.0019	0.00167	0.00145	0.00123	0.00105
4×10^5	0.00215	0.00192	0.0017	0.0015	0.0013	0.0011	0.00094
10^6	0.0019	0.00169	0.00149	0.0013	0.00112	0.00098	0.00084

Note: (1) Cyclic strain rate: 1×10^{-3} in./in./sec. (1×10^{-3} m/m/s).

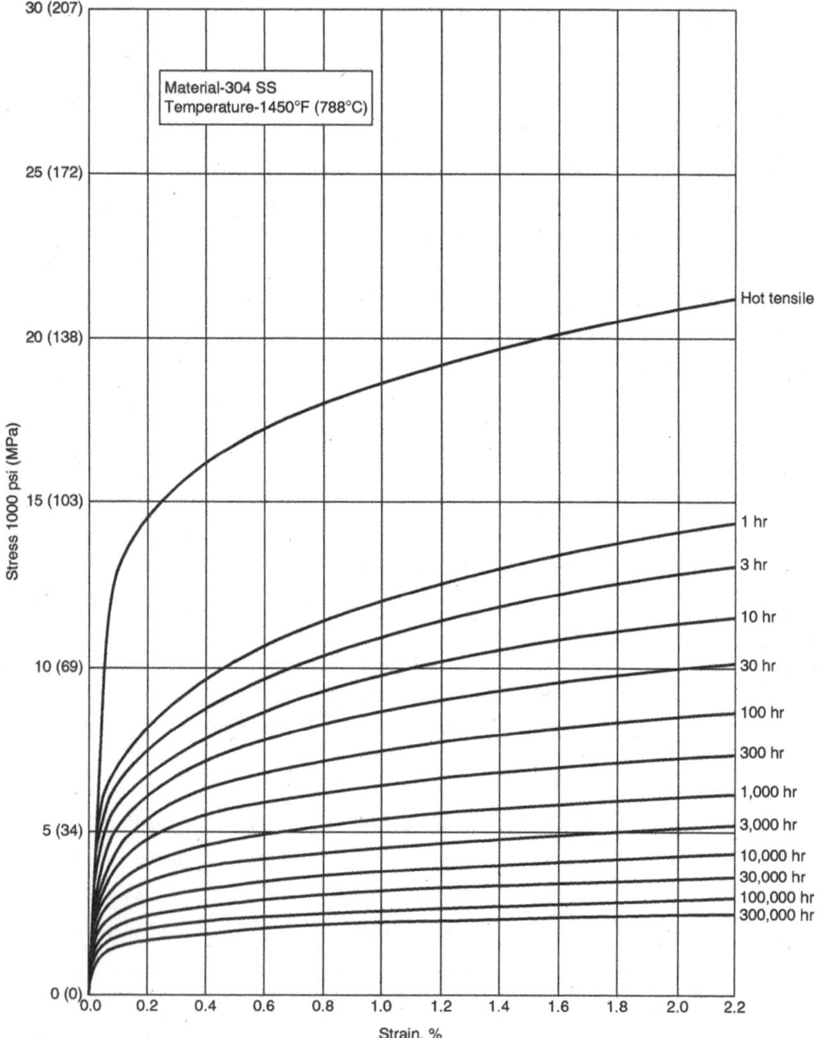

Figure 2.5 Average isochronous stress-strain curves for type 304 stainless steel at 1450°F (788°C) [ASME II-D].

6) Calculate the rupture ratio

$$\sum \left(t_r \,/\, t_m \right)_k \tag{2.24}$$

where

t_r = number of required time for the part.

t_m = time to rupture.

Table 2.3 Expected minimum stress-to-rupture values, 1000 psi, Type 304 stainless steel [ASME II-D].

US customary units

Temp., °F	1 h	10 h	30 h	10^2 h	3 × 10^2 h	10^3 h	3 × 10^3 h	10^4 h	3 × 10^4 h	10^5 h	3 × 10^5 h
800	57.0	57.0	57.0	57.0	57.0	57.0	57.0	57.0	51.0	44.3	39.0
850	56.5	56.5	56.5	56.5	56.5	56.5	50.2	45.4	40.0	34.7	30.5
900	55.5	55.5	55.5	55.5	51.5	46.9	41.2	36.1	31.5	27.2	24.0
950	54.2	54.2	51.0	48.1	43.0	38.0	33.5	28.8	24.9	21.2	18.3
1000	52.5	50.0	44.5	39.8	35.0	30.9	26.5	22.9	19.7	16.6	14.9
1050	50.0	41.9	37.0	32.9	28.9	25.0	21.6	18.2	15.5	13.0	11.0
1100	45.0	35.2	31.0	27.2	23.9	20.3	17.3	14.5	12.3	10.2	8.6
1150	38.0	29.5	26.0	22.5	19.3	16.5	13.9	11.6	9.6	8.0	6.6
1200	32.0	24.7	21.5	18.6	15.9	13.4	11.1	9.2	7.6	6.2	5.0
1250	27.0	20.7	17.9	15.4	13.0	10.8	8.9	7.3	6.0	4.9	4.0
1300	23.0	17.4	15.0	12.7	10.5	8.8	7.2	5.8	4.8	3.8	3.1
1350	19.5	14.6	12.6	10.6	8.8	7.2	5.8	4.6	3.8	3.0	2.4
1400	16.5	12.1	10.3	8.8	7.2	5.8	4.7	3.7	3.0	2.3	1.9
1450	14.0	10.2	8.8	7.3	5.8	4.6	3.8	2.9	2.3	1.8	1.4
1500	12.0	8.6	7.2	6.0	4.9	3.8	3.0	2.4	1.8	1.4	1.1

7) The fatigue ratio calculated in Step 2 and the rupture ratio calculated in Step 6 are combined in accordance with the following creep-fatigue equation

$$\left[\sum\left(n/N_f\right)_j+\sum\left(t_r/t_m\right)_k\right]\le D_{\text{cf}}.$$

where

D_{cf} = total creep fatigue damage factor obtained from Figure 2.6.

The curves in Figure 2.6, also shown in ASME VIII-2, can be represented by the following equations

For 304 stainless steel, 316 stainless steel, and 9Cr-1Mo-V steel

$$\sum(\Delta t/t_d)=-2.333\left[\sum\left(n/N_d\right)\right]+1 \text{ for } 0<\sum\left(n/N_d\right)\le 0.3. \qquad (2.26)$$

$$\sum(\Delta t/t_d)=-0.429\left[\sum\left(n/N_d\right)\right]+0.429 \text{ for } 0.3<\sum\left(n/N_d\right)\le 1.0. \qquad (2.27)$$

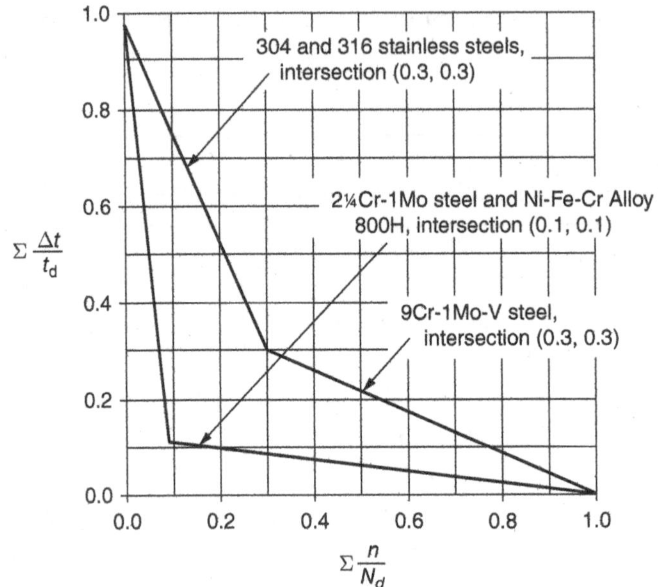

Figure 2.6 Creep-fatigue damage envelope [ASME VIII-2].

For Nickel alloy 800H and 2.25Cr-1Mo steel

$$\sum(\Delta t/t_d) = -9\left[\sum\left(n/N_d\right)\right] + 1 \text{ for } 0 < \sum\left(n/N_d\right) \leq 0.1 \tag{2.28}$$

$$\sum(\Delta t/t_d) = -0.111\left[\sum\left(n/N_d\right)\right] + 0.111 \text{ for } 0.1 < \sum\left(n/N_d\right) \leq 1.0. \tag{2.29}$$

It should be pointed out that the intermediate coordinates for the 9Cr-1Mo-V curve in the nuclear code ASME III-5 are 0.1 and 0.01 rather than 0.3 and 0.3. These lower code coordinates are based on load relaxation. It is assumed in the above analysis that thermal stresses diminish relatively quickly when calculating rupture life t_m. It is also assumed that stress concentration factors due to holes, lugs, etc., are included in the strain values when entering Table 2.2, but are excluded when entering the stress values in Table 2.3. Stress concentration factors may be obtained from various publications, such as Pilkey and Pilkey (2008), or by experimental or theoretical methods.

Example 2.4 The support system in Figure 2.4 is subjected to the cycles shown in Figure 2.7. The system temperature increases from ambient to 1450°F. It stays at 1450°F for about three years (30,000 h). The temperature is then dropped down to ambient temperature where the system is inspected

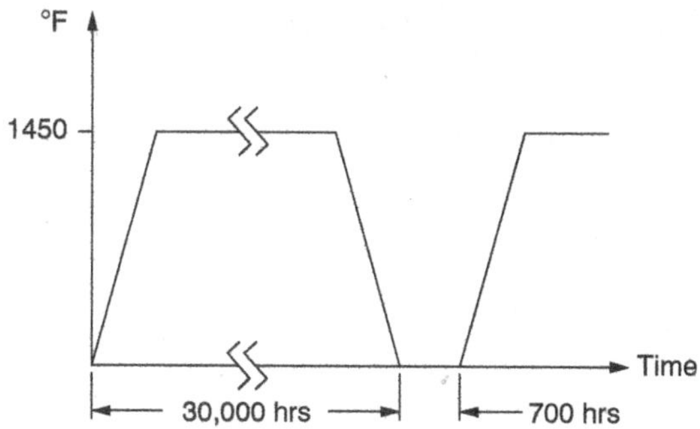

Figure 2.7 Operating cycles.

before it starts up again. The process is repeated 10 times during the expected life of the system (300,000 h). Evaluate the creep fatigue property of the support system based on the mechanical stress obtained from Example 2.2 plus the thermal stress obtained from Example 2.3. Let $E = 18{,}700$ ksi at 1450°F.

Solution

First trial

1) Calculate the strain corresponding to the maximum stress in the member due to mechanical and thermal stresses in accordance with Eq. (2.22).
   A summary of the stresses from Examples 2.2 and 2.3 are shown in Table 2.4. From Eq. (2.22), the maximum strain is

$$\Delta\varepsilon = (|3000|)/18{,}700{,}000 = 1.60\times10^{-4}\,\text{in./in.}$$

2) Calculate the fatigue ratio (n/N_f).

   Total number of cycles, $n = 10$

Table 2.2 is limited to 1300°F. At this temperature, a life of 1 million cycles is obtained for a strain level of 0.00084. The strain level calculated above is 0.00016, which is about five times smaller. Accordingly, we can use the strain value of 0.00084 in./in. as a starting number for 1 million cycles at 1300°F. This

Table 2.4 Stress summary of trial one in members of Figure 2.4, psi.

Member	Mechanical stress, P_L, from Example 2.2	Thermal stress, Q, from Example 2.3	Mechanical plus thermal stress
1	1350	1650	3000
2	1800	−2850	−1050
3	1350	1650	3000

strain value is much larger than the calculated value of 0.00016 and is thus very conservative. From Eq. (1.3), the Larson-Miller parameter is

$$\begin{aligned} P_{\mathrm{LM}} &= (460+1300)(20+\log_{10}1{,}000{,}000) \\ &= 45{,}760. \end{aligned}$$

Using this parameter at 1450°F results in

$$\begin{aligned} &45{,}760 = (460+1450)(20+\log_{10}N_f) \\ &\log_{10}N_f = 3.96 \quad \text{and} \quad N_f = 9080 \text{ cycles.} \end{aligned}$$

From Eq. (2.23),

$$\sum (n/N_f) = 10/9080 \approx 0.$$

3) Calculate the mechanical stress P_{L}.
   From Table 2.4, maximum $P_{\mathrm{L}} = 1800$ psi.

4) Check isochronous curve
   From Figure 2.5 with $P_{\mathrm{L}} = 1800$, a value of $\varepsilon = 0.4\%$ is obtained. This value is well below the limiting value of 1%.

5) Obtain the number of hours, t_{m}, from the rupture curve.
   From Table 2.3, the 1450°F line gives a life of 100,000 h for a stress of 1800 psi.

6) Calculate the rupture ratio from Eq. (2.24)

$$\sum \left(t_r/t_m\right)_k = (300{,}000/100{,}000) = 3.0.$$

Because this ratio is substantially greater than 1.0, the test fails and the diameter of 1.75 in. (area = 2.42 in.^{2}) for the rods is not adequate for 300,000 h.

Second trial

From Table 2.3, the maximum stress P_{L} cannot exceed 1400 psi at 1450°F in order to obtain 300,000 h. Thus, the area of the rods must be increased to

$$A = (1800/1400)(2.42) = 3.1 \text{ in.}^2.$$

Try a rod diameter of 2.0 in. ($A = 3.14$ in.^{2}).

The new stresses are shown in Table 2.5.

Steps 1 to 4 are acceptable.

7) Obtain the number of hours, t_{m}, from the rupture curve.
   From Table 2.3, the 1450°F line gives a life of 300,000 h at a stress of 1400 psi.

$$t_{\mathrm{m}} = 300{,}000 \text{ h}.$$

8) Calculate the rupture ratio from Eq. (2.24)

$$\sum \left(t_{\mathrm{r}}/t_{\mathrm{m}}\right)_{\mathrm{k}} = (300{,}000/300{,}000) = 1.0.$$

Table 2.5 Stress summary of trial 2 in members of Figure 2.4, psi.

Member	Mechanical stress, P_L, from Example 2.2	Thermal stress, Q, from Example 2.3	Mechanical plus thermal stress
1	1040	1650	2690
2	1385	−2850	−1465
3	1040	1650	2690

9) Calculate Eq. (2.25).
From Figure 2.6 with a value of $(n/N_d) \approx 0.0$ and a value of $(t_r/t_m) = 1.0$, it can be shown that D is within the envelope for 304 stainless steel. Hence, the structural members are adequate for 10 cycles and an expected life of 300,000 h. Use 2.0-in. diameter rods.

2.6 Reference Stress Method

Many structural components in a boiler and pressure vessel operating in the creep range can be analyzed, as a limiting case, by plastic analysis, as demonstrated by the results in Table 2.1. The reference stress method, discussed in Section 1.5.3.2, is based on the following equation to determine the equivalent creep stress in a structure

$$\sigma_R = (P / P_L)\sigma_y \tag{2.30}$$

where

P = applied load on the structure
P_L = the limit value of applied load based on plastic analysis
σ_R = reference stress
σ_y = yield stress of the material.

Plastic analysis results in a larger load-carrying capacity of the indeterminate structure than an elastic analysis. In plastic analysis Wang (1970), the structure is loaded until the member with the highest stress reaches a specified limiting stress. Additional increment in the applied loading is assumed possible subsequent to removing the member with the limiting stress from further consideration. The analysis continues through various cycles. At the end of each cycle, members that have reached their limiting stress are removed from further analysis and the process continues by increasing the load until the structure becomes unstable. The structure is assumed to support the entire load accumulated through the various cycles. In many instances this load is substantially larger than that used in elastic analysis and is the limiting case of creep analysis.

It should be noted that this methodology works for systems where the members are of relatively similar stiffness and adequate ductility. If one member is substantially stiffer than the others, it will carry most of the initial load and can rupture before it sheds load to the other members.

Example 2.5 In Example 2.2, let the cross-sectional area of all members = 2.42 in.^{2}. The total weight of the cyclone and its contents is 10,000 lb. Determine the forces in the braces using the reference stress method. The material of the braces is 304 stainless steel. The design temperature is 1450°F and the modulus of elasticity is 18,700 ksi. Assume an allowable stress of 1800 psi and average yield stress of 20,300 psi.

Solution

The maximum load capacity per brace using yield stress as a criterion is

$$\text{Load / brace} = (20{,}300)(2.42) = 49{,}126 \text{ lb}$$

The limit value of the applied load, P_{lim}, based on plastic analysis, is calculated as

$$P_{\text{lim}} = 134{,}200 \text{ lb}$$

The stress in all of the truss members in the creep range is then calculated from Eq. (2.19) as

$$\sigma_R = (10{,}000/134{,}200)(20{,}300) = 1510 \text{ psi.}$$

This stress value is smaller than the maximum allowable stress value of 1800 psi obtained elastically for Member 2 in Example 2.2. Thus, the designer has two choices at this time. One is to use the new lower stress of 1510 psi in calculating creep-fatigue cycles. The other choice is to reduce the size of the members based on the lower stress. In this case, the required brace area is (1510/1800)(2.42) = 2.03 in.^{2}.

$$\text{Use 1.625-in. rods.} \qquad A = 2.07 \text{ in.}^2.$$

Thus, in this case, the reference stress method results in a 14% reduction of the required area [(2.42 – 2.07)/2.42] compared to elastic analysis using the same factor of safety.

2.7 Elastic Follow-up

Elastic follow-up is a condition that occurs at elevated temperature where additional strain accumulates in a more highly stressed local region adjacent to a lower stressed region due to unbalanced stress relaxation when subjected to strain-induced loading. A brief discussion of elastic follow-up was presented in Section 1.6.1. The following example illustrates the stress patterns due to elastic follow-up in a two-bar system.

Example 2.6 The two-element system shown in Figure 2.8a is subjected to a 200°F reduction in temperature. Determine stresses and strain (elastic follow-up) from $t = 0$ h to $t = 10{,}000$ h for the following two conditions.

Case 1: $A_1 = 0.40$ in.^{2} $A_2 = 0.10$ in.^{2} ratio of area = 4
Case 2: $A_1 = 0.15$ in.^{2} $A_2 = 0.10$ in.^{2} ratio of area = 1.5

Data:

Material: Type 304 SS	Operating temperature = 1050°F
$\Delta T = -200$°F	$\alpha = 10.4 \times 10^{-6}$ in./in.-F
$E = 22{,}400{,}000$ psi	$L = 20.0$ in.

Solution

Strain due to ΔT, Figure 2.8b, is

$$\varepsilon_{T1} = \varepsilon_{T2} = \alpha\Delta T = -0.00208 \text{ in./in.} \tag{1}$$

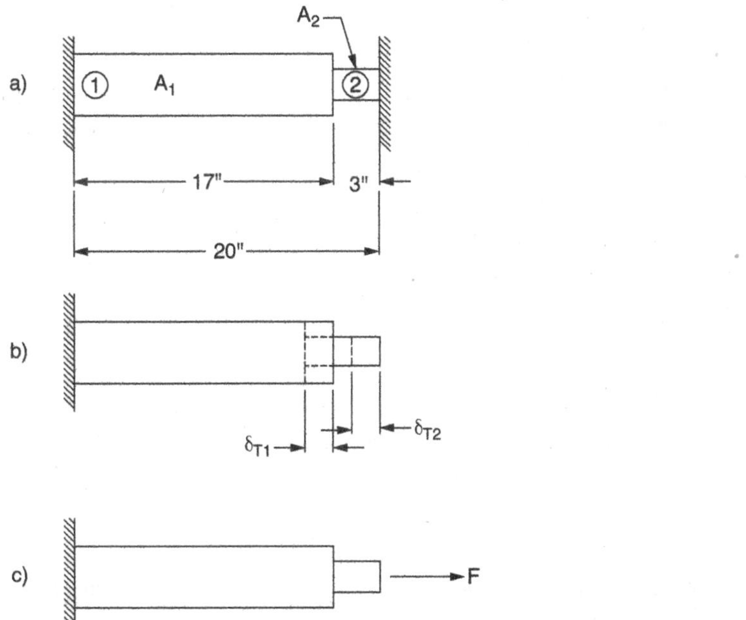

Figure 2.8 Two-element system.

Deflection due to ΔT, Figure 2.8b, is

$$\delta_{T1} = \alpha \Delta T\ L_1 = \varepsilon_{T1} L_1 \tag{2a}$$

$$= (-0.00208)(17)$$
$$= -0.0354 \text{ in.}$$

$$\delta_{T2} = \alpha \Delta T\ L_2 = \varepsilon_{T2} L_2 \tag{2b}$$

$$= (-0.00208)(3)$$
$$= -0.00624 \text{ in.}$$

$$\text{total } \delta = -0.0416 \text{ in.}$$

CASE 1
$A_1 = 0.40$ in.^{2} $A_2 = 0.10$ in.^{2} ratio of area = 4

Case 1 – Elastic analysis:
Assume both Members 1 and 2 to be elastic.

The force required to reduce the thermal deflection to zero is given by:

Deflection due to temperature = deflection in Element 1 due to force F + deflection in Element 2 due to force F.

$$\delta_T = \delta_1 + \delta_2. \tag{3}$$

This equation can be written as

$$\alpha \Delta T\ L = (FL_1 / A_1 E_1) + (FL_2 / A_2 E_2). \tag{4}$$

From Eq. (4),

$$(10.4x10^{-6})(200)(20)(22.4 \times 10^6) = F(17/0.4 + 3/0.1)$$
$$F = 12{,}853 \text{ lbs}$$
$$S_1 = 12{,}853/0.4 = 32{,}133 \text{ psi}$$
$$S_2 = 12{,}853/0.1 = 128{,}530 \text{ psi}$$
$$\varepsilon_1 = 12{,}853/(0.4)(22{,}400{,}000) = 0.0014 \text{ in/in.}$$
$$\varepsilon_2 = 12{,}853/(0.1)(22{,}400{,}000) = 0.0057 \text{ in/in.}$$

A summary of the strains, together with the results of a nonlinear relaxation analysis [performed by Mark Messner at Argonne National Laboratory], are shown in Table 2.6 and Figure 2.9.

Table 2.6 Strain values for area ratio of 4:1.

ratio 4:1	Member 1 ratio 4:1	Member 2 ratio 4:1
Elastic	0.0014	0.0057
Relaxation	0.00023	0.0126

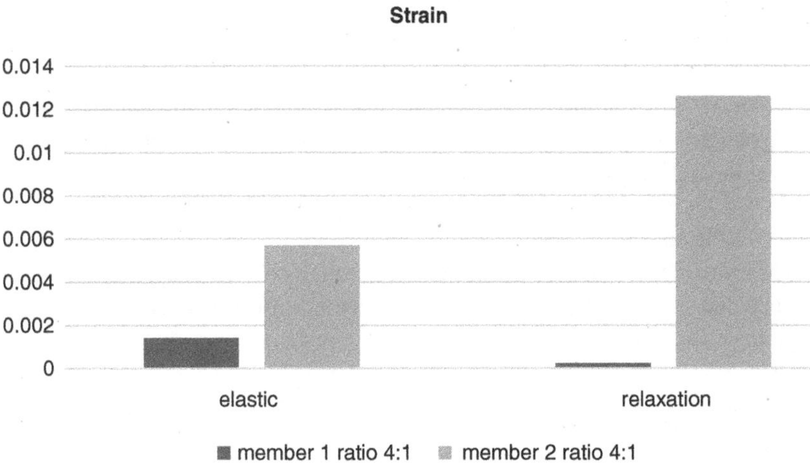

Figure 2.9 Strain values for area ratio of 4:1.

The strain in localized high stress Member 2 increased by a factor of two over that calculated elastically using the more sophisticated inelastic relaxation method.

CASE 2

$A_1 = 0.15$ in.^{2} $A_2 = 0.10$ in.^{2} ratio of area = 1.5

Case 2 – Elastic analysis:

Assume both Members 1 and 2 to be elastic.

The force required to reduce the thermal deflection to zero is given by:

Deflection due to temperature = deflection in Element 1 due to force F + deflection in Element 2 due to force F.

$$\delta_T = \delta_1 + \delta_2. \tag{5}$$

This equation can be written as

$$\alpha \Delta T\ L = \left(FL_1 / A_1E_1\right) + \left(FL_2 / A_2E_2\right). \tag{6}$$

From Eq. (6),

$$\left(10.4x10^{-6}\right)\left(200\right)\left(20\right)\left(22.4x10^{6}\right)=F\left(17/0.15+3/0.10\right)$$
$$F=6501\text{ lbs}$$
$$S_1=6501/0.15=43{,}340\text{ psi}$$
$$S_2=6501/0.1=65{,}010\text{ psi}$$
$$\varepsilon_1=6501/\left(0.15\right)\left(22{,}400{,}000\right)=0.0019\text{ in./in.}$$
$$\varepsilon_2=6501/\left(0.1\right)\left(22{,}400{,}000\right)=0.0029\text{ in./in.}$$

A summary of the strains, together with the results of a nonlinear relaxation analysis [performed by Mark Messner at Argonne National Laboratory], are shown in Table 2.7 and Figure 2.10.

Here the strain in the higher stressed bar increased a factor of three compared to the elastically calculated distribution.

These examples illustrate the potentially significant impact of strain redistribution due to plasticity and creep in a system with a localized high stress region.

Table 2.7 Strain values for area ratio of 1.5:1.

Ratio 1.5:1	Member 1 ratio 1.5:1	Member 2 ratio 1.5:1
elastic	0.0019	0.0029
relaxation	0.00071	0.0098

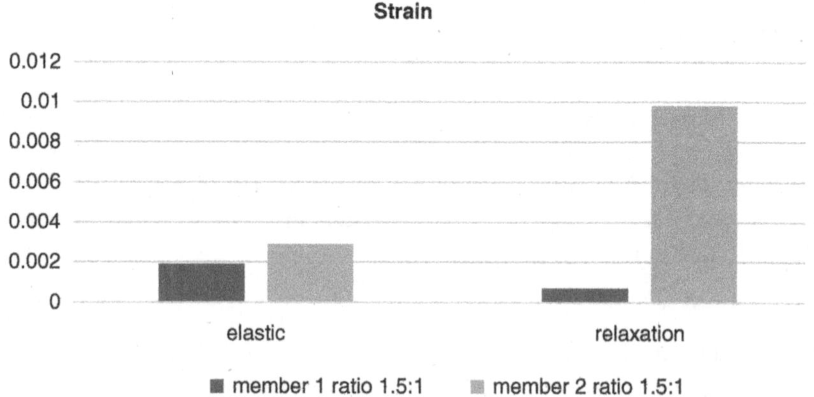

Figure 2.10 Strain values for area ratio of 1.5:1.

Problems

2.1 An internal cyclone is supported by three braces as shown. All braces have the same cross-sectional area of 2.0 in.^{2}. The total weight of the cyclone and its contents is 15,000 lb. The material of the braces is 304 stainless steel. The design temperature is 1450°F and the modulus of elasticity is 18 700 ksi. The allowable stress in accordance with ASME VIII-1 is 1800 psi at 1450°F and 20,000 psi at room temperature. Member 3 is subjected to a 25°F increase due to insulation problems. Let $\alpha = 10.8 \times 10^{-6}$ in./in.°F. The yield stress is 30,000 psi at room temperature and 10,600 psi at 1450°F. Check the following:

a) Stresses due to mechanical loads

b) Stresses due to mechanical and thermal loads in accordance with VIII-2 criteria.

2.2 Use the stresses obtained from Problem 2.1 to evaluate the creep-rupture property for 30,000 h and one cycle. Use the properties given in Example 2.4.

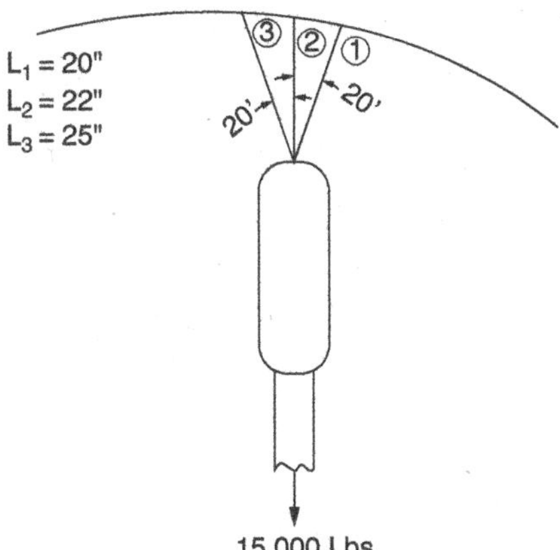

3

Structural Members in Bending

Heat treating of a nickel alloy vessel at 1950°F (1065°C) [Nooter Construction, Inc.].

Analysis of ASME Boiler, Pressure Vessel, and Nuclear Components in the Creep Range, Second Edition. Maan H. Jawad and Robert I. Jetter.
© 2022 John Wiley & Sons Ltd. Published 2022 by John Wiley & Sons Ltd.

3.1 Introduction

Many components in pressurized equipment operating in the creep range consist of beams and flat plates in bending. These include piping components, internal catalyst supports, nozzle covers, and tube sheets. In this chapter, the characteristics of beams in bending are studied first, followed by an evaluation of flat circular plates in bending.

The characteristics of beams in bending operating in the creep range have been studied by many engineers and researchers. Theoretically, the analysis is based on Norton's relationship, which correlates the stress and strain rate in the creep regime, as discussed in Chapters 1 and 2, and is given by

$$d\varepsilon / dt = k'\sigma^n \tag{3.1}$$

where

k' = constant
n = creep exponent, which is a function of material property and temperature
$d\varepsilon/dt$ = strain rate
σ = stress.

This equation, however, is impracticable for most problems encountered by an engineer, as discussed in Section 2.1. A simpler method, referred to as the stationary stress or elastic analog method, is normally used to solve beam-bending problems and is given by

$$\varepsilon = K'\sigma^n \tag{3.2}$$

where

K' = constant
ε = strain.

It should be noted that when $n = 1.0$ in Eq. (3.2), then $K' = 1/E$, where E is the elastic modulus. The application of Eq. (3.2) to regular engineering problems, although simpler than Eq. (3.1), is still very complicated due to the nonlinear relationship between stress and strain. This difficulty is overcome by performing plastic analysis to approximate the results obtained by Eq. (3.2), as explained later in the chapter.

3.2 Bending of Beams

The equations for the bending of a beam in the creep regime are based on assumptions Finnie and Heller (1959) that are similar to those for beams in elastic analysis; i.e.,

- The length of the beam is much larger than its cross section.
- Plane cross sections before bending remain plane after bending.
- Bending deflection is small compared to the length of the beam.
- Stress-strain diagrams are the same for both the tensile as well as the compressive sides of the beam.
- The plane of bending is a plane of symmetry.

The strain at a given point in a beam due to bending is

$$\varepsilon = z/\rho \tag{3.3}$$

where

z = location of a point measured from the neutral axis
ρ = radius of curvature.

The beam is assumed to have achieved a stationary stress condition. This assumption is realistic for components in pressure vessel applications where temperature and loading are constant for extended periods of operation during the cycle. Combining Eqs. (3.2) and (3.3) results in

$$\sigma = K'^{(-1/n)}(z/\rho)^{1/n} \tag{3.4}$$

Equating the internal moment in a cross section to the applied external moment gives

$$M = \int \sigma z dA \tag{3.5}$$

where

dA = unit cross-sectional area of the beam
M = applied external moment.
Combining Eqs. (3.4) and (3.5) yields

$$M = \int K'^{(-1/n)}\rho^{(-1/n)}(z)^{(1+1/n)} dA. \tag{3.6}$$

Define the creep moment of inertia as

$$I_n = \int (z)^{1+1/n} dA \tag{3.7}$$

where

I_n = creep moment of inertia.
And combining Eqs. (3.4) and (3.6) gives

$$\sigma = M z^{1/n} / I_n. \tag{3.8}$$

3.2.1 Rectangular Cross Sections

For a rectangular cross section with $B =$ width and $H =$ height, Eq. (3.7) gives

$$I_n = B \int_{-H/2}^{H/2} (z)^{1+1/n} dz$$

or

$$I_n = \frac{2nB}{1+2n}(H/2)^2 (H/2)^{1/n}. \tag{3.9}$$

Substituting this expression into Eq. (3.8), and defining the elastic moment of inertia as

$$I = BH^3/12, \tag{3.10}$$

$$\sigma = \left(\frac{MH}{2I}\right)\left(\frac{1+2n}{3n}\right)\left(\frac{z}{H/2}\right)^{1/n}. \tag{3.11}$$

This equation reduces to a maximum value of

$$\sigma = 6M/\left(BH^2\right) \quad \text{for } n = 1.0. \tag{3.12}$$

The first bracketed term in Eq. (3.11) is the conventional (Mc/I) expression in elastic analysis. The second and third bracketed terms are the stress modifiers due to creep. When $n = 1$, Eq. (3.11) gives the conventional elastic bending equation. Figure 3.1 shows the stress distribution in the beam cross section when $n = 1$, 3, 6, and 10. Notice the reduction in maximum stress due to the creep function n.

3.2.2 Circular Cross Sections

For tubes having circular cross sections with $R_o =$ outside radius and $R_i =$ inside radius, Eq. (3.7) gives Finnie and Heller (1959)

$$I_n = \frac{8I}{[3+(1/n)](\pi)^{1/2}} \left[\frac{1-\left(R_i/R_o\right)^{3+(1/n)}}{1-\left(R_i/R_o\right)^4}\right] \frac{\Gamma[1+(1/2n)]}{\Gamma[1.5+(1/2n)]} \tag{3.13}$$

where

$I =$ moment of inertia of a tubular cross section
$= \pi\left(R_o^4 - R_i^4\right)/4$

and

$\Gamma =$ gamma function tabulated in Table 3.1.
The bending equation becomes

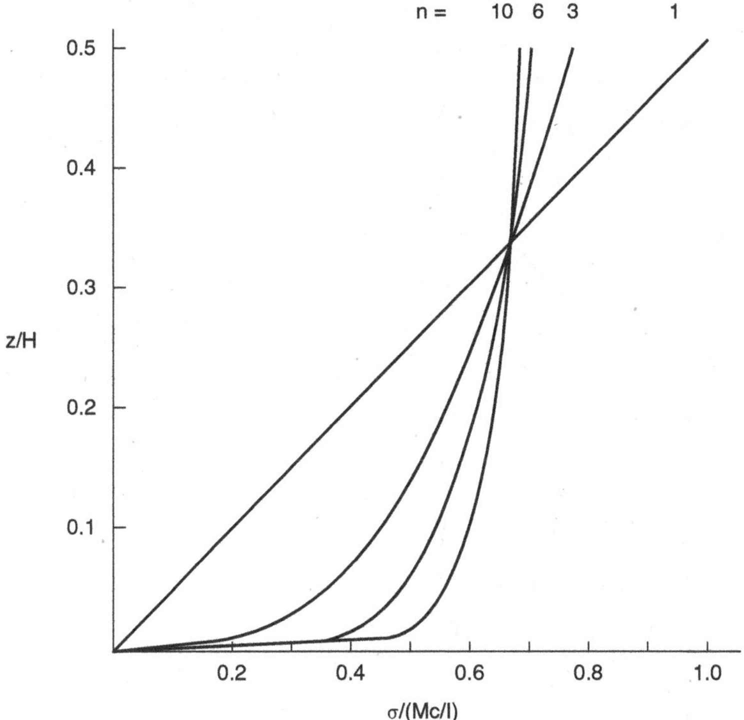

Figure 3.1 Stress distribution in a rectangular beam as a function of *n*.

Table 3.1 Some tabulated values[1] of Γ(n).

(n)	Γ(n)
1.0	1.0000
1.1	0.9514
1.2	0.9182
1.3	0.8975
1.4	0.8873
1.5	0.8862
1.6	0.8935
1.7	0.9086
1.8	0.9314
1.9	0.9618
2.0	1.0000

[1] Note: $\Gamma(n) = \int_0^\infty \left(e^{-x} x^{n-1}\right) dx$

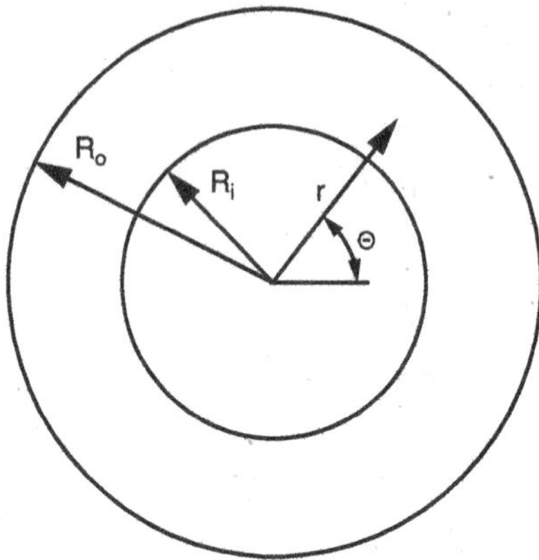

Figure 3.2 Thick cylindrical shell.

$$\sigma = (MR_o / I_n)(r \sin\theta / R_o)^{1/n} \tag{3.14}$$

where r is as defined in Figure 3.2.

For thin shells and pipes, this equation reduces to

$$\sigma = M / (\pi t R_o^2) \quad \text{for } n = 1.0. \tag{3.15}$$

Example 3.1 A beam in a high temperature application is subjected to a 1000 ft-lb maximum bending moment. The allowable stress is 2000 psi. Determine

(a) The required cross section of a rectangular beam if an elastic analysis was performed where n = 1.0.
(b) The required cross section of a rectangular beam if a creep analysis was performed where n = 4.7.
(c) The required cross section of a pipe beam if an elastic analysis was performed where n = 1.0.
(d) The required cross section of a pipe beam if a creep analysis was performed where $n = 4.7$.

Solution

(a) From Eq. (3.12)

$$\begin{aligned} BH^2 &= 6(1000 \times 12)/2000. \\ &= 36 \text{ in.}^3 \end{aligned}$$

Use 6 × 1 in. rectangular cross section.

(b) From Eq. (3.11) with $z = H/2$

$$2000 = \left(\frac{(1000\times 12)H}{2I}\right)\left(\frac{1+2(4.7)}{3(4.7)}\right).$$

$$BH^2 = 26.55 \text{ in.}^3$$

Use 5.0 × 1 – 1/16 in. rectangular cross section.

(c) From Eq. (3.15),

$$tR_o^2 = 1000\times 12/(2000\pi).$$
$$= 1.91 \text{ in.}^3$$

Use an 8-in. Sch 5S pipe (OD = 8.625 in., $t = 0.109$ in., and $tR_o^2 = 2.03$ in.^{3}).

(d) Assume an 8-in. Sch 5S pipe.

$OD = 8.625$ in. $\quad t = 0.109$ in.

$R_o = 4.3125$ in. $\quad R_i = 4.2035$ in.

$I = \pi\,(4.3125^4 - 4.2035^4)\,/4 = 26.44$ in.^{4}

$R_i/R_o = 0.9747 \quad 1/n = 0.2128 \quad 1/2n = 0.1064.$

From Eq. (3.13)

$$I_n = \frac{8(26.44)}{(3.2128)(\pi)^{1/2}}\left[\frac{1-(0.9747)^{3.2128}}{1-(0.9747)^4}\right]\frac{\Gamma\,[1.1064]}{\Gamma\,[1.6064]}$$
$$= (37.145)(0.8112)(1.065)$$
$$= 32.09 \text{ in.}^4.$$

From Eq. (3.14)

$$\sigma = (1000\times 12)(4.3125)/\,32.09$$
$$= 1615 \text{ psi} < 2000 \text{ psi}.$$

Use an 8-in. Sch 5S pipe

3.3 Shape Factors

The shape factor is defined as the ratio of the moment that a cross section can withstand when analyzed using creep or plastic analysis to the moment that the cross section can withstand using elastic analysis. In this section, the shape factors for rectangular and hollow circular cross sections are derived based on creep and plastic analysis. These shape factors will be used later in Chapters 4 and 5 when analyzing shells.

3.3.1 Rectangular Cross Sections

For rectangular cross sections, the relationship between maximum stress and moment in the elastic range is

$$\sigma = 6M / BH^2 \tag{3.16}$$

where

B = width of beam
H = height of beam
M = applied bending moment
σ = maximum bending stress.

The relationship between maximum stress and moment in the creep range is obtained from Eq. (3.11). Substituting $z = H/2$ and $I = BH^3/12$ results in

$$\sigma = \left(\frac{6M}{BH^2}\right)\left(\frac{1+2n}{3n}\right) \tag{3.17}$$

or

$$\sigma = \left(\frac{1}{\text{SF}}\right)\left(\frac{6M}{BH^2}\right) \tag{3.18}$$

where

$$\text{SF} = \text{shape factor}$$
$$= \frac{3n}{1+2n} \tag{3.19}$$

For $n = 1$, the shape factor is 1.0, and Eqs. (3.16) and (3.18) become the same. In the creep range, n varies for different materials and temperatures. Values of n are published for a number of ferrous and non-ferrous alloys at different temperatures Odqvist (1966). The n values range between 2.5 for annealed carbon steel at 1020°F (550°C) and 10 for aluminum at 300°F (150°C). Hult (1966) has published similar data. Variation in SF with respect to n is shown in Table 3.2.

The ASME uses an SF of 1.25 for calculations made in the creep range. This is based on a conservative value of $n = 2.5$.

Table 3.2 shows that the value of the shape factor is equal to 1.5 as n approaches infinity. It so happens that the 1.5 shape factor also corresponds to

Table 3.2 Variation in shape factor for rectangular cross sections with respect to n.

n	1	2.5	5	10	∞
SF	1	1.25	1.36	1.43	1.50

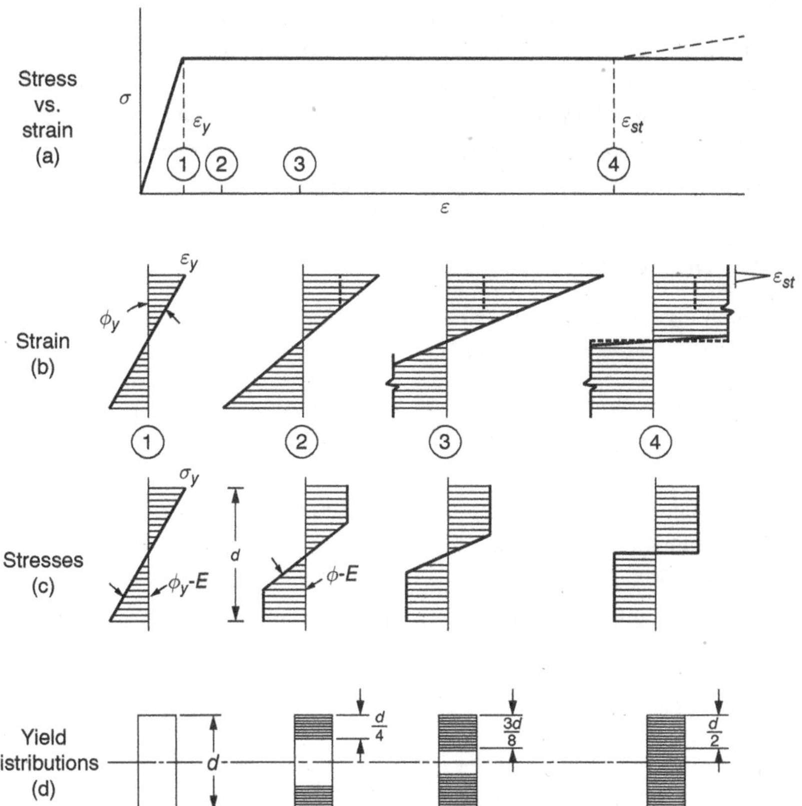

Figure 3.3 Stress and strain distribution [Based on Beedle, 1966].

that obtained from plastic analysis of beams. This is illustrated in the stress-strain diagram of Figure 3.3a for an elastic perfectly plastic material. The stress and strain distribution increases proportionately with an increase in the external moment as shown in Figure 3.3b, Points 1 through 4. The stress distribution across the thickness gradually changes from triangular to rectangular in shape as the strain increases from Points 1 to 2, and finally to Point 4. In the fully plastic region at Point 3, the stress expression is obtained by equating internal and external moments

$$\sigma = 4M \,/\, BH^2. \tag{3.20}$$

A comparison of Eqs. (3.16) and (3.20) shows that, for the same bending moment, the stress obtained by using Eq. (3.20) from plastic analysis is 50% less than that obtained by using Eq. (3.16) for elastic analysis. In other words, there is a stress reduction factor of 1.5 when using plastic analysis compared to elastic analysis. This fact is utilized by the ASME code as well as other international codes to increase the allowable elastic stress values by a factor of 1.5 when analyzing beams of rectangular cross section, bending of plates, and local bending of shells. The ASME code uses a conservative shape factor of 1.25 for rectangular cross sections in the creep range.

3.3.2 Circular Cross Sections

The relationship between maximum stress and moment for pipes of circular cross section in the elastic range is

$$\sigma = MR_o / I \tag{3.21}$$

where

I = moment of inertia of a tubular cross section $= \pi\left(R_o^4 - R_i^4\right)/4$
R_i = inside radius of beams
R_o = outside radius of beams.

The relationship between maximum stress and moment in the creep range is obtained from Eq. (3.14) as

$$\sigma = MR_o / I_n \tag{3.22}$$

or

$$\sigma = \left(\frac{1}{\mathrm{SF}}\right)\left(\frac{MR_o}{I}\right) \tag{3.23}$$

where SF is the shape factor defined by

$$\mathrm{SF} = \frac{8}{[3+(1/n)](\pi)^{1/2}}\left[\frac{1-\left(R_i/R_o\right)^{3+(1/n)}}{1-\left(R_i/R_o\right)^4}\right]\frac{\Gamma[1+(1/2n)]}{\Gamma[1.5+(1/2n)]}. \tag{3.24}$$

For $n = 1$ the shape factor is 1.0, and Eq. (3.21) for elastic analysis and Eq. (3.23) for creep analysis become the same. Variation in SF with respect to n is shown in Table 3.3 for hollow thin circular cross sections ($R_i \gg t$).

Table 3.3 Variation in shape factor for hollow circular cross sections with respect to n.

n	1	2.5	5	10	∞
SF	1	1.14	1.20	1.243	1.27

3.4 Deflection of Beams

The approximate relationship between the deflection of a beam and the radius of curvature, ρ, is given by the mathematical equation

$$\frac{d^2w}{dx^2} = \frac{1}{\rho} \tag{3.25}$$

where

w = deflection of beam.

Combining this equation with the stationary stress Eq. (3.2), as well as Eqs. (3.3) and (3.8), yields the following expression for the deflection of a beam

$$\frac{d^2w}{dx^2} = K'M^n / I_n^n \tag{3.26}$$

where

I_n = moment of inertia as defined by Eq. (3.7)
M = applied moment on beam.

Equation (3.26) is nonlinear for all values of n, except $n = 1.0$. Thus, its application is very cumbersome for all but the simplest cases. Some of these simple cases Faupel and Fisher (1981) are shown below.

- Cantilever beam (Figure 3.4a) with an end load F

$$\delta_A = \frac{K'F^n\left(L^{n+2}\right)}{I_n^n(n+2)} \tag{3.27}$$

$$\theta_A = \frac{-K'F^n\left(L^{n+1}\right)}{I_n^n(n+1)} \tag{3.28}$$

- Cantilever beam (Figure 3.4b) with a uniform load q

$$\delta_A = \frac{K'q^n\left(L^{2n+2}\right)}{I_n^n 2^n(2n+2)} \tag{3.29}$$

$$\theta_A = \frac{-K'q^n\left(L^{2n+1}\right)}{I_n^n 2^n(2n+1)} \tag{3.30}$$

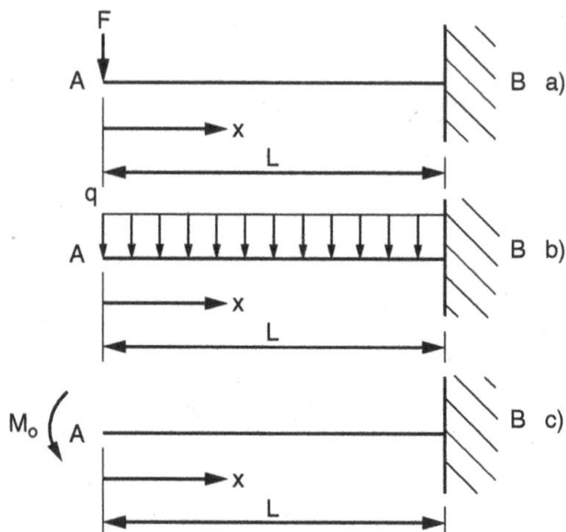

Figure 3.4 Cantilever beam.

- Cantilever beam (Figure 3.4c) with a counterclockwise end moment *MA*

$$\delta_A = \frac{K'M_A^n L^2}{2I_n^n} \tag{3.31}$$

$$\theta_A = \frac{-K'M_A^n L}{I_n^n}. \tag{3.32}$$

The following example illustrates the application of Eq. (3.26) to a two-span piping system.

Example 3.2 Determine the maximum bending moments in the piping system shown in Figure 3.5a. Assume a uniform load $q = 0.1$ kips/ft due to piping weight and contents and $L = 20$ ft. Let $n = 1$ and 4, and compare the results with plastic analysis.

Solution

Due to symmetry, the structure is reduced to that shown in Figure 3.5b. The bending moment in the beam is

$$M = R_A x - qx^2/2. \tag{a}$$

Substituting this equation into Eq. (3.26) gives

$$w'' = \left(K'/I_n^n\right)\left(R_A x - qx^2/2\right)^n. \tag{b}$$

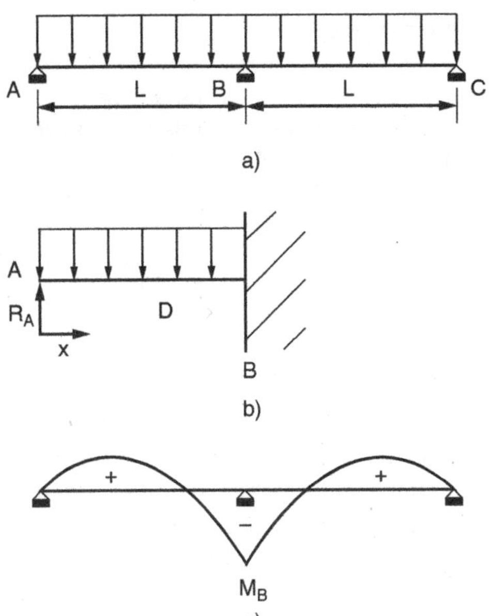

Figure 3.5 Two-span beam.

- n = 1

The value of R_A is obtained by solving Eq. (b) for the three boundary conditions

$$w = 0 \text{ at } x = 0, L \text{ and } w' = 0 \text{ at } x = L$$

This gives

$$R_A = 0.375qL \text{ and } M = 0.375qLx - qx^2/2. \tag{c}$$

- n = 4

The value of R_A is obtained by solving Eq. (b) for the three boundary conditions

$$w = 0 \text{ at } x = 0, L \text{ and } w' = 0 \text{ at } x = L.$$

The expression for the reaction obtained from the boundary conditions is a fourth-order equation. A numerical solution gives

$$R_A = 0.40qL \text{ and } M = 0.40qLx - qx^2/2. \tag{d}$$

Table 3.4 Solution of Eqs. (c) and (d) in Example 3.2.

Moment	Maximum negative moment at Point B	Maximum positive moment at Point D	Location of maximum positive moment
Elastic analysis ($n = 1.0$)	$qL^2/8$	$qL^2/14.22$	$x = 0.375L$
Creep analysis ($n = 4.0$)	$qL^2/10$	$qL^2/12.5$	$x = 0.40L$
Plastic analysis	$qL^2/11.63$	$qL^2/11.63$	$x = 0.414L$

Solutions of Eqs. (c) and (d) are shown in Table 3.4. When $n = 1$, Eq. (c) yields the moment expression for an elastic analysis. It has a maximum negative value at support B of $M_B = qL^2/8$ and a maximum positive value at $x = 0.375L$ of $M = qL^2/14.22$. For $n = 4.0$, the maximum negative value at support B is obtained from Eq. (d) as $M_B = qL^2/10$ and a maximum positive value at $x = 0.40L$ of $M = qL^2/12.5$. Thus, for $n = 4.0$, the moments in the beam have redistributed where the negative and positive moments are closer in magnitude. It is of interest to note that a plastic analysis for this structure gives a bending moment with a maximum negative value at support B of $M_B = qL^2/11.63$ and a maximum positive value at $x = 0.414L$ of $M = qL^2/11.63$. Thus, the stationary creep analysis of this structure can be approximated by performing the much easier plastic analysis even when the values of n are relatively low. One method of plastic analysis is illustrated in the next section.

The stationary stress and plastic stress analyses discussed in the above example are fairly tedious to perform for complex piping loops. The elastic analysis, however, gives conservative values as shown in Table 3.4 and is much easier to perform. Accordingly, ASME B31.1 (Power Piping) and B31.3 (Process Piping) allow elastic analysis of piping loops in high-temperature applications. The allowable stress values at the high temperatures are based on creep and rupture criteria. This elastic analysis, and some of the requirements of ASME B31.3, are discussed in Section 3.7.

3.5 Stress Analysis

One method of analyzing a piping system in the creep region is to perform a pseudo-plastic analysis. This is accomplished by running an incremental elastic analysis where the load is increased until the yield stress is reached at one location in the system. A hinge is then placed at this location and the stiffness matrix is modified accordingly. A second elastic cycle is then performed until a second location reaches the yield stress and a hinge is applied at the second location. The procedure continues until the structure becomes unstable, similarly to the procedure described in Chapter 2. The elastic formulation of the governing equations for this type of analysis is developed in this section for a two-dimensional system. Generally, a three-dimensional frame analysis is performed for piping loops. Such an analysis is beyond the scope of this book. Accordingly, in this chapter the moments M_i are obtained from a simple two-dimensional structural analysis to demonstrate the procedure.

The general elastic relationship between the applied joint loads and forces in a frame, Figure 3.6, is given by

$$[P]=[K][X]. \tag{3.33}$$

where

$[P]$ = applied loads at frame joints
$[K]$ = global stiffness matrix of frame given in Eq. (3.34)
$[X]$ = deformation of frame joints.

Figure 3.6 shows a typical member in bending. Positive direction of forces, applied loads, joint deflections, and angles are shown in Figure 3.6. Member 1, shown in the figure, starts at Node 1 and ends at Node 2. At Node 1, forces

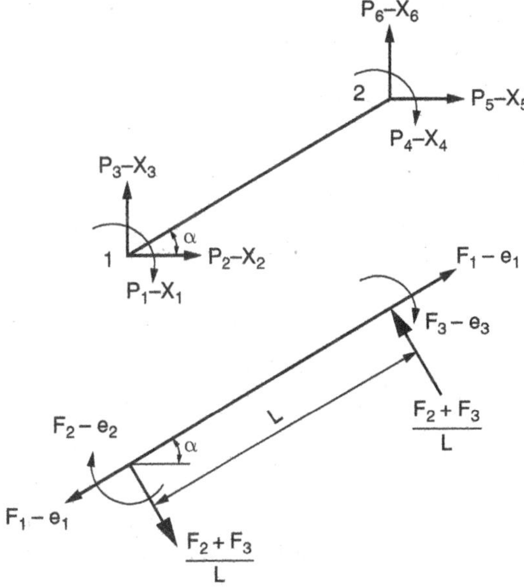

Figure 3.6 Sign convention for beams.

P_1, P_2, and moment M_1, or deflections X_1, X_2, and rotation θ_1,may be applied as shown in the figure. Similarly, forces P_1, P_2, and moment M_1, or deflections X_1, X_2, and rotation θ_1, can also be applied at Node 2. The local member rotation angle α, with respect to the global coordinate system of the member, is measured positive counterclockwise from the positive x direction at the beginning of the member. Tensile force is as shown in the figure.

The global stiffness matrix of the frame is assembled from the individual frame member stiffnesses [k]. The stiffness matrix, [k], of any member in a frame Weaver and Gere (1990) is expressed as

$$[K]=\begin{bmatrix} k_{11} & k_{12} & k_{13} & k_{14} & k_{15} & k_{16} \\ & k_{22} & k_{23} & k_{24} & k_{25} & k_{26} \\ & & k_{33} & k_{34} & k_{35} & k_{36} \\ & & & k_{44} & k_{45} & k_{46} \\ \text{symmetric} & & & & k_{55} & k_{56} \\ & & & & & k_{66} \end{bmatrix} \tag{3.34}$$

where

$$\begin{aligned}
k_{11} &= 4EI/L, & k_{12} &= \left(6EI/L^2\right)\sin\alpha \\
k_{13} &= -\left(6EI/L^2\right)\cos\alpha, & k_{14} &= 2\mathrm{EI}/\mathrm{L} \\
k_{15} &= -k_{12}, & k_{16} &= -k_{13} \\
k_{22} &= (EA/L)\cos^2\alpha + \left(12EI/L^3\right)\sin^2\alpha \\
k_{23} &= (EA/L)\sin\alpha\cos\alpha - \left(12EI/L^3\right)\sin\alpha\cos\alpha \\
k_{24} &= k_{12},\ k_{25} = -k_{22},\ k_{26} = -k_{23} \\
k_{33} &= (EA/L)\sin^2\alpha + \left(12EI/L^3\right)\cos^2\alpha \\
k_{34} &= k_{13},\ k_{35} = k_{26},\ k_{36} = -k_{33} \\
k_{44} &= k_{11},\ k_{45} = -k_{12},\ k_{46} = -k_{13} \\
k_{55} &= k_{22},\ k_{56} = k_{23},\ k_{66} = k_{33}
\end{aligned}$$

Equation (3.34) is a 6 × 6 matrix because there is a total of six degrees of freedom at each end of the member. They are X_1through X_6 or P_1 through P_6. The stiffness matrix of the frame is assembled by combining the stiffness matrices of all the individual members into one matrix, called the global matrix. Once the global stiffness matrix [K] is developed, Eq. (3.33) can be solved for the unknown joint deflections [X]. This assumes that the load matrix [P] at the nodal points is known.

$$[X]=[K]^{-1}[P] \tag{3.35}$$

The forces in the individual members of the frame are then calculated from the equation

$$[F]=[D][X_m] \tag{3.36}$$

where

[F] = force in member
[D] = force-deflection matrix of a member
$[X_m]$ = matrix for deflection at ends of individual members.
The matrix [D]is defined as

$$[D]=\begin{bmatrix} 0 & D_{12} & D_{13} & 0 & D_{15} & D_{16} \\ D_{21} & D_{22} & D_{23} & D_{24} & D_{25} & D_{26} \\ D_{31} & D_{32} & D_{33} & D_{34} & D_{35} & D_{36} \end{bmatrix} \tag{3.37}$$

where

$$D_{12}=-(AE/L)\cos\alpha \qquad D_{13}=-(AE/L)\sin\alpha \qquad D_{15}=-D_{12}$$
$$D_{16}=-D_{13} \qquad D_{21}=4EI/L \qquad D_{22}=\left(6IE/L^2\right)\sin\alpha$$
$$D_{23}=-\left(6IE/L^2\right)\cos\alpha \qquad D_{24}=2EI/L \qquad D_{25}=-D_{22}$$
$$D_{26}=-D_{23},\ D_{31}=D_{24} \qquad D_{32}=D_{22},\ D_{33}=D_{23}, \qquad D_{34}=D_{21}$$
$$D_{35}=D_{25},\quad D_{36}=D_{26}$$

The application of these equations to a simple piping loop is illustrated in the following example.

Example 3.3 The pipe loop shown in Figure 3.7a has a 4-in. STD weight pipe with an OD = 4.5 in., thickness = 0.237 in., and weight = 10.79 lb/ft. The pipe carries fluid with density of 62.4 lb/ft^3. Determine the maximum stress due to pipe and fluid weight using elastic analysis. The pipe hanger CE prevents vertical deflection at that location. Let the elastic modulus = 27 × 10^6 psi. The elbows at Points B and C have a bend radius of 6.0 in.

Solution

(a) Elastic analysis
From Figure 3.7b,
Member AB:

$$A=3.175\text{ in.}^2,\quad I=7.210\text{ in.}^4,\quad L=240\text{ in.}$$
$$\alpha=0\text{ deg.},\quad \sin\alpha=0.0,\quad \cos\alpha=1.0$$

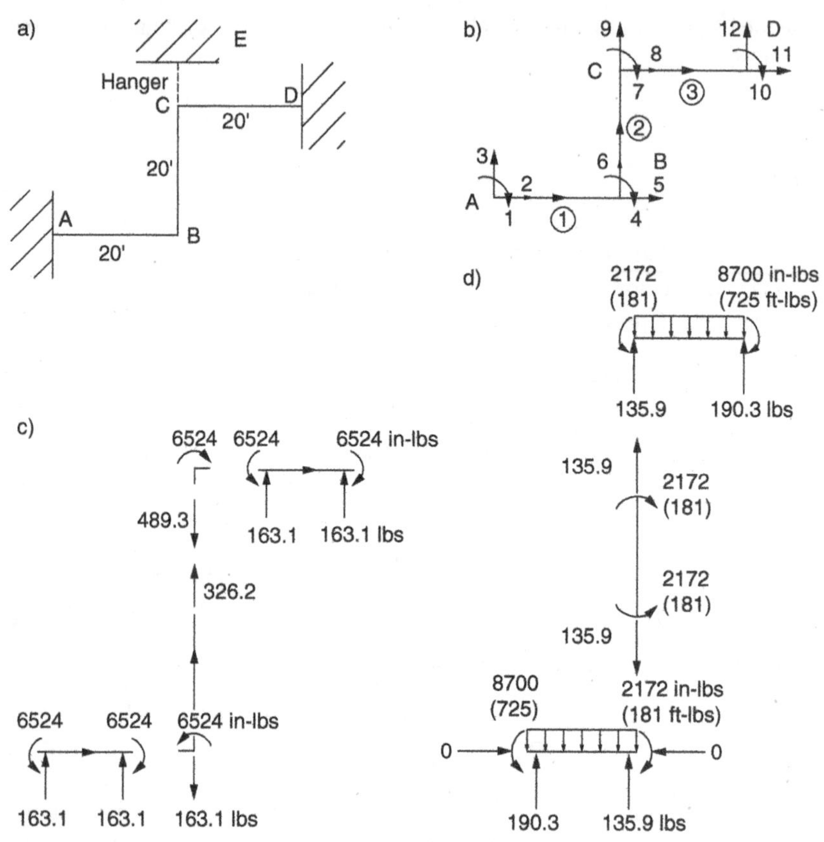

Figure 3.7 Pipe loop.

From Eq. (3.34), $[k]_{AB}$ =

1	2	3	4	5	6	
3,244,639	0	−20,279	1,622,319	0	20,279	1
	357,080	0	0	−357,080	0	2
		168.9916	−20,279	0	−168.9916	3
Symmetric			3,244,639	0	20,279	4
				357,080	0	5
					168.9916	6

Member BC:

$$A = 3.174 \text{ in.}^2, \quad I = 7.210 \text{ in.}^4, \quad L = 240 \text{ in}$$
$$\alpha = 90 \text{ deg}, \quad \sin\alpha = 1.0, \quad \cos\alpha = 0.0$$

From Eq. (3.34), $[k]_{BC}$ =

4	5	6	7	8	9	
3,244,639	20,279	0	1,622,319	−20,279	0	4
	168.9916	0	20,279	−168.9916	0	5
		357,080	0	0	−357,080	6
Symmetric			3,244,639	−20,279	0	7
				168.9916	0	8
					357,080	9

Member CD:

$$A = 3.174\ \text{in.}^2,\quad I = 7.210\ \text{in.}^4,\quad L = 240\ \text{in.}$$
$$\alpha = 0\ \text{deg},\quad \sin\alpha = 0.0,\quad \cos\alpha = 1.0$$

From Eq. (3.34), $[k]_{CD}$ =

7	8	9	10	11	12	
3,244,639	0	−20,279	1,622,319	0	20,279	7
	357,080	0	0	−357,080	0	8
		168.9916	−20,279	0	−168.9916	9
Symmetric			3,244,639	0	20,279	10
				357,080	0	11
					168.9916	12

From Figures 3.6 and 3.7, it is seen that the degrees of freedom 1, 2, 3, 9, 10, 11, and 12 must be set to zero because the pipe loop is fixed at these nodal points. This reduces the stiffness matrix k_{AB} to a 3 × 3 matrix corresponding to Displacements 4, 5, and 6. It also reduces stiffness matrix k_{BC} to a 5 × 5 matrix corresponding to Displacements 4, 5, 6, 7, and 8, and matrix k_{CD} to a 2 × 2 matrix corresponding to Displacements 7 and 8. The total global matrix is then a 5 × 5 matrix with magnitude $[K]_{\text{global}}$ =

4	5	6	7	8	
6,489,278	20,279	20,279	1,622,319	−20,279	4
	357,249	0	20,279	−169.0	5
		357,249	0	0	6
Symmetric			6,489,278	−20,279	7
				357,249	8

The applied uniform load consists of dead weight plus contents

$$q = 10.79 + 62.4\pi[4.5 - 2(0.237)]^2 / [4(144)] = 10.79 + 5.52 = 16.31 \text{ lb/ft}$$

$$\text{fixed end moments} = 16.31(20)^2 / 12 = 543.7 \text{ft} - \text{lb} = 6524 \text{ in.} - \text{lb}$$

$$\text{reactions} = 16.31(20)/2 = 163.1 \text{ lb}$$

From Figure 3.7c, the forces on Joints 4, 5, 6, 7, and 8 are

$$[F] = \begin{bmatrix} -6524 \\ 0 \\ -163.1 \\ 6524 \\ 0 \end{bmatrix}.$$

The joint deflections are obtained from Eq. (3.35) as

$$\begin{bmatrix} X_4 \\ X_5 \\ X_6 \\ X_7 \\ X_9 \end{bmatrix} = \begin{bmatrix} 13.3920 \times 10^{-4} \\ 0 \\ -38.0525 \times 10^{-5} \\ 13.4015 \times 10^{-4} \\ 0 \end{bmatrix} \begin{matrix} \text{rad; clockwise rotation} \\ \text{in.} \\ \text{in.; downward deflection.} \\ \text{rad; clockwise rotation} \\ \text{in.} \end{matrix}$$

The member forces are obtained from Eq. (3.36) as

1. Member 1

$$\begin{bmatrix} F \\ M_A \\ M_B \end{bmatrix} = \begin{bmatrix} 0 \\ -8700 \\ 2172 \end{bmatrix} \begin{matrix} \text{lb} \\ \text{in.} - \text{lb} \\ \text{in.} - \text{lb} \end{matrix} = \begin{bmatrix} 0 \\ -725 \\ 181 \end{bmatrix} \begin{matrix} \text{lb} \\ \text{ft} - \text{lb} \\ \text{ft} - \text{lb} \end{matrix}$$

2. Member 2

$$\begin{bmatrix} F \\ M_B \\ M_C \end{bmatrix} = \begin{bmatrix} 135.9 \\ -2172 \\ 2172 \end{bmatrix} \begin{matrix} \text{lb} \\ \text{in.} - \text{lb} \\ \text{in.} - \text{lb} \end{matrix} = \begin{bmatrix} 135.9 \\ -181 \\ 181 \end{bmatrix} \begin{matrix} \text{lb} \\ \text{ft} - \text{lb} \\ \text{ft} - \text{lb} \end{matrix}$$

3. Member 3

$$\begin{bmatrix} F \\ M_C \\ M_D \end{bmatrix} = \begin{bmatrix} 0 \\ -2172 \\ 8700 \end{bmatrix} \begin{matrix} \text{lb} \\ \text{in.} - \text{lb} \\ \text{in.} - \text{lb} \end{matrix} = \begin{bmatrix} 0 \\ -181 \\ 725 \end{bmatrix} \begin{matrix} \text{lb} \\ \text{ft} - \text{lb.} \\ \text{ft} - \text{lb} \end{matrix}$$

Figure 3.7c shows the final forces and bending moments in Members AB, BC, and CD.

- Maximum bending stress at the supports

$$S = \frac{(8700)(2.25)}{7.21} = 2715 \text{ psi}$$

  This value must be less than the allowable stress at the given temperature.
- Maximum stress at the elbows
  From Table 3.5,

$$h = 0.237(6) / (2.13)^2 = 0.313$$
$$i_1 = 0.9 / 0.313^{2/3} = 1.952$$
$$M = 181(1.952) = 353.3 \text{ ft} - \text{lb}$$
$$S = \frac{(353.3)(12)(2.25)}{7.21} = 1323 \text{ psi}$$

  This value must be less than the allowable stress at the given temperature.

3.5.1 Commercial Programs

Numerous computer programs are available for determining moments and forces in piping loops, such as AutoPIPE, CAEPIPE, CAESAR II, and PipePak. These sophisticated programs take into consideration such items as pressure, thermal expansion, dead weight, fluid weight, support movements, and hanger locations. Most of the programs also take into account the reduction in moments and forces in the piping loop due to elbow radii and other curved members. They also check stresses in accordance with different codes and standards such as ASME B31.1 and B31.3.

Example 3.4 The final forces, moments, and stress calculated in Example 3.3 were compared with results obtained from the commercial piping program AutoPIPE. The program was run with the same loads and geometry of Example 3.3. The final forces and bending moments calculated by the program are shown in Figure 3.8. A comparison between the bending moments obtained manually (Figure 3.7c), and those obtained from AutoPIPE (Figure 3.8) show very good agreement at the supports and 15% variance at the elbows, with the manual method being on the conservative side. The 15% difference is because the manual method assumes a sharp corner at Points B and C, whereas AutoPIPE takes the elbow curvature into consideration.

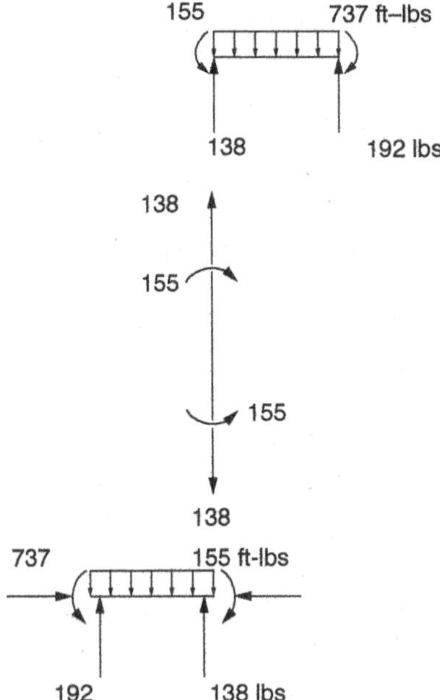

Figure 3.8 Moment and force obtained from AutoPIPE.

The stress at the supports and elbows calculated by AutoPIPE are

- Stress at supports = 2753 psi
- Stress at elbows = 1169 psi

The stress at the elbow takes into consideration the stress intensification factor calculated from ASME B31.1.

3.6 Reference Stress Method

The reference stress method, discussed in Section 1.5.3.2, is based on the following equation to determine the equivalent creep stress in a structure

$$\sigma_R = \left(P / P_L\right)\sigma_y \tag{3.38}$$

where

P = applied load on the structure
P_L = the limit value of applied load based on plastic analysis
σ_R = reference stress
σ_y = yield stress of the material.

There are many methods of performing plastic analysis in a piping system. One such method is to use elastic equations with incremental applied load conditions until one of the joints reaches its plastic limit. Then the structure is modified and a second run is performed until a second joint reaches its limit. The analysis continues until the structure becomes unstable, at which point the analysis is terminated and the maximum load carrying capacity of the structure is obtained. Another method, for simple piping systems, is to use plastic structural analysis. This is demonstrated in the following example.

Example 3.5 Solve Example 3.3 using plastic structural analysis. Use a shape factor of 1.27 from Table 3.3 and assume $\sigma_y = 20{,}000$ psi.

Solution

The plastic moments in the pipe loop are shown in Figure 3.9. For reasons of symmetry, Member AB will be used.

$$\sum \text{of internal energy} = \sum \text{of external energy}$$

$$M_\text{p}(\theta) + M_\text{p}(2\theta) + M_\text{p}(\theta) = P_\text{L}(L)(\delta)/2. \quad \text{(a)}$$

Let $\tan\theta \approx \theta \approx \delta/(L/2)$. Then Eq. (a) becomes

$$M_\text{p} = P_\text{L} L^2/16.$$

Substituting this quantity into the expression

$$\sigma_\text{y} = \frac{M_\text{p}(R/2)}{(I)(\text{SF})}$$

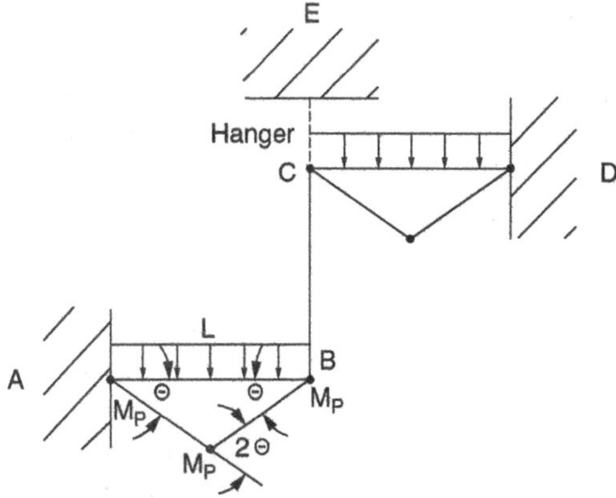

Figure 3.9 Plastic moments in a pipe loop.

and solving for P_L gives

$$P_L = 272.8 \text{ lb / ft}.$$

Solving Eq. (3.38) for the reference stress gives

$$\sigma_R = (16.31 / 272.8)(20{,}000) = 1195 \text{ psi}.$$

This value is substantially lower than the stress value of 2716 psi obtained from elastic analysis in Example 3.3.

3.7 Piping Analysis – ASME B31.1 and B31.3

3.7.1 Introduction

A piping system is one of the more complex structural configurations a component designer or analyst is likely to encounter. Although the straight sections are relatively simple, the connecting components, principally elbows and branch connections, have complex stress distributions even under simple loading conditions. In addition, the piping system frequently contains other components, such as flanged joints and valves. Added to this complexity is the fact that all components of the piping system interact with each other such that the loading on any single component is a function of the response of the whole piping system to applied loading conditions.

There are many references for the design and evaluation of piping systems and it is not the purpose of this section to replicate those sources. Rather, the purpose is to (1) acquaint the designer/analyst with the overall approach to piping design, as embodied in ASME B31.1 Power Piping (ASME B31.1 Power piping 2020) and ASME B31.3 Process Piping (ASME B31.3 Process Piping 2020), (2) discuss some of the issues impacting piping design in the creep regime, and (3) provide a hand evaluation of a simple pipeline to illustrate the application of ASME B31.3. Although they differ in some details, the overall approach in ASME B31.1 and ASME B31.3 is quite similar. The following discussion is based largely on Chapters 16 and 17 by Becht in the Companion Guide to the "ASME Boiler and Pressure Vessel Code" Rao (2018) and Chapter 38 by Rodabaugh in the same reference.

3.7.2 Design Categories and Allowable Stresses

There are three general categories of design requirement in ASME B31.1 and B31.3: pressure design, sustained and occasional loading, and thermal expansion.

3.7.2.1 Pressure Design

There are four basic approaches to pressure design: (1) components for which there are tabulated pressure/temperature ratings in their applicable standard; (2) components that are specified to have the same pressure rating as the attached piping; (3) components such as straight pipe and branch connections for which design equations are provided to determine minimum required thickness; and (4) non-standard components for which experimental methods, for example, may be used. The design equations for minimum wall thickness are nominally based on Lame's hoop stress equations, which take into account the variation of pressure-induced stress through the wall thickness. The allowable stress for pressure loading is based on criteria similar to those used for ASME I and VIII Div 1. In the creep regime, this is the lower of 67% of the average creep rupture strength in 100,000 h, 80% of the minimum creep strength or the average stress to cause a minimum creep rate of 0.01% per 1000 h.

3.7.2.2 Sustained and Occasional Loading

Whereas the pressure stress criteria are nominally based on circumferential stress, the sustained and occasional load limits are based on longitudinal bending stresses. The limit for sustained loading (e.g., pressure and weight) is the same as for pressure design. However, because the longitudinal stress due to pressure is nominally half the hoop stress, there is a margin left for longitudinal bending and pressure stresses. The limit for occasional loading, pressure, and weight plus seismic and wind, is a factor that is higher depending on the loading duration.

Both ASME B31.1 and B31.3 specify the use of 0.75 times the basic factor stress intensification factors for sustained and occasional loading; however, Becht in Rao (2018) notes that some piping analysis computer programs use the full intensification factor for the sustained and occasional load limit. Considering also stress redistribution effects due to creep as discussed earlier in this chapter, it is recommended that the full intensification factor be used in evaluation of sustained and occasional load stresses for ASME B31.1 and B31.3 applications in the creep regime.

3.7.2.3 Thermal Expansion

Not all piping systems require an analytical evaluation of piping system stresses due to restrained thermal expansion. Guidelines for the exceptions are provided in ASME B31.1 and B31.3. When an analytical evaluation is made, it is based on elastic analysis using classical global stiffness matrices based on superposition principals. However, the flexibility of piping components, such as elbows and miter joints, is modified by flexibility factors that approximate the additional flexibility of these components as compared to a straight pipe with the same length as the center-line length of the curved pipe. Formulas for these flexibility factors are provided in the respective codes.

In addition to enhanced flexibility, stress levels in components such as elbows and branch connections are also higher than would be the case for a straight pipe under the same loads. To account for these higher stresses, stress intensification factors are also provided. Component stresses are then calculated from the equation

$$S_b = \frac{\left[\left(i_i M_i\right)^2 + \left(i_o M_o\right)^2\right]^{1/2}}{z} \tag{3.39}$$

where

i_i = in-plane stress intensification factor (see Table 3.5)
i_o = out-of-plane stress intensification factor (see Table 3.5)
M_i = in-plane bending moment
M_o = out-of-plane bending moment
S_b = resultant bending stress
z = section modulus of pipe.

The allowable stress range for restrained thermal expansion is designed to achieve shakedown in a few cycles. The background on shakedown concepts is discussed in more detail in Section 1.6.3. The basic limit Severud (1975, 1980)

Table 3.5 Flexibility and stress intensification factors [ASME B31.3].

		Stress Intensification			
Description	**Flexibility factor, *k***	**Out-of-plane, *Io***	**In-plane, *Ii***	**Flexibility characteristic, *h***	**Sketch**
Welding elbow or pipe bend	$\frac{1.65}{h}$	$\frac{0.75}{h^{2/3}}$	$\frac{0.9}{h^{2/3}}$	$\frac{\bar{T}R_1}{r_2^2}$	$\bar{T}$, r_2, R_1 = bend radius
Closely spaced miter bend $s < r_2 (1 + \tan\theta)$	$\frac{1.52}{h^{5/6}}$	$\frac{0.9}{h^{2/3}}$	$\frac{0.9}{h^{2/3}}$	$\frac{\cot\theta}{2}\left(\frac{s\bar{T}}{r_2^2}\right)$	$\bar{T}$, r_2, s, θ, $R_1 = \frac{s\cot\theta}{2}$

In the above equations, the stress range reduction factor, *f*, is based on a fatigue curve for butt-welded carbon steel pipe developed by Markl (1960).

for the allowable range of restrained thermal expansion stresses in ASME B31.1 and 31.3 is given by the equation

$$S_A = f\left(1.25S_c + 0.25S_h\right) \tag{3.40}$$

where

S_A = allowable displacement stress range
S_c = allowable stress at the cold end of the thermal cycle
S_h = allowable stress at the hot end of the thermal cycle
f = stress range reduction factor used to account for fatigue effects when the number of equivalent full-range thermal cycles exceeds 7000 (about once a day for 20 years).

The above equation does not give credit for the unused portion of the allowable longitudinal stress, S_L, discussed in Section 3.5.2.2. If such credit is taken, Eq. (3.40) becomes

$$S_A = f\left[1.25\left(S_c + S_h\right) - S_L\right]. \tag{3.41}$$

Although Eq. (3.41) provides a somewhat higher allowable stress, the difference becomes small as the temperature moves into the creep regime and the allowable stress is dependent on creep properties. Equation (3.40) has the advantage that thermal expansion stresses may be evaluated separately from the local values of longitudinal stresses, S_L.

3.7.3 Creep Effects

There are several areas where creep effects play an important role in piping design, among them are weld strength reduction factors, elastic follow-up, and the effects of creep on cyclic life.

3.7.3.1 Weld Strength Reduction Factors

Weld strength reduction factors were introduced in the 2004 edition of ASME B31.3 to account for the reduced creep rupture strength of weld metal as compared to base metal. Although the designer may use factors based on specific data, a general factor is provided that is applicable to all materials in lieu of specific weld metal data. This factor varies linearly from 1.0 at 950°F (510°C) to 0.5 at 1500°F (815°C).

3.7.3.2 Elastic Follow-Up

The use of elastic analysis procedures to determine the stress/strain distribution due to restrained thermal expansion is based on the assumption that the piping system is balanced without localized high stress areas. Examples of piping configurations that are unbalanced are provided in ASME B31.3, but there are no quantitative criteria provided. An unbalanced system operating in the creep regime can be particularly susceptible to elastic follow-up considerations, as discussed in Section 1.6.1. Under severe follow-up conditions, with localized stress/strain concentrations, the resultant localized stress due to restrained thermal expansion may not relax, thus behaving more like a sustained load. Under these circumstances – i.e., a significantly unbalanced system – it may be appropriate to treat these stresses similarly to a sustained load with an allowable stress of S_h.

3.7.3.3 Cyclic Life Degradation

The tests on which the *f* factors are based were performed on carbon and stainless steel components below the creep regime. Operating in the creep regime can have several deleterious effects. First, the basic continuous cycling fatigue curve will be lower at higher temperatures. Second, as previously discussed in Section 1.6.4, there is a hold-time effect on cyclic life due to the accumulation of creep damage as restrained thermal expansion stresses relax. And, third, due to Neuber-like effects, the strain at structural discontinuities will be greater than predicted by elastic analyses. Taking direct account of these phenomena goes well beyond the scope of normal piping analyses. There is, however, a conservative approximation included in ASME III-5 for Class B nuclear components at elevated temperature. The changes in the *f* factor, as described in Article HCB-1-2000 of ASME III-5, are an indication of the potential significance of the effects of creep on cyclic life. For carbon steel, instead of a reduction starting at 7000 cycles below the creep regime, at 750°F (400°C) the reduction starts at 50 cycles and decreases to five cycles at 900°F (480°C). For 304 SS, the reduction starts at 850°F (455°C), reducing to 50 cycles at 1000°F (540°C) and five cycles at 1200°F (650°C).

The above discussions should not be taken as direct recommendations for reduction in allowable stress levels for ASME B31.1 and B31.3 in the creep regime. Application of these criteria has, after all, led to a long history of successful service experience. It is, however, a cautionary recommendation against pushing the criteria to their limits – particularly when dealing with unfavorable geometries and a large number of service cycles.

3.8 Circular Plates

The differential equations for the bending of circular plates in the creep regime have no closed-form solution. Numerical solutions obtained by Odqvist (1966), Kraus (1980), and other authors have shown that the stationary stress solution of circular plate equations in the creep range approach the results obtained by plastic theory as the value of n increases. For $n = 1$, the elastic and creep results are the same.

Table 3.6 Maximum moment in a solid circular plate[1] [Jawad (2018)].

Loading condition	Simply supported edge[2]	Fixed edge[3]
Uniform load, P (elastic solution)	$PR^2/4.85$	$PR^2/8$
Uniform load, P (plastic solution)	$PR^2/6$	$PR^2/12$
Center load, P (plastic solution)	$F/(2\pi)$	$F/(4\pi)$

[1] Poisson's ratio is taken as 0.30.
[2] Maximum moment is at the center of the plate.
[3] Maximum moment is at the edge of the plate.F, concentrated load at center of plate; P, uniform applied pressure; R_o, radius of plate.

Table 3.6 shows various values of the maximum bending moment in a circular plate using elastic and plastic analysis. For a simply supported plate, the maximum bending moment due to lateral pressure and based on elastic analysis is

$$M = PR^2 / 4.85 \tag{3.42}$$

where

M = bending moment
P = applied pressure
R = radius of plate.

The required thickness is obtained from the equation $S = 6\,M/t^2$ as

$$t = R\,[1.237P / S]^{1/2} \tag{3.43}$$

where

S = allowable stress
t = thickness.

Equation (3.43) can be rewritten in a slightly different form by substituting SE for S and d for $2R_o$. The result is

$$t = d\left[CP / SE_0\right]^{1/2} \tag{3.44}$$

where

d = diameter of the plate
E_o = joint efficiency
C = constant (0.309 for simply supported plate 0.188 for fixed plate).

The ASME I and VIII-1 codes use Eq. (3.44) for a variety of unstayed flat heads. Numerous values of C are given for the various head configurations and shell attachment details. ASME I and VIII-1 use Eq. (3.44) for all temperature ranges, including those in the creep range.

It is of interest to note that a plastic analysis, performed as a limit case for creep analysis, results in a maximum moment for the simply supported case $M = PR^2/6$, as shown in Table 3.6. From the expression $S = 6\ M/t^2$, and using a shape factor of 1.25 from Table 3.2, we have

$$t = d[0.2P / S]^{1/2}. \tag{3.45}$$

A comparison of Eqs. (3.44) and (3.45) shows that a thickness reduction of 20% is achieved by performing plastic analysis compared to elastic analysis for the same allowable stress.

Problem

3.1 The 12-in. diameter Standard Schedule pipe loop is as shown in the figure. The pipe loop and the vessels undergo a temperature increase from ambient to 1000°F. Calculate the thermal stress in the pipe loop and compare it with the allowable stress values. The allowable stress at ambient temperature is 17,100 psi and at 1000°F it is 8000 psi. Let $f = 1.0$, $E = 20.4 \times 10^6$ psi, $\alpha = 8.2 \times 10^6$ in./in.-°F, $A = 14.6$ in.^{2}, $I = 279$ in.^{4}, $t = 0.375$ in.

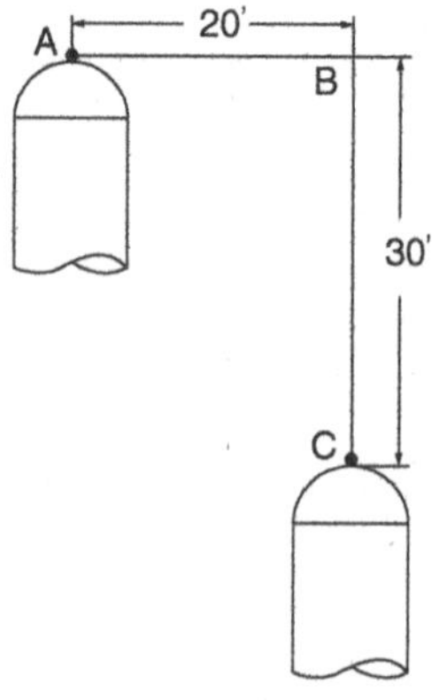

4

Analysis of ASME Pressure Vessel Components

Load-Controlled Limits

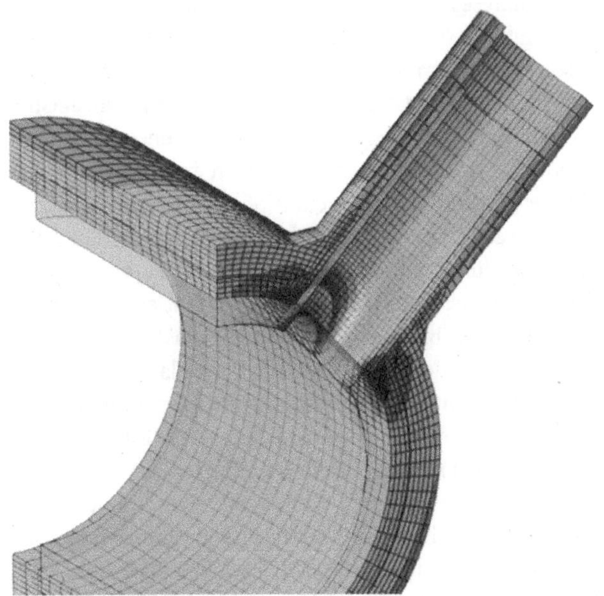

Finite element model [Chithranjan Nadarajah]

4.1 Introduction

The ability of pressure vessel shells to perform properly in the creep range depends on many factors, such as their stress level, material properties, temperature range, as well as operating temperature and pressure cycles. ASME Sections I and VIII-1, which are based on Design by Rules rather than Design

Analysis of ASME Boiler, Pressure Vessel, and Nuclear Components in the Creep Range, Second Edition. Maan H. Jawad and Robert I. Jetter.
© 2022 John Wiley & Sons Ltd. Published 2022 by John Wiley & Sons Ltd.

by Analysis, allow components to be designed elastically at elevated temperature service as long as the allowable stress is based on creep-rupture data obtained from ASME II-D. In many cases, however, additional analyses are required to ascertain the adequacy of such components at elevated temperatures. The effect of creep on long-time exposure of various components in power boilers to high temperatures is of interest. The effect of continuous startup and shutdown of heat recovery steam generators (HRSG) on life expectancy at elevated temperatures is one of the design issues. Another safety concern is the effect of long-time exposure of equipment used in the petrochemical industry to high temperatures.

These additional analyses include evaluation of bending and thermal stresses as well as creep and fatigue. The sophistication of the analyses could range from a simple elastic solution to a more complicated plastic or a very complex creep analysis. The preference of most engineers is to perform an elastic analysis because of its simplicity, expediency, and cost effectiveness. The approximate elastic analysis is not as accurate as plastic or creep analysis, which reflects more accurately the stress-strain interaction of the material. However, elastic analysis is sufficiently accurate for most design applications. The ASME's elastic procedure compensates for this approximation by separating the stresses obtained from elastic analysis into various stress categories in order to simulate, as closely as practically possible, the more accurate plastic or creep analysis.

ASME I and VIII-1 provide Design-by-Rule formulae used in conjunction with implicit time-dependent allowable stress values for component sizing and local design details. The allowable stress values for ASME I and VIII-1 are tabulated in Tables 1A and B in ASME II-D. In the high temperature regime where creep effects prevail, the time-dependent allowable stress values are a function of creep response at 100,000 h and are not intended to represent specification of an explicit design life. Also, neither ASME I nor VIII-1 provide explicit rules or methods for evaluation of deformation-controlled stress limits, such as for thermal stress or fatigue/creep-fatigue. However, both provide nonspecific guidance if these loads need consideration. Thus, to paraphrase for ASME-I, PG-16 notes that the Section does not have rules for all possible details of design and when details are not given it is intended that the Manufacturer use appropriate details done by appropriate analytical methods or rules from other design codes. For ASME VIII-1, UG-20 recommends design details per WRC Bulletin 470, "Recommendations for Design of Vessels for Elevated Temperature Service" but notes that the manufacturer is not relieved of the design responsibility with regard to consideration of stresses associated with both steady-state conditions and transient events, such as startup, shutdown, intermittent operation, thermal cycling, etc., as defined by the user.

The 2007 edition of ASME VIII-2 and subsequent editions cover temperatures in the creep regime above the previous limits of 700°F (370°C) and 800°F (425°C) for ferritic and austenitic materials, respectively. ASME VIII-2 (Part 4, Class 1 and Class 2) and (Part 5) also require either meeting the requirements for exemption from fatigue analysis, or, if that requirement is not satisfied, meeting the requirements in ASME-2 (Part 5) for fatigue analysis. However, above the 700/800°F (370/425°C) limit, the only available option for either Part 4 or Part 5 is to satisfy the exemption from fatigue analysis requirements based on comparable equipment in comparable service because the fatigue curves required for a full fatigue analysis are limited to 700°F (370°C) and 800°F (425°C) and the cyclic design evaluation methodology above this temperature doesn't account for creep damage; for example, creep-fatigue.

ASME VIII-2's supplemental creep rules are detailed rules for analyzing boiler and pressure vessel components in the creep regime that supplement the Design-by-Analysis rules in ASME VIII-2 Part 5. These rules are a simplification of the rules given in ASME III-5 for Class A components operating the creep regime. The features of ASME III-5 are discussed in more detail in Chapter 8. As in ASME III-5, ASME VIII-2's supplemental creep rules have explicit time-dependent allowable stress values for a limited number of materials and procedures for evaluating strain limits and creep-fatigue. The focus of this chapter is on the evaluation of load-controlled, primary stress limits. The evaluation of strain limits is addressed in Chapter 5 and creep-fatigue damage in Chapter 6.

4.2 Design Thickness

The required shell thickness for ASME I and VIII vessels at any temperature is obtained from the corresponding equations given in these two sections. When an additional stress analysis is specified or required at elevated temperatures, then the procedure established in ASME VIII-2 for creep analysis, Figure 4.1, is used as a guideline. The procedure is based on first calculating a thickness of the component.

The thickness is based on design conditions and is obtained from applicable design equations in ASME I or VIII-2 Part 4. Stresses due to operating conditions are then calculated and maintained below specified allowable limits obtained from ASME VIII-2 Part 5.

For illustration purposes, the following equations for cylindrical shells are given and will be used throughout the next two chapters in solving various problems.

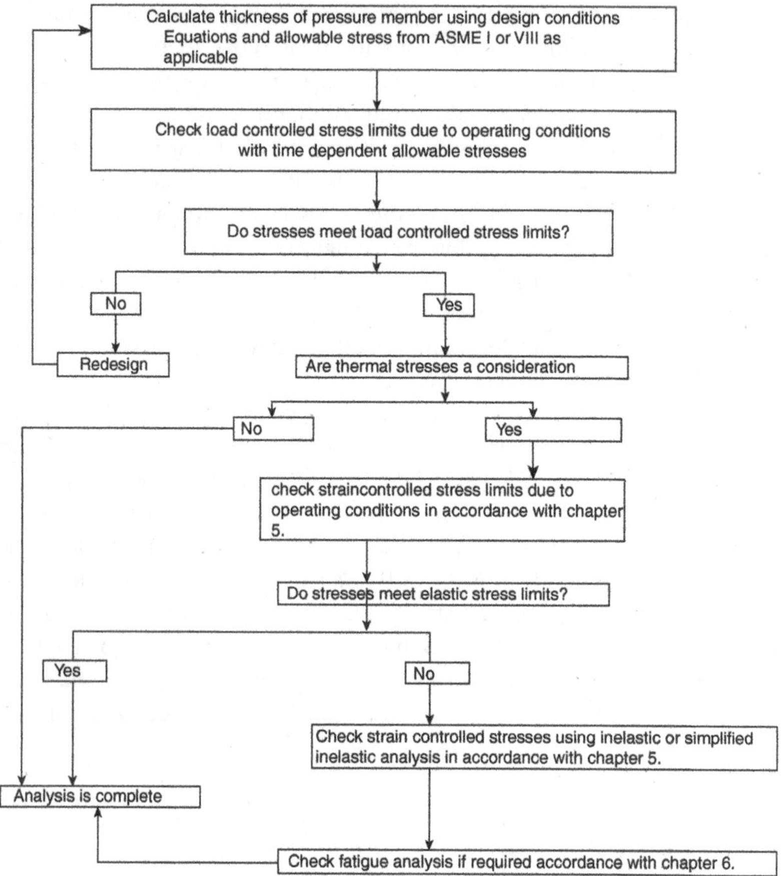

Figure 4.1 Sequence in creep analysis for design and operating conditions.

4.2.1 ASME I

The design shell thickness when thickness does not exceed one-half the inside radius is given by

$$t = \frac{PD_0}{2SE_0 + 2yP} + c \tag{4.1}$$

or

$$t = \frac{PR_i}{SE_o - (1-y)P} + c \tag{4.2}$$

where

c = corrosion allowance
D_o = outside diameter
E_o = joint efficiency obtained from ASME I
P = internal pressure
R_i = inside radius
S = allowable stress obtained from ASME II-D
t = thickness
y = temperature coefficient as given in Table 4.1.

The y factor in the above two equations was introduced in ASME I in the mid-1950s Winston et al. (1954) to take into account the reduction of stress due to redistribution when the temperature is in the creep region.

The design shell thickness, when thickness exceeds one-half the inside radius, is

$$t = \left[(Z_1)^{1/2} - 1\right] R_i \tag{4.3}$$

where $Z_1 = (SE_o + P)/(SE_o - P)$.

This equation is based on Lame's Eq. (4.9), discussed later in this chapter. ASME I recently added a new shell design equation in Appendix A of ASME I and is given by

$$t = D_i\left(e^{(P/SE_o)} - 1\right)/2 + c + f' \tag{4.4}$$

where

f' = thickness factor for expanded tube ends.

Equation (4.4) is based on limit analysis. It is applicable at low as well as high temperatures. However, caution must be exercised when using this equation at elevated temperatures for thick-wall cylinders or material embrittled with aging because cracking on the surface will cause loss of load-carrying capacity.

4.2.2 ASME VIII

The design shell thickness in ASME VIII-1 when thickness does not exceed one-half the inside radius or pressure does not exceed 0.385 SE_o is given by

$$t = \frac{PR_i}{SE_o - 0.6P} + c. \tag{4.5}$$

The design shell thickness in ASME VIII-1 when thickness exceeds one-half the inside radius or pressure exceeds 0.385 SE_o is

$$t = \left(Z^{1/2} - 1\right)R_i + c \tag{4.6}$$

where $Z = (SE_o + P)/(SE_o - P)$.

E_o = joint efficiency obtained from ASME VIII

Table 4.1 The y coefficients in Eqs. (4.1) and (4.2) of ASME I.

	Temperature[1]							
	900 (480) and below	950 (510)	1000 (540)	1050 (565)	1100 (595)	1150 (620)	1200 (650)	1250 (675) and above
Ferritic	0.4	0.5	0.7	0.7	0.7	0.7	0.7	0.7
Austenitic	0.4	0.4	0.4	0.4	0.5	0.7	0.7	0.7
Alloy 800	0.4	0.4	0.4	0.4	0.4	0.4	0.5	0.7
800H, 800 HT	0.4	0.4	0.4	0.4	0.4	0.4	0.5	0.7
825	0.4	0.4	0.4	-	-	-	-	-
230 Alloy	0.4	0.4	0.4	0.4	0.4	0.4	0.5	0.7
N06045	0.4	0.4	0.4	0.4	0.5	0.7	0.7	0.7
N06690	0.4	0.4	0.4	0.4	0.5	0.7	0.7	-
Alloy 617	0.4	0.4	0.4	0.4	0.4	0.4	0.5	0.7

[1]Values are in °F, those in parentheses are in °C.

The design shell thickness in ASME VIII-2, Part 4 (Design by Formula), is

$$t = D_i\left(e^{P/SE_o} - 1\right)/2. \tag{4.7}$$

This equation is based on limit analysis.

Example 4.1 A pressure vessel is constructed of alloy 2.25Cr-1Mo steel and has an inside radius of 5 in. The design temperature is 1000°F, the design pressure is 3000 psi, and the allowable stress is 7800 psi. The joint efficiency, E_o, is equal to 1.0 and the corrosion allowance, c, is zero. What is the required thickness

(a) in accordance with ASME VIII-1?
(b) in accordance with ASME I?
(c) in accordance with ASME VIII-2?

Solution
(a) From Eq. (4.5)

$$t = \frac{(3000)(5)}{(7800)(1.0) - 0.6(3000)} + 0.0$$
$$= 2.5 \text{ in.}$$

(b) From Table 4.1, $y = 0.7$ and from Eq. (4.2)

$$t = \frac{(3000)(5)}{(7800)(1.0) - (1 - 0.7)(3000)} + 0.0$$
$$= 2.17 \text{ in.}$$

(c) From Eq. (4.7)

$$t = 10\left(e^{3000/7800(1.0)} - 1\right)/2$$
$$= 2.35 \text{ in.}$$

Notice that the thickness obtained from ASME I is smaller than that obtained from ASME VIII-1 due to the y factor adjustment for elevated temperature. However, the obtained thicknesses from ASME I and VIII-1 would be identical at lower temperatures, where the y factor in Eq. (4.2) is taken as 0.4.

The thickness of an ASME I shell at 1000°F, obtained from Eq. (4.4), which has recently been added to ASME I, is 2.35 in. Thus, at this high temperature it is more economical to use the 2.17 in. obtained from Eq. (4.2). However, at temperatures below the creep value, Eqs. (4.2) and (4.4) yield about the same results.

The thickness obtained from the ASME VIII-2 Part 4 equation is less than that obtained from the ASME VIII-1 equation for the same allowable stress. This is due to the stress redistribution taken into account in Eq. (4.7) as a result of limit analysis.

All cylindrical shell equations in ASME I and VIII are based on Lame's equation, or an approximation of it. Lame's equation, which is based on elastic analysis, is expressed as

$$\sigma = P\frac{\left[\left(R_0 / R\right)^2 + 1\right]}{\gamma^2 - 1} \tag{4.8}$$

where

R = radius
R_o = outside radius
$\gamma = R_o/R_i$.

This equation has a maximum stress value at the inner surface where $R = R_i$ of

$$\sigma = P\left(\frac{\gamma^2 + 1}{\gamma^2 - 1}\right). \tag{4.9}$$

The outside stress with $R = R_o$ is given by the equation

$$\sigma = 2P / \left(\gamma^2 - 1\right). \tag{4.10}$$

Example 4.2 Find the inside and outside stress values in the ASME VIII-1 and ASME I vessels of Example 4.1 using Lame's Eqs. (4.9) and (4.10).

Solution

(a) For ASME VIII-1 with $t = 2.5$ in.

$$\gamma = 7.5 / 5 = 1.5.$$

The inside stress is obtained from Eq. (4.9)

$$\begin{aligned}\sigma_i &= 3000\frac{\left[1.5^2 + 1\right]}{1.5^2 - 1} \\ &= 7800 \text{ psi}\end{aligned}$$

which is the same as the allowable stress.

The outside stress is obtained from Eq. (4.10)

$$\begin{aligned}\sigma_0 &= 2(3000) / \left(1.5^2 - 1\right). \\ &= 4800 \text{ psi}\end{aligned}$$

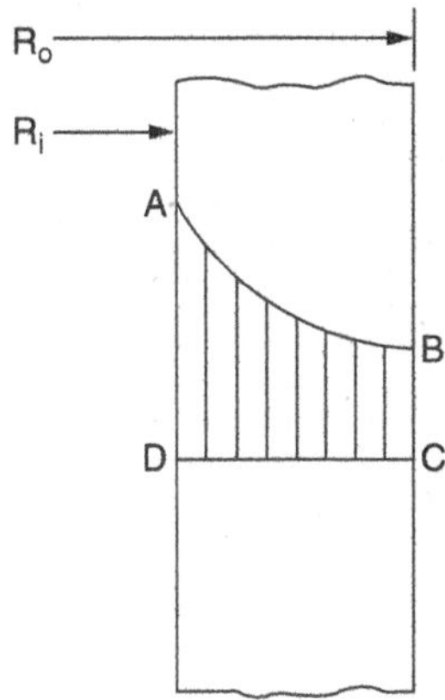

Figure 4.2 Circumferential stress distribution through the thickness of a cylindrical shell.

(b) For ASME I with $t = 2.17$ in.

$$\gamma = 7.17/5 = 1.434.$$

The inside stress is

$$\begin{aligned}\sigma_i &= 3000\frac{\left[1.434^2 + 1\right]}{1.434^2 - 1}.\\ &= 8680 \text{ psi}\end{aligned}$$

This value, which is based on elastic analysis, is larger than the allowable stress at elevated temperatures. However, at a sustained elevated temperature, this stress relaxes to a lower value due to creep (as discussed later in this chapter).

The outside stress is

$$\begin{aligned}\sigma_0 &= 2(3000)/\left(1.434^2 - 1\right).\\ &= 5680 \text{ psi}\end{aligned}$$

A plot of Eq. (4.9) for a thick shell is shown as line AB in Figure 4.2. This line shows a nonlinear stress distribution across the thickness. ASME VIII-2 and III, as well as some other international codes, splits this nonlinear calculated stress into different components when an elastic stress analysis is performed. Methods of splitting this stress into various categories, and the definitions of these categories, are discussed next.

4.3 Stress Categories

The evaluation for the load-controlled stress limits is made in accordance with Figure 4.1. The procedure for creep analysis in ASME VIII-2 consists of first calculating a trial thickness of the member and then performing an elastic stress

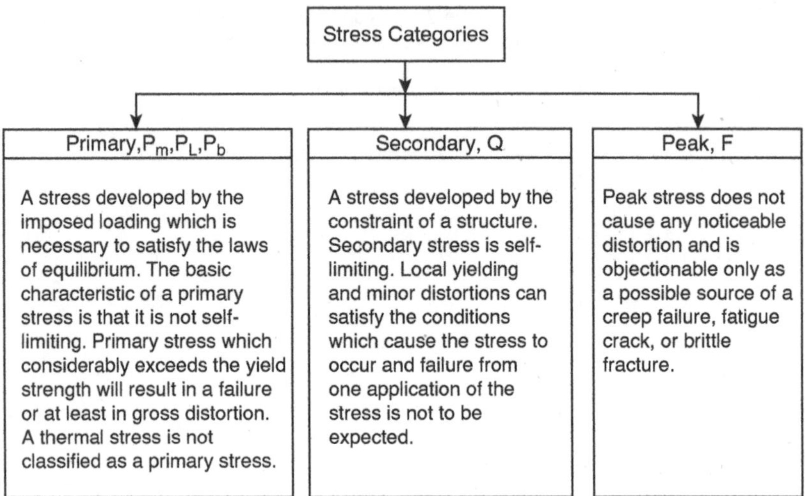

Figure 4.3 Stress categories.

analysis to ascertain its adequacy in the creep regime. The stress in the member is normally due to various operating conditions, such as pressure, temperature, and other loads. These loads may result in a complex stress profile across the thickness. This complex stress profile is separated into various categories to simplify the elastic analysis. Each of the stress categories is then compared to an allowable stress or equivalent strain limit. To demonstrate this procedure, it is necessary to define the various stress categories in accordance with the ASME code. ASME VIII-2 and III define three different stress categories as shown in Figure 4.3. They are referred to as the primary stress, secondary stress, and peak stress. The ASME description of these stress categories follows.

4.3.1 Primary Stress

Primary stress is any normal or shear stress developed by an imposed loading, which is necessary to satisfy the laws of equilibrium of external and internal forces and moments.

The basic characteristic of a primary stress is that it is not self-limiting. Primary stress is further subdivided into two categories, as shown in Figure 4.4. The first category is primary membrane stress, which could either be general P_m or local P_L whereas the second category is primary bending stress, P_b.

4.3.1.1 General Primary Membrane Stress (P_m)

This stress is so distributed in the structure that no redistribution of load occurs as a result of yielding. An example is the stress in a circular cylindrical or spherical shell,

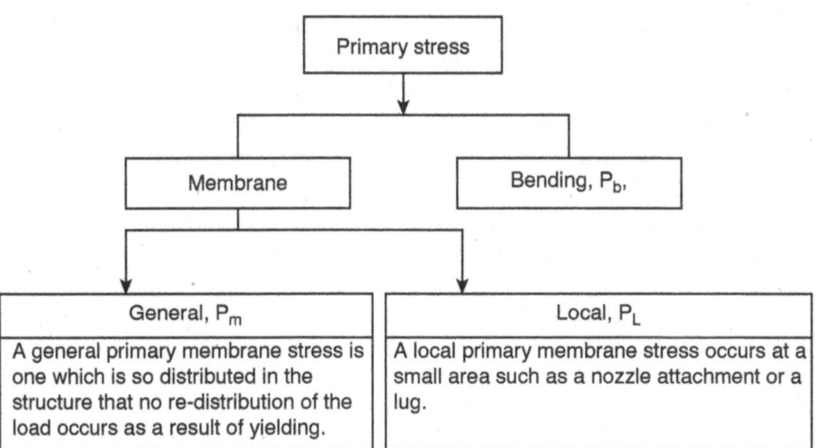

Figure 4.4 Primary stress.

away from discontinuities, due to internal pressure or to distributed live loads.

4.3.1.2 Local Primary Membrane Stress (P_L)

Cases arise in which a membrane stress, produced by pressure or other mechanical loading and associated with a primary or a discontinuity effect, produces excessive distortion in the transfer of load to other portions of the structure. Conservatism requires that such a stress be classified as local primary membrane stress even though it shows some characteristics of a secondary stress. Examples include the membrane stress in a shell produced by external loads and moment at a permanent support or at a nozzle connection.

It is important to emphasize that the above definition includes pressure-induced loads as well as those due to mechanical loading, and that the definition is equally applicable at any location in the structure.

4.3.1.3 Primary Bending Stress (P_b)

This stress is the variable component of normal stress in a cross section. An example is the bending stress in the central portion of a flat head due to pressure.

4.3.2 Secondary Stress, Q

Secondary stress is a normal stress or a shear stress developed by the constraint of adjacent material or by self-constraint of the structure, and thus it is normally associated with deformation-controlled quantity at elevated temperatures. The basic characteristic of a secondary stress is that it is self-limiting. Local yielding and minor distortions can satisfy the conditions that cause the stress to occur and failure from one application of the stress is not to be expected. Examples of secondary stresses are bending stress at a gross structural discontinuity, bending stress due to a linear radial thermal strain profile through the thickness of section, and stress produced by an axial temperature distribution in a cylindrical shell.

The designer must keep in mind that thermally induced membrane stresses are considered primary in the paragraph for creep-fatigue evaluations of ASME VIII-2. Similarly, pressure-induced primary plus secondary stresses are considered primary stress in the creep-fatigue evaluation of ASME VIII-2.

4.3.3 Peak Stress, F

Peak stress is that increment of stress that is additive to the primary plus secondary stresses by reason of local discontinuities of local thermal stress, including the effects of stress concentrations. The basic characteristic of a peak stress is that it does not cause any noticeable distortion and is objectionable only as a possible source of a fatigue crack or a brittle fracture, and, at elevated temperatures, as a possible source of localized rupture or creep fatigue failure. Some examples of peak stress are thermal stress at local discontinuities and cladding, and stress at a local discontinuity.

4.3.4 Separation of Stresses

A complex stress pattern in a cross section must be separated into primary, secondary, and peak stress components in accordance with ASME. Table 4.2 lists the stress categories of some commonly encountered load cases. The table shows that internal pressure for cylindrical shells must be divided into primary membrane, P_m, and secondary, Q, stresses. Accordingly, the parabolic stress distribution given by Eq. (4.9) and shown as area ABCD in Figure 4.2 must be decomposed into three parts as shown in Figure 4.5. The membrane stress CDEF is obtained by integrating area ABCD over the thickness and dividing by the total thickness. This results in the simple equation

$$P_m = \frac{P}{\gamma - 1} = (PR_i)/t. \tag{4.11}$$

The secondary bending stress component, Q, is given by coordinates EG and FH. The value of Q is determined by integrating area BJF times its moment

Table 4.2 Classification of stresses for some typical cases [ASME VIII-2].

Examples of Stress Classification				
Vessel Component	**Location**	**Origin of Stress**	**Type of Stress**	**Classification**
Any shell including cylinders, cones, spheres, and formed heads	Shell plate remote from discontinuities	Internal pressure	General membrane	P_m
			Gradient through plate thickness	Q
		Axial thermal gradient	Membrane	Q
			Bending	Q
	Near nozzle or other opening	Net-section axial force and/or bending moment applied to the nozzle, and/or internal pressure	Local membrane	P_L
			Bending	Q
			Peak (fillet or corner)	F
	Any location	Temperature difference between shell and head	Membrane	Q
			Bending	Q
	Shell distortions such as out-of-roundness and dents	Internal pressure	Membrane	P_m
			Bending	Q
Cylindrical or conical shell	Any section across entire vessel	Net-section axial force, bending moment applied to the cylinder or cone, and/or internal pressure	Membrane stress averaged through the thickness, remote from discontinuities; stress component perpendicular to cross section	P_m
			Bending stress through the thickness; stress component perpendicular to cross section	P_b
	junction with head or flange	Internal pressure	Membrane	P_L
			Bending	Q

(*Continued*)

Table 4.2 (Continued)

Examples of Stress Classification				
Vessel Component	**Location**	**Origin of Stress**	**Type of Stress**	**Classification**
Dished head or conical head	Crown	Internal pressure	Membrane	P_m
			Bending	P_b
	Knuckle or junction to shell	Internal pressure	Membrane	P_L [Note (1)]
			Bending	Q
Flat head	Center region	Internal pressure	Membrane	P_m
			Bending	P_b
	Junction to shell	Internal pressure	Membrane	P_L
			Bending	Q [Note (2)]
Perforated head or shell	Typical ligament in a uniform pattern	Pressure	Membrane (averaged through cross section)	P_m
			Bending (averaged through width of ligament., but gradient through plate)	P_b
			Peak	F
	Isolated or atypical ligament	Pressure	Membrane	Q
			Bending	F
			Peak	F
Nozzle	Within the limits of reinforcement	Pressure and external loads and moments, including those attributable to restrained free end displacements of attached piping	General membrane	P_m
			Bending (other than gross structural discontinuity stresses) averaged through nozzle thickness	P_m

Table 4.2 (Continued)

Examples of Stress Classification				
Vessel Component	**Location**	**Origin of Stress**	**Type of Stress**	**Classification**
	Outside the limits of reinforcement	Pressure and external axial, shear, and torsional loads, including those attributable to restrained free end displacements of attached piping	General membrane	P_m
		Pressure and external loads and moments, excluding those attributable to restrained free end displacements of attached piping	Membrane Bending	P_L P_b
		Pressure and all external loads and moments	Membrane Bending Peak	P_L Q F
	Nozzle wall	Gross structural discontinuities	Membrane Bending Peak	P_L Q F
		Differential expansion	Membrane Bending Peak	Q Q F

(Continued)

Table 4.2 (Continued)

Examples of Stress Classification				
Vessel Component	**Location**	**Origin of Stress**	**Type of Stress**	**Classification**
Cladding	Any	Differential expansion	Membrane	F
			Bending	F
Any	Any	Radial temperature distribution [Note (3)]	Equivalent linear stress [Note (4)]	$\boldsymbol{Q}$
			Nonlinear portion of stress distribution	$\boldsymbol{F}$
Any	Any	Any	Stress concentration {notch effect}	$\boldsymbol{F}$

Notes
1 Consideration shall be given to the possibility of wrinkling and excessive deformation in vessels with large diameter-to-thickness ratio.
2 If the bending moment at the edge is required to maintain the bending stress in the center region within acceptable limits, the edge bending is classified as P_b; otherwise, it is classified as Q.
3 Consider possibility of thermal stress ratchet.
4 Equivalcnt linear stress is defined as the linear stress distribution that has the same net bending moment as the actual stress distribution.

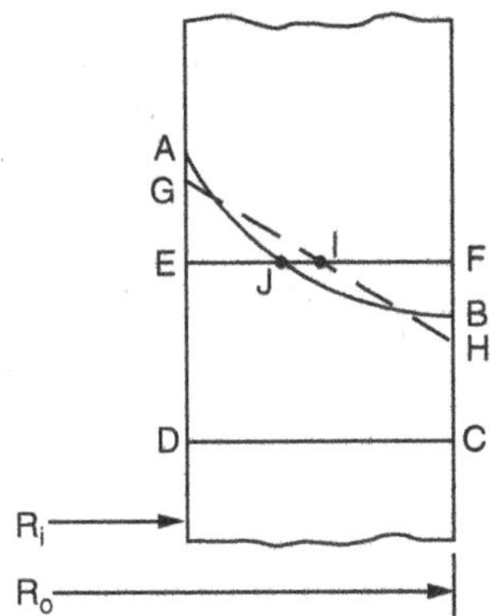

Figure 4.5 Separation of stresses in a cylindrical shell.

arm around Point *J* plus area *EAJ* times its moment arm around *J* and then multiplying the total sum by the quantity $6/t^2$. This results in a *Q* value of

$$Q = \frac{6P\gamma}{(\gamma-1)^2}\left[\frac{1}{2} - \frac{\gamma}{\gamma^2-1}(\ln\gamma)\right]. \tag{4.12}$$

The peak stress component, *F*, is obtained by subtracting the quantities P_m plus *Q* from the total stress at a given location. Hence,

$$F_{max} = P\left(\frac{\gamma^2+1}{\gamma^2-1}\right) - P_m - Q. \tag{4.13}$$

Example 4.3 A steam drum has an inside diameter of 30 in. and outside diameter of 46 in. The applied pressure is 3300 psi. Determine the following:
(a) Maximum and minimum stresses
(b) Membrane stress P_m
(c) Secondary stress *Q*
(d) Peak stress *F*.

Solution
(a) From Lame's Eq. (4.9) with $\gamma = 23/15 = 1.533$, the maximum and minimum stresses at the inside and outside surfaces are

$$\sigma_i = 3300\frac{\left[1.533^2+1\right]}{1.533^2-1}$$
$$= 8190 \text{ psi}$$
$$\sigma_o = 2(3300)/\left(1.533^2-1\right)$$
$$= 4890 \text{ psi}$$

(b) From Eq. (4.11),

$$\begin{aligned} P_m &= \frac{3300}{1.533-1} \\ &= 6190 \text{ psi} \end{aligned}$$

(c) From Eq. (4.12),

$$\begin{aligned} Q &= \frac{6(3300)(1.533)}{(1.533-1)^2}\left[\frac{1}{2} - \frac{1.533}{1.533^2-1}(\ln 1.533)\right] \\ &= 106{,}845(0.5-0.485) \\ &= 1590 \text{ psi} \end{aligned}$$

(d) The maximum peak stress is obtained from Eq. (4.13)

$$\begin{aligned} F &= \sigma_{\text{inside}} - (P_m + Q) \\ &= 8190 - (6180 + 1590) \\ &= 410 \text{ psi} \end{aligned}$$

4.3.5 Thermal Stress

Most vessels operating at elevated temperatures are subjected to thermal gradients and stress. If there are rapid thermal transients in thick-walled vessels, the thermal stress may be nonlinear through the thickness of the component and thus require separation into membrane, secondary, and peak stresses. In Table 4.2, the thermal stress in vessel components is generally classified as either secondary or peak depending on whether the thermal stress is general or local as shown in Figure 4.6. There are some conditions where the membrane stress due to thermal conditions is classified as primary stress in the creep range. This condition will be discussed in later chapters.

Generally, thermal stress affects the cycle life of a component; the analysis is discussed in Chapter 5. However, some thermal stress is also evaluated at steady-state conditions at such areas as skirt and lug attachments.

4.4 Equivalent Stress Limits for Design and Operating Conditions

ASME I and VIII provide design rules for common components such as shells, heads, nozzles, and covers. These rules are intended to keep the design stresses P_m, P_L, and P_b in the components within allowable stress limits. The design pressure and temperature are assumed as slightly higher than the operational pressure and temperature and are taken at a point in the operational cycle where they are at maximum. Much pressurized equipment, such as power boilers and hydrocrackers in refineries, operates in essentially a steady-state condition. However, the combination of mechanical and thermal loading may necessitate additional creep analysis. Such analysis is discussed in Chapters 5 and 6.

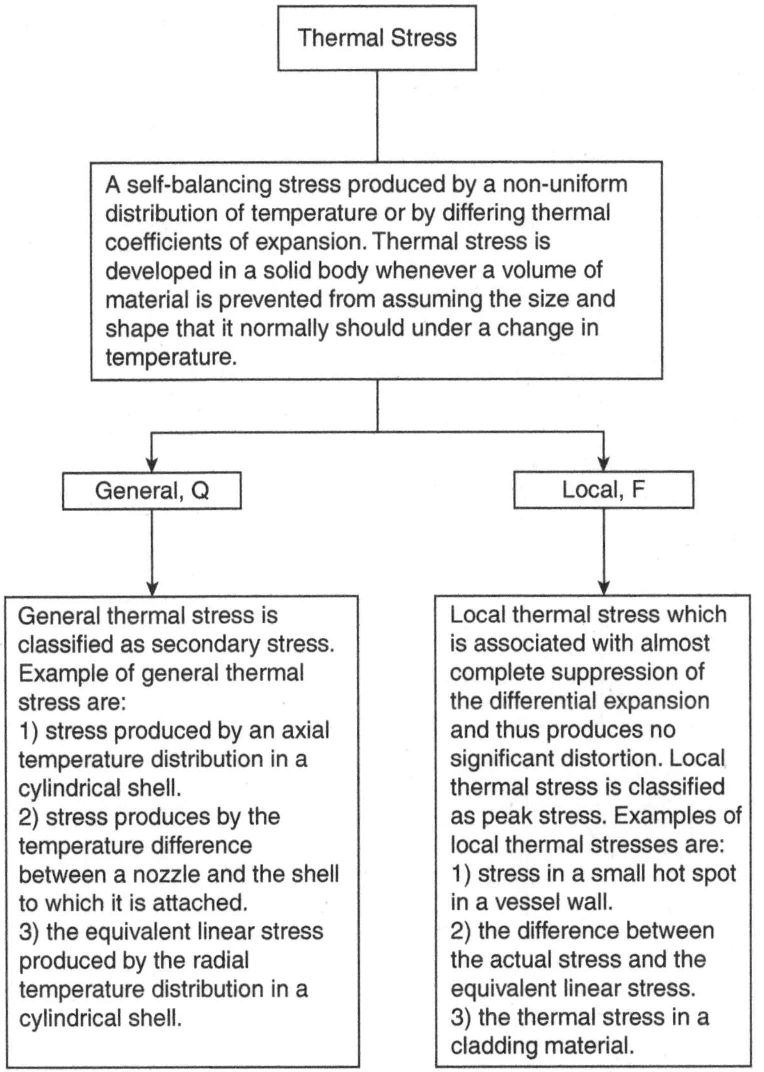

Figure 4.6 Thermal stress categories.

ASME VIII-2 has specific procedure for setting limits on various combinations of the stresses $P_{m,}$ P_L, P_b, and Q. The maximum values of these combinations are called equivalent stress values and are designated by P_m, P_L, P_b, and Q, as shown in Figure 4.7. The procedure for calculating equivalent stress values from stress values is given by the following steps.

1) At a given time during the operation (such as a steady-state condition), separate the calculated stress at a given point in a vessel into (P_m), ($P_L + P_b$), and ($P_L + P_b + Q$). The quantity P_L is either general or local primary membrane stress.
2) Each of these three bracketed quantities is actually six quantities acting as normal and shear stresses on an infinitesimal cube at the point selected.
3) Repeat Steps 1 through 2 at a different time frame, such as startup.
4) Find the algebraic sum of the stresses from Steps 1 through 3. This sum is the stress range of the quantities (P_m), ($P_L + P_b$), and ($P_L + P_b + Q$).

Stress Categories and Limitis of Equivalent Stress

Stress Category	Primary			Secondary Membrane plus Bending	Peak
	General Membrane	Local Membrane	Bending		
Description	Average primary stress across solid section. Excludes discontinuities and concentrations. Produced only by mechanical loads.	Average stress across any solid section. Considers discontinuities but not concentrations. Produced only by mechanical loads.	Component of primary stress proportional to distance from centroid of solid section. Excludes discontinuities and concentrations. Produced only by mechanical loads.	Self-equilibrating stress necessary to satisfy continuity of structure.Occurs at structural discontinuities. Can be caused by mechanical load or by differential thermal expansion. Excludes local stress concentrations.	1. Increment added to primary or secondary stress by a concentration {notch}. 2. Certain thermal stresses that may cause fatigue but not distortion of vessel shape.
Symbol	P_m	P_L	P_b	Q	F

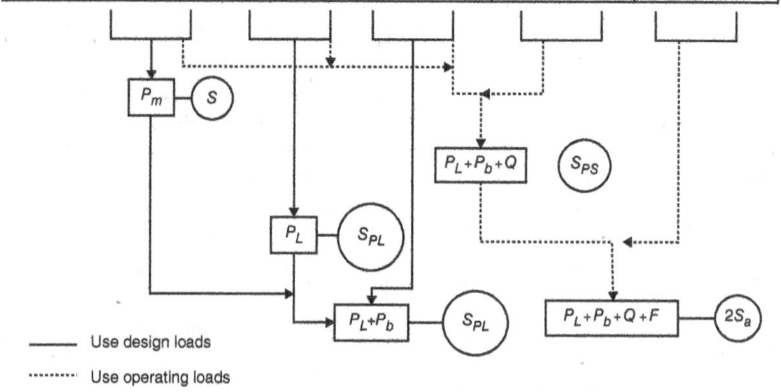

P_m — S

P_L+P_b+Q — S_{PS}

P_L — S_{PL}

P_L+P_b — S_{PL}

P_L+P_b+Q+F — $2S_a$

——— Use design loads

········· Use operating loads

Figure 4.7 Stress categories and limits of equivalent stress [ASME VIII-2].

5) From Step 4, determine the three principal stresses S_1, S_2, and S_3 for each of the stress categories (P_m), ($P_L + P_b$), and ($P_L + P_b + Q$).
6) From Step 5, determine the equivalent stress, S_e, for each of the stress categories (P_m), ($P_L + P_b$), and ($P_L + P_b + Q$) in accordance with the strain energy (von Mises) equation

$$S_e = \frac{1}{(2)^{1/2}}\left[(S_1 - S_2)^2 + (S_2 - S_3)^2 + (S_3 - S_1)^2\right]^{1/2}. \quad (4.14)$$

The maximum absolute values obtained in Step 6 are called equivalent stresses (P_m), ($P_L + P_b$), and $\left(P_L + P_b + Q\right)$. The von Mises equation (Eq. 4.14) is for ductile materials and is not applicable to brittle materials or materials that become brittle during operation. It should also be noted that, for shells subjected to internal pressure, the equivalent stress values obtained from the von Mises theory are about 14% smaller than those obtained from the shear theory used in ASME VIII-2 prior to 2007.

7) The equivalent stress values in Step 6 are analyzed in accordance with Figure 4.7 for temperatures below the creep range and Figure 4.8 for temperatures in the creep range. This chapter covers load-controlled stress limits as shown in Figure 4.8. Strain and deformation limits are covered in Chapter 5 and creep fatigue is covered in Chapter 6.

The design equivalent stress limits for P_m, P_L, and P_b at temperatures below the creep range are shown by the solid lines in Figure 4.7. The design equivalent stress is generally limited to

$P_m \leq S$ for general membrane stress
$P_L \leq S_{PL}$ for local membrane stress
$(P_L + P_b) \leq S_{PL}$ for local membrane and bending primary stresses
where S_{PL} is the larger of $1.5S$ or S_y.

Many of the construction details given in ASME I and VIII for major components meet these design stress limits, and in general there is no need to run a stress check due to design conditions. The situation in the creep range is different in that the $1.5S$ limit on P_L and P_b does not apply when loading conditions are of sufficient duration for creep effects to redistribute elastically calculated stress and strain. This is reflected by the use of the K_t factor in the stress evaluation in ASME VIII-2 and III-5. And, although many construction details given in ASME I and VIII for major components meet the design stress limits in the creep range as given by the first row in Figure 4.8, other components such as nozzle reinforcement, lugs, and attachments, may require additional analysis in the creep range in accordance with

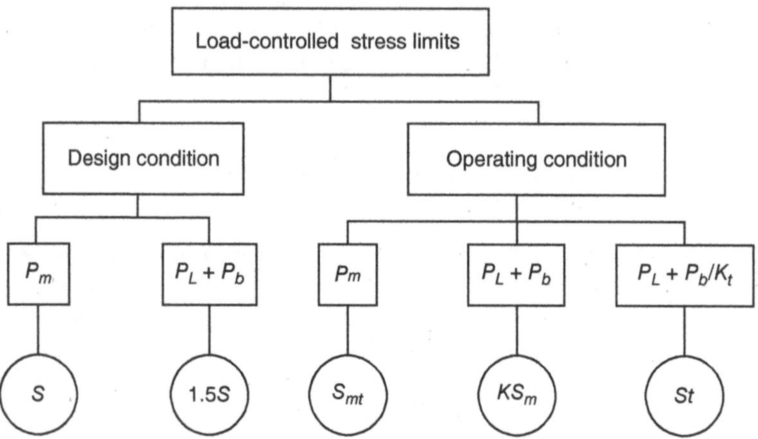

Figure 4.8 Flow Diagram for Load-Controlled Stress Limits [ASME VIII-2].

Figure 4.8 for steady load conditions when the design specifications require it or when the designer deems it necessary.

For the operating condition, it is necessary to determine the equivalent stress values of P_m, P_L, P_b, and Q. The ASME requirements for defining equivalent stress limits differ substantially for temperatures below the creep zone from those in the creep zone. The criteria below the creep zone for ASME VIII-2 are given by the dotted lines in Figure 4.7 and show that primary plus secondary equivalent stress values are limited by the quantity S_{PS}. This quantity is essentially limited by the larger of $3S$ or $2S_y$, where S and S_y are the average value for the highest and lowest temperature of the cycle. The criteria for stress in the creep range in accordance with ASME VIII-2 are shown in Figure 4.8. It involves checking load-controlled then strain-controlled stresses. A detailed description of the ASME VIII-2 methodology is given in this chapter for load-controlled stress and in Chapter 5 for strain-controlled stress.

The above procedure for the separation of stresses is illustrated as follows.

For calculating the stresses on the inside surface of a thick cylindrical shell subjected to internal pressure.

- Step 1
  (a) Circumferential stress
  From Eq. (4.11),

$$P_{mH} = \frac{P}{\gamma - 1}. \tag{4.15}$$

Due to internal pressure away from discontinuities, the quantity $P_{mH} = P_{LH}$ in this case and thus,

$$P_{LH} = \frac{P}{\gamma - 1},\ P_{bH} = 0$$

$$P_{LH} + P_{bH} = \frac{P}{\gamma - 1}. \tag{4.16}$$

From Eq. (4.12),

$$Q_H = \frac{6P\gamma}{(\gamma-1)^2}\left[\frac{1}{2} - \frac{\gamma}{\gamma^2 - 1}(\ln\gamma)\right]$$

and

$$P_{LH} + P_{bH} + Q_H = \frac{P}{(\gamma-1)^2}\left[(4\gamma - 1) - \frac{6\gamma^2}{\left(\gamma^2 - 1\right)}(\ln\gamma)\right]. \tag{4.17}$$

(b) Longitudinal stress
From Jawad and Farr (2019),

$$P_{LL} = \frac{P}{\gamma^2 - 1}. \tag{4.18}$$

Due to internal pressure away from discontinuities

$$P_{bL} = 0 \quad Q_L = 0.$$

Hence,

$$P_{LL} + P_{bL} = \frac{P}{\gamma^2 - 1} \tag{4.19}$$

and

$$P_{LL} + P_{bL} + Q = \frac{P}{\gamma^2 - 1}. \tag{4.20}$$

(c) Radial stress

$$P_{LR} = -P/2 \tag{4.21}$$

$$P_{bR} = 0, \quad Q_R = -P/2.$$

Hence,

$$P_{LR} + P_{bR} = -P/2 \tag{4.22}$$

and

$$P_{LR} + P_{bR} + Q_R = -P. \tag{4.23}$$

- Step 2
  All shearing stress is zero in a cylindrical cylinder subjected to internal pressure.
- Step 3
  It will be assumed that the initial stresses before application of pressure are all zero.
- Step 4
  Because the stresses in Step 3 are zero, the total sum of the equivalent stress values is given in Step 1.
- Step 5
  (P_m), which is the same as P_L in this case, are given by Eqs. (4.15), (4.18), and (4.21)

$$S_1 = \frac{P}{\gamma - 1}, \quad S_2 = \frac{P}{\gamma^2 - 1}, \quad S_3 = -P/2. \tag{4.24}$$

The principal stresses for the stress value ($P_L + P_b$) are given by Eqs. (4.16), (4.19), and (4.22)

$$S_1 = \frac{P}{\gamma - 1}, \quad S_2 = \frac{P}{\gamma^2 - 1}, \quad S_3 = -P/2. \tag{4.25}$$

The principal stresses for the stress value ($P_L + P_b + Q$) are given by Eqs. (4.17), (4.20), and (4.23)

$$\left.\begin{array}{l} S_1 = \dfrac{P}{(\gamma - 1)^2}\left[(4\gamma - 1) - \dfrac{6\gamma^2}{\gamma^2 - 1}(\ln \gamma)\right] \\ S_2 = \dfrac{P}{\gamma^2 - 1}, S_3 = -P \end{array}\right. . \tag{4.26}$$

- Step 6
  The stress differences for P_m are obtained from Eq. (4.24) as

$$\left.\begin{array}{l} S_1 - S_2 = \dfrac{P}{\gamma^2 - 1} \\ S_2 - S_3 = \dfrac{P\left(\gamma^2 + 1\right)}{2\left(\gamma^2 - 1\right)} \\ S_3 - S_1 = \dfrac{-P(\gamma + 1)}{2(\gamma - 1)} \end{array}\right| . \tag{4.27}$$

The stress differences for $(P_L + P_b)$ are obtained from Eq. (4.25) as

$$\left.\begin{aligned} S_1 - S_2 &= \frac{P\gamma}{\gamma^2 - 1} \\ S_2 - S_3 &= \frac{P\left(\gamma^2 + 1\right)}{2\left(\gamma^2 - 1\right)} \\ S_3 - S_1 &= \frac{-P(\gamma + 1)}{2(\gamma - 1)} \end{aligned}\right|. \tag{4.28}$$

The stress differences for $(P_L + P_b + Q)$ are obtained from Eq. (4.26) as

$$\begin{aligned} S_1 - S_2 &= \frac{2P\gamma}{(\gamma - 1)^2}\left[\frac{(2\gamma + 1)}{\gamma + 1} - \frac{3\gamma}{\gamma^2 - 1}(\ln\gamma)\right] \\ S_2 - S_3 &= \frac{P\gamma^2}{\gamma^2 - 1} \\ S_3 - S_1 &= \frac{-P}{(\gamma - 1)^2}\left[\left(\gamma^2 + 2\gamma\right) - \frac{6\gamma^2}{\gamma^2 - 1}(\ln\gamma)\right]. \end{aligned} \tag{4.29}$$

Equations (4.27), (4.28), and (4.29) are substituted into Eq. (4.14) to obtain the equivalent stress values of P_m, $(P_L + P_b)$, and $(P_L + P_b + Q)$.

- Step 7
  The equivalent stress values P_m, $(P_L + P_b)$, and $(P_L + P_b + Q)$ obtained in Step 6 are analyzed in accordance with Figure 4.7 for temperatures below the creep range and Figure 4.8 for temperatures in the creep range, as explained in this section.

4.5 Load-Controlled Limits for Components Operating in the Creep Range

Pressure vessel and boiler components operating in the creep range may require, in some instances, special analysis in accordance with load- and strain-controlled limits. These instances include sudden upset or regeneration conditions, a few operating cycles with temperature spikes, and design conditions based on life expectancy greater than the assumed 100,000 h. Moreover, some construction components, such as nozzles, support and jacket attachments, and tubes, may require such analysis in equipment operating under steady-state conditions. The procedure for such analysis is given in this section.

In the creep range, the criteria for equivalent stress limits shown in Figure 4.8 are based on limiting the stresses for the design loads and then for the operating loads.

The design limits in Figure 4.8 can be considered satisfied for head and shell components when the designer uses ASME I and VIII details of construction. The reason is that the requirements of $P_m < S_o$ and $P_L + P_b < 1.5S_o$ are fulfilled when using the details of ASME I and VIII. Other details, such as nozzle reinforcement and jacket attachments, are based, in part, on Design by Rules and may require analysis at the discretion of the designer.

In the creep range, the ASME VIII-2 operating limits in the second row of Figure 4.8 must be complied with in a similar fashion as the ASME VIII-2 limits shown in Figure 4.7 designated by the dotted lines for temperatures below the creep range. The limits are based on (1) load-controlled stresses and (2) strain- and deformation-controlled stresses. The load-controlled stresses are discussed in this chapter and the strain-controlled stresses are discussed in Chapter 5.

The load-controlled limits are given by

$$P_m < S_{mt} \tag{4.30}$$

$$\left(P_L + P_b\right) < KS_m \tag{4.31}$$

and

$$\left[P_L + \left(P_b / K_t\right)\right] < S_t \tag{4.32}$$

where K is given in Table 4.3.

For welded construction, the values of S_{mt} and S_t listed above are further defined as

S_{mt} = lower value of S_{mt} or $0.8(R_w)(S_r)$

S_t = lower value of St or $0.8(R_w)(S_r)$.

R_w = ratio of weld metal creep rupture strength to base metal creep rupture strength. Actual values are given in ASME II-D.

Example 4.4 The vessel shown in Figure 4.9a is constructed of SA-387-22 Cl.1. The shell is assumed to be thin such that the stress distribution across the thickness is uniform due to applied pressure. The design and operating conditions are:

Design pressure = 0 psi to 500 psi
Design temperature = 70°F to 825°F
Operating pressure = 0 psi to 450 psi
Operating temperature = 70°F to 810°F
R_i = 16.00 in.
$\mu = 0.30$
Joint efficiency = 1.0
Use 100,000 h service life.

Table 4.3 Shape factors.

Shape	I/c[1]	Z[2]	Shape factor, K
B, t	$Bt^2/6$	$Bt^2/4$	1.5
t, R	$\pi R^2 t$	$4 R^2 t$	1.27
R	$\pi R^3/4$	$4 R^3/3$	1.7
R_o, R_i	$(\pi/4)R_0^3\left(1-\gamma_1^4\right)$	$(4/3)R_0^3\left(1-\gamma_1^3\right)$	$(16/3\pi)\left(1-\gamma_1^3\right)/\left(1-\gamma_1^4\right)$ Ranges between 1.27 and 1.70

[1] $\sigma = Mc/I$ for elastic analysis.
[2] $\sigma = M/Z$ for plastic analysis.
$\gamma_1 = R_i/R_o$.

(a) Calculate the required thicknesses in accordance with VIII-1

The designer of an ASME VIII-1 vessel is required, occasionally, to check the stresses at various components subsequent to the design. However, ASME VIII-1 does not provide rules for such detailed stress analysis. Accordingly, the designer uses the rules of ASME VIII-2, Part 5, when the temperature is below

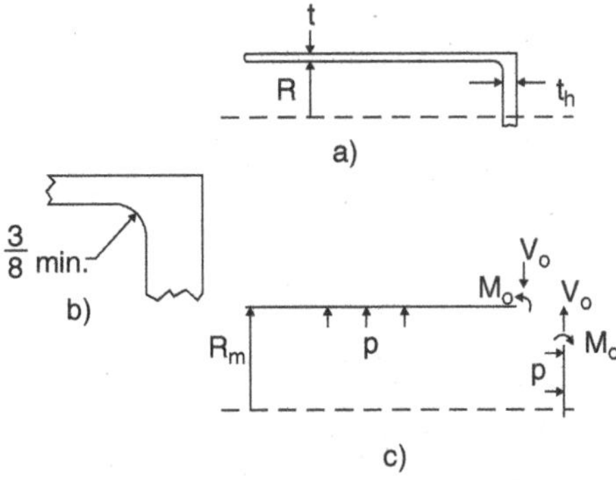

Figure 4.9 Head-to-shell junction.

the creep range, and the creep rules of ASME VIII-2 when the temperature is in the creep range. This procedure is demonstrated in (b) and (c) below.

(b) Evaluate the operating stress values at the inside surface of the head-to-shell junction using the criteria of ASME VIII-2, Part 5, as a general guide.

(c) Evaluate the operating stress values at the inside surface of the head-to-shell junction in accordance with the criteria for load-controlled stress in ASME VIII-2, as a general guide.

In item (c) only the load-controlled stresses will be checked. The strain and deformation-controlled stresses, which must also be simultaneously evaluated in accordance with ASME VIII-2, will be discussed in Chapter 5.

Solution

(a) Design Condition

From ASME II-D, the allowable stress, S, at 825°F is 16,600 psi. This temperature is below the 900°F set by ASME II-D for this material, where creep and rupture criteria control.

The required shell thickness based on ASME VIII-1's equation is

$$\begin{aligned} t &= PR_i / (SE_o - 0.6P) \\ &= (500)(16) / [16{,}600(1.0) - 0.6(500)] \\ &= 0.49 \text{ in.} \end{aligned}$$

Use $t = ½$ in.

The required head thickness is obtained from the following equations.

$$\begin{aligned} m &= \text{ calculated thickness of shell/actual thickness of shell} \\ &= 0.49 / 0.5 \\ &= 0.98. \end{aligned}$$

Bending factor C is given by

$$\begin{aligned} C &= 0.44\text{m} \\ &= 0.43. \end{aligned}$$

The required flat head thickness is obtained from

$$\begin{aligned} t_h &= D_i(CP / S)^{1/2} \\ &= (32)[(0.43)(500) / 16{,}600]^{1/2}. \\ &= 3.64 \text{ in.} \end{aligned}$$

Use $t_h = 3.75$ in.

Details of a shell-to-head junction in accordance with ASME VIII-1 are shown in Figure 4.9b.

(b) Evaluation of Operating Stresses in Accordance with VIII-2, Part 5

Operating pressure = 450 psi

Operating temperature = 810°F

The 810°F operating temperature is below the 850°F limit set by ASME VIII-2, where the allowable stress values are controlled by creep for this material. Hence, creep is not a consideration in accordance with the rules of ASME VIII-2.

From ASME II-D for VIII-2 values, S_{Av} = 18,790 psi. (S = 20,000 psi at room temperature and S = 17,580 psi at 810°F).

From Table Y-1 of ASME II-D, av. S_y = 28,260 psi. (S_y = 30,000 psi at room temperature and S_y = 26,520 psi at 810°F).

From Table TM-1 of ASME II-D, E = 26,200,000 psi at 810°F.

The unknown moment, M_o, and shear force, V_o, at the junction are shown in Figure 4.9c. Their values are obtained from the following two compatibility equations

$$\text{Deflection of shell due to pressure} + M_0 + V_0 = \text{Deflection of plate due to } V_0 \quad (1)$$

and

$$\text{Rotation of shell due to } M_0 + V_0 = \text{Rotation of plate due to } M_0 + \text{pressure} \quad (2)$$

Deflection of shell

Due to pressure = $0.85PR_m^2/Et$

Due to moment = $M_o/2D\beta^2$

Due to radial load = $-V_o/2D\beta^3$

Rotation of shell

Due to pressure = 0

Due to moment = $-M_o/D\beta$

Due to radial load = $V_o/2D\beta^2$

Radial deflection of flat head

Due to pressure = 0

Due to moment = 0

Due to radial load = $R_m V_o/Et_h$

Rotation of flat head

Due to pressure = $-3(1-\mu)PR_m^3/2Et_h^3$

Due to moment = $12(1-\mu)R_m M_o/Et_h^3$

Due to radial load = 0

where

$D = Et^3/12(1-\mu^2)$

E = modulus of elasticity

R_m = mean radius
t = thickness of shell
t_h = thickness of flat head

$\beta = \left[3\left(1-\mu^2\right)/R_m^2t^2\right]^{0.25}$

μ = Poisson's ratio.

Solving the two compatibility equations (Eqs. 1 and 2) for M_o and V_o gives $M_o = 1283 \text{ in.} - \text{lb/in.}$ and $V_o = 994$ lbs./in..

It is assumed that the initial stresses at the beginning of the operating cycle are zero. Hence, the following calculations can be treated as stress-range values rather than stress values. Also, only the equivalent stress quantity ($P_L + P_b + Q$) is required to be checked for the operating conditions in accordance with Figure 4.7.

- Shell stress-range calculations

Axial stress

$$P_L = PR_i/2 = 7200 \text{ psi} \tag{1}$$

$$P_b = 0 \tag{2}$$

$$Q = 6M/t^2 = 30{,}800 \text{ psi} \tag{3}$$

$$\left(P_L + P_b + Q\right) = 38{,}000 \text{ psi} \tag{4}$$

Circumferential stress

$$P_L = 0 \text{ psi} \tag{5}$$

$$P_b = 0 \text{ psi} \tag{6}$$

$$Q = 0.3(30{,}800) = 9240 \text{ psi} \tag{7}$$

$$\left(P_L + P_b + Q\right) = 9240 \text{ psi} \tag{8}$$

Radial stress

$$P_L = -225 \text{ psi} \tag{9}$$

$$P_b = 0 \text{ psi} \tag{10}$$

$$Q = -225 \text{ psi} \tag{11}$$

$$\left(P_L + P_b + Q\right) = -450 \text{ psi.} \tag{12}$$

The three principal stresses for $(P_L + P_b + Q)$ at the inside surface of the shell at the head-to-shell junction are then given by Eqs. (4), (8), and (12):

$$S_1 = 38{,}000 \text{ psi}, \quad S_2 = 9240 \text{ psi}, \quad S_3 = -450 \text{ psi}.$$

Thus, the maximum equivalent stress value of $(P_L + P_b + Q)$ is obtained from Eq. (4.14) as

$$\begin{aligned} P_L + P_b + Q &= 0.707\left[(28{,}760)^2 + (9690)^2 + (38{,}450)^2\right]^{1/2}. \\ &= 34{,}600 \text{ psi} \end{aligned}$$

The allowable stress is the larger of

$$2S_y = 2(28{,}260) = 56{,}520 \text{ psi}$$

or

$$3S = 3(18{,}790) = 56{,}370 \text{ psi}.$$

Thus, the calculated value of 34,600 psi is less than the allowable stress of 56,520 psi. Hence, based on an elastic analysis below the ASME VIII-2 creep range, the stress at the inside surface of the shell at the shell-to-head junction is adequate.

The above calculations satisfy the stress requirements of ASME VIII-2.

The following calculation for secondary stress, Q, is detailed here but is not required in the ASME VIII-2 load-controlled stress evaluations. It is performed here for expediency for use in Chapter 5 in the strain and deformation limit calculations.

(Q)calculations

From Eqs. (3), (7), (11), and (4.14), The maximum equivalent stress value of Q is 27,540 psi

- Head stress calculations

  Radial stress

$$P_L = V / t_h = 235 \text{ psi} \tag{13}$$

$$P_b = 6M / t_h^2 = 550 \text{ psi} \tag{14}$$

$$Q = 0 \text{ psi} \tag{15}$$

$$\left(P_L + P_b + Q\right) = 815 \text{ psi} \tag{16}$$

Tangential stress

$$P_L = 0\,\text{psi} \tag{17}$$

$$P_b = 0.3(550) = 165\text{ psi} \tag{18}$$

$$Q = 0\,\text{psi} \tag{19}$$

$$(P_L + P_b + Q) = 165\text{ psi} \tag{20}$$

Through thickness stress

$$P_L = -225\,\text{psi} \tag{21}$$

$$P_b = 0\,\text{psi} \tag{22}$$

$$Q = -225\,\text{psi} \tag{23}$$

$$(P_L + P_b + Q) = -450\text{ psi}. \tag{24}$$

The three principal stresses at the inside surface of the shell at the head-to-shell junction are then given by Eqs. (16), (20), and (24)

$$S_1 = 815\text{ psi},\quad S_2 = 165\text{ psi},\quad S_3 = -450\text{ psi}.$$

Thus, the maximum equivalent stress value of ($P_L + P_b + Q$) is obtained from Eq. (4.14) as

$$\begin{aligned}(P_L + P_b + Q) &= 0.707\left\{(815-165)^2 + [165-(-450)]^2 + (-450-815)^2\right\}^{1/2}\\ &= 1095\text{ psi}\end{aligned}$$

By inspection, these equivalent stress values are well below the allowable stress limits.

The stress at the middle of the head is equal to the stress assuming a simply supported head minus the stress caused by the edge moment due to the shell restraint

$$\begin{aligned}P_b &= \left[3(3+\mu)PR_m^2/8t_h^2\right] - 550\\ &= 10{,}460 - 550\\ &= 9910\text{ psi} < 1.5S_m\end{aligned}$$

The edge stress σ can be classified as either primary bending stress, P_b, or secondary stress, Q, depending on the design of the head. If the edge moment is used to reduce the moment in the middle of the head in order to meet the allowable stress criterion, 1.5S, then it must be classified as P_b. If the stress in the middle of the head is based on a simply supported head without taking the edge stress into consideration, then the edge stress is classified as secondary.

The following calculation for secondary stress, Q, is detailed here but is not required in the ASME VIII-2 load-controlled stress evaluations. It is performed here for expediency for use in Chapter 5 in the strain and deformation limit calculations.

From Eqs. (15), (19), (23), and (4.14),

Maximum equivalent stress value of $Q = 225$ psi.

(c) Evaluation of Operating Condition in Accordance with Load-Controlled Stress
From ASME II-D, S_o = 16,600 psi at 810°F.
From ASME II-D, Figure E-100.4-4, S_{mt} = 17,100 psi at 810°F for 100,000 h.
From ASME II-D, Figure E-100.5-4, S_t = 17,200 psi at 810°F for 100,000 h.
From ASME II-D, Figure E-100.11-1, R_w = 1.0 at 810°F.
From ASME II-D, Figure E-100.7-4, S_r = 25,800 psi at 810°F for 100,000 h.

$$\begin{aligned} S_{mt} &= \text{lower value of } S_{mt} \text{ or } 0.8(R)(S_r) \\ &= 17{,}100 \text{ psi, or} \\ &= 0.8(1.0)(25{,}800) = 20{,}640 \text{ psi} \end{aligned}$$

Use S_m = 17,600 psi.

$$S_t = \text{lower value of } S_t \text{ or } 0.8(R_w)(S_r).$$

Use S_t = 17,200 psi.

It should be noted that in Part (b) of this example, only the equivalent stress ($P_L + P_b + Q$) is required to be checked for the operating conditions in accordance with Figure 4.7. However, in the creep range, the quantities P_m, ($P_L + P_b$), ($P_L + P_b/K_t$), Q, and ($P_L + P_b + Q$) need to be checked in accordance with the requirements of Figure 4.8. In this chapter, only the quantities P_m, ($P_L + P_b$), and ($P_L + P_b/K_t$) will be evaluated. The remaining quantities, Q, and ($P_L + P_b + Q$) will be calculated in this example but discussed in Chapter 5.

- Shell stress calculations
  In the shell and head calculations, values of P_m and P_L will be conservatively assumed to be similar. These values are obtained from the von Mises equation given in ASME VIII-2.

(P_m)

From Eqs. (1), (5), (9), and (4.14), maximum equivalent stress value of $P_m = 7315$ psi

($P_L + P_b$)

From Eqs. (1 + 2), (5 + 6), (9 + 10), and (4.14), maximum equivalent stress value of ($P_L + P_b$) = 7315 psi

($P_L + P_b/K_t$)

From Table 4.3, $K = 1.5$

$$K_t = (1 + K)/2 = 1.25.$$

From (Eq.1 + Eq.2/K_t), (Eq.5+ Eq.6/K_t), (Eq.9 + Eq.10/K_t), and Eq. (4.14), maximum equivalent stress value of ($P_L + P_{b/}K_t$) = 7375 psi.

Load-controlled stress Limits

From Eq. (4.30)

$$P_m < S_{mt}$$
$$7315 < 17{,}100 \quad \text{ok.}$$

From Eq. (4.31)

$$(P_L + P_b) < KS_m$$
$$7315 < 1.5(16{,}600) \quad \text{ok.}$$

From Eq. (4.32)

($P_L + P_{b/}K_t$) < St 7315 < 17,200 ok.

- Head equivalent stress calculations

In the shell and head calculations, values of P_m and P_L will be conservatively assumed to be similar.

(P_m)

From Eqs. (13), (17), (21), and (4.14) maximum equivalent stress value of $P_m = 425$ psi.

($P_L + P_b$)

From Eqs. (13 + 14), (17 + 18), (21 + 22), and (4.14) maximum equivalent stress value of ($P_L + P_b$) = 910 psi.

($P_L + P_b/K_t$)

From Table 4.3, $K = 1.5$

$$K_t = (1 + K)/2 = 1.25.$$

From (Eq. 13 + Eq. 14/K_t), (Eq. 17 + Eq. 18/K_t), (Eq. 21 + Eq. 22/K_t), and Eq. (4.14) maximum equivalent stress value of ($P_L + P_b/K_t$) = 810 psi.

4.6 Reference Stress Method

The separation of stresses, at a given point in a pressure vessel part, into primary and secondary components is a very tedious task, as illustrated in Section 4.4. Whenever possible, many designers rely on other approximate approaches to design components. One such approach is the reference stress method, which is a convenient means of characterizing the stress in a component operating in the creep range without having to separate the stresses into primary and secondary components. The method is based on limit analysis and is an asymptotic solution to obtaining stress in a component.

The rationale for using limit analysis in determining stress in pressure components operating in the creep range was discussed in Chapters 1 and 2. The relationship between strain rate and stress in the creep range is normally taken as

$$d\varepsilon / dt = k'\sigma^n \tag{4.33}$$

where

$d\varepsilon/dt$ = strain rate

k' = constant

n = creep exponent, which is a function of material property and temperature

σ = stress.

It was shown in Chapter 1 that a stationary stress (elastic analog) condition is reached after a few hours of operating a component in the creep range under a constant strain rate. The relationship between strain and stress can then be taken as

$$\varepsilon = K'\sigma^n \tag{4.34}$$

where

ε = strain

K' = constant.

4.6.1 Cylindrical Shells

The governing equation for the circumferential stress in a cylindrical shell due to internal pressure in the creep range under a stationary stress condition [Finnie and Heller (1959)] is derived from Eq. (4.34) as

$$S = P\left[\left(\frac{2-n}{n}\right)(R_0 / R)^{2/n} + 1\right] / \left(\gamma^{2/n} - 1\right) \tag{4.35}$$

where

P = internal pressure
R = radius at any point in the shell wall
R_i = inside radius
R_o = outside radius
S = circumferential stress
$\gamma = R_o/R_i$.

A plot of this equation for $n > 2$ is shown by line EF in Figure 4.10. This line shows that, when the component is operating in the creep range, the stress at the inside surface decreases and the stress at the outside surface increases compared to the stress in the non-creep region, as illustrated by line AB from Eq. (4.8).

A comparison of Eqs. (4.8) and (4.35) is given in Table 4.4. For $n = 1$, the creep (Eq. 4.35) and the elastic (Eq. 4.8) equations yield the same answer, as they should. For n values between one and two, the maximum stress given by Eq. (4.35) is on the inside surface of the shell similar to the elastic equation. However, for n values greater than two the maximum stress given by Eq. (4.35) is on the outside surface of the shell. Hence, creep fatigue could occur on either surface depending on the material value of n. However, for most materials used in pressure vessel construction, the n value is greater than two and the maximum stress in the creep regime is on the outside surface as shown in Table 4.4. The designer should be aware of the stress reversal from the inside to the outside surface when evaluating creep and fatigue stress. Table 4.4 also shows that the coefficients obtained from Eq. (4.35) reach an asymptotic value for large values of n. This fact is crucial when the designer decides to substitute inelastic analysis for creep analysis.

The following example illustrates the application of Eq. (4.35).

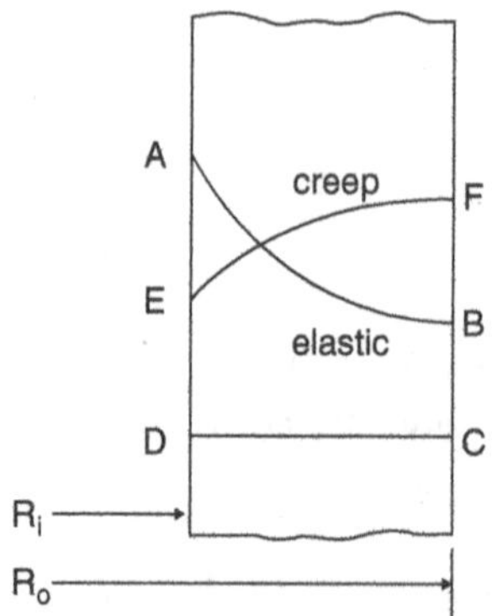

Figure 4.10 Elastic versus stationary creep stress distribution.

Table 4.4 Stress coefficients, S/P, for shell equations and thicknesses.

R_o/R_i		1.67		1.25		1.1	
Shell type		**Thick**		**Intermediate**		**Thin**	
Location		**Inside**	**Outside**	**Inside**	**Outside**	**Inside**	**Outside**
	n						
Lame's equation (Eq. 4.8)		2.12	1.12	4.56	3.56	10.52	9.52
Creep equation (Eq. 4.35)	1	2.12	1.12	4.56	3.56	10.52	9.52
	1.5	1.69	1.36	4.18	3.85	10.17	9.84
	2	1.49	1.49	4.00	4.00	10.00	10.00
	5	1.16	1.76	3.68	4.28	9.69	10.29
	10	1.05	1.85	3.58	4.38	9.59	10.39
	50	0.97	1.93	3.50	4.46	9.51	10.47
	100	0.96	1.94	3.49	4.47	9.50	10.48
Inelastic equation (Eq. 4.22)			1.95		4.48		10.49

Example 4.5 A cylindrical shell is constructed from annealed 2.25Cr-1Mo steel and has an inside diameter of 48 in. The internal pressure is 4000 psi and the design temperature is 1000°F. Let $E_o = 1.0$. The isochronous curves for this material are shown in Figure 4.11.

a) Determine the required thickness using Section VIII-1 criteria.
b) Determine the required thickness using Eq. (4.20) and a 100,000-h life.

Solution

(a) Division VIII-1

The allowable stress from Section II-D is 7800 psi. Since $P > 0.385S$, thick shell equations must be used.

$$\begin{aligned} Z &= \frac{SE_0 + P}{SE_0 - P} \\ &= (7800 + 4000) / (7800 - 4000) \\ &= 3.11 \end{aligned}$$

$$\begin{aligned} t &= R_i\left(Z^{1/2} - 1\right) \\ &= 24(1.7635 - 1). \\ &= 18.32 \text{ in.} \end{aligned}$$

It should be noted that the allowable stress of 7800 psi is based, in part, on rupture at 100,000 h.

a) Equation (4.35)

A value of $n = 6.26$ is obtained from Points A and B in Figure 4.11 and Eq. (4.34). From Eq. (4.35) with $R = R_o$,

$$\frac{7800}{4000} = \frac{((2-n)/n)\left(R_0/R\right)^{2/n} + 1}{\left(R_0/R_i\right)^{2/n} - 1}$$

$$1.95 = \frac{(-0.6805)(1.0)^{2/6.26} + 1.0}{\left(R_0/24.0\right)^{2/6.26} - 1.0}$$

or

$$\begin{aligned} R_0 &= 38.56 \text{ in.} \\ t &= R_0 - R \\ &= 14.56 \text{ in.} \end{aligned}$$

The thickness, which is based on stationary creep considerations, is about 25% less than the thickness obtained from ASME VIII-1. This is because ASME VIII-1 uses Lame's Eq. (4.9), which gives a high stress value at the inner

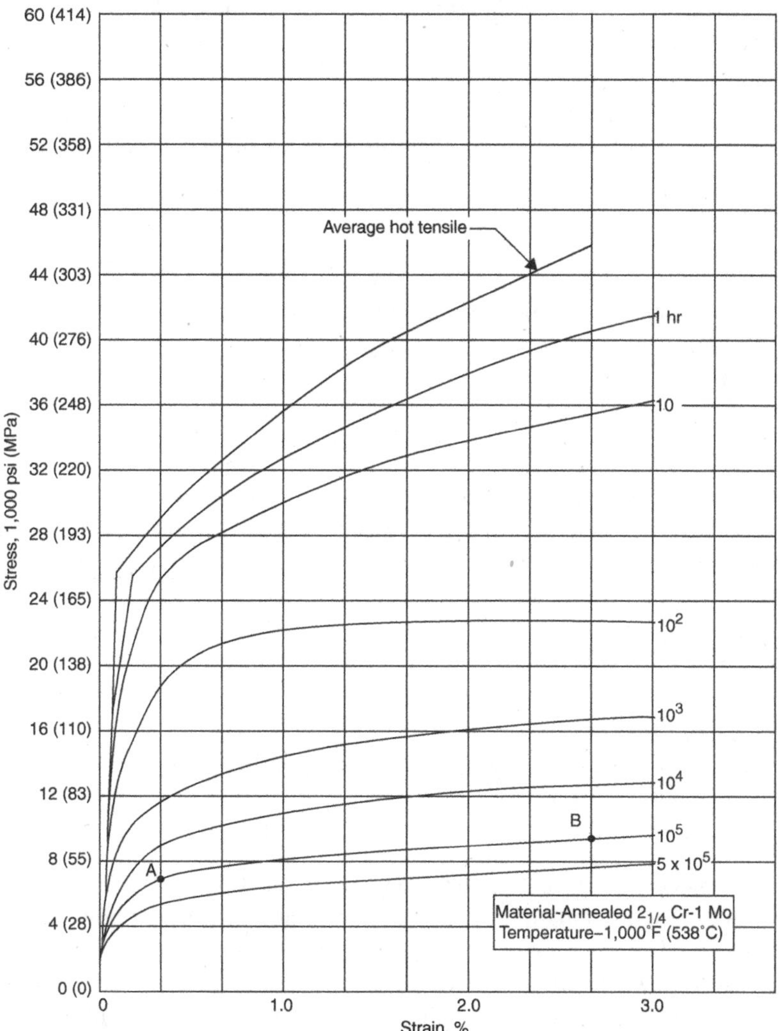

Figure 4.11 Average isochronous stress strain curves for 2.25Cr-1Mo annealed steel at 1000°F (538°C) [ASME II-D].

surface, whereas Eq. (4.35) redistributes the stresses resulting in a maximum stress at the outside surface that is lower than Lame's high stress at the inner surface as shown in Figure 4.10.

The last entry in Table 4.4 is based on the maximum pressure allowed in a cylinder based on limit analysis. The stress on the inside surface of a cylinder due to pressure when the cylinder begins to yield on the inside surface Chen and Zhang (1991) is given by

$$S = P^{*}\left(\gamma^{2}+1\right)/\left(\gamma^{2}-1\right) \tag{4.36}$$

whereas the maximum stress on the outside of a cylinder when the wall is completely yielded is given by

$$S = P^{**}/\ln(\gamma) \tag{4.37}$$

where

P^* = pressure when cylinder wall begins to yield
P^{**} = pressure when cylinder wall completely yields
S = stress
$\gamma = R_o/R_i$.

The last entry in Table 4.4 is based on the limit analysis Eq. (4.37). Results are almost identical to those obtained from Eq. (4.35) with large values of n. This conclusion is used to justify the use of limit analysis for creep evaluations and is the basis for the Reference Stress method used in industry Larsson (1992). The general equation for the reference stress is given by

$$S_R = \left(P/P_u\right)S_y \tag{4.38}$$

where

P = applied load
P_u = ultimate load assuming rigid perfectly plastic material
S_R = reference stress
S_y = yield stress.

The procedure for using the stress reference method consists of first assuming a thickness or configuration of a given part and then performing a limit analysis. A reference stress is then obtained from the limit analysis and compared to an allowable stress for the component. The procedure is based on trial-and-error to obtain a configuration that is stressed within the allowable values.

The main method of obtaining reference stress is by using a numerical procedure such as a finite element analysis. The reason for this is that closed-form solutions are only available for a handful of components such as shells, flat covers, and beams.

It is noteworthy that, at present, there is no general consensus among engineers regarding a stress reference equation for thermal loads Goodall (2003) or one that combines pressure and thermal loads.

Example 4.6 What is the required thickness of the heat exchanger channel shown in Figure 4.12? The material is 304 stainless steel. The design temperature is 1150°F and the design pressure is 2500 psi. The thickness is to be calculated in accordance with

a) ASME VIII-1
b) creep Eq. (4.35)
c) limit Eq. (4.37)
d) reference stress Eq. (4.38)

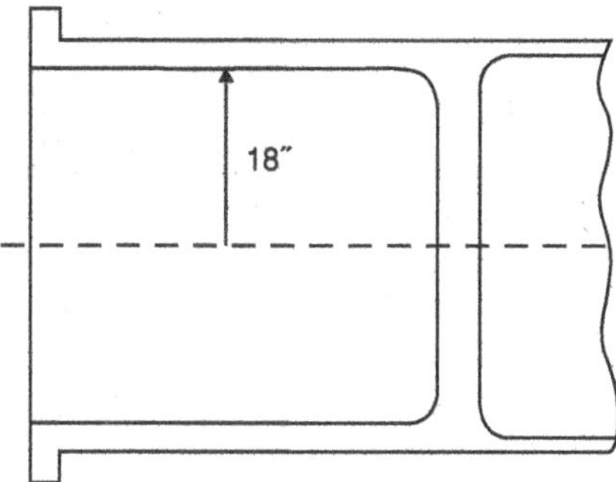

Figure 4.12 Heat Exchanger channel.

Solution

a) ASME VIII-1
The allowable stress, *S*, from ASME II-D is 7700 psi.
The required thickness from ASME VIII-1 is

$$t = PR_i/(SE_o - 0.6P)$$
$$= 2500 \times 18/(7700 \times 1.0 - 0.6 \times 2500)$$
$$= 7.26 \text{ in.}$$
$$\gamma = (18 + 7.26)/18 = 1.4033.$$

From Eq. (4.8), the stress at the inside and outside surfaces is

$$\sigma_i = 2500\left[\left(1.4033^2 + 1\right)/\left(1.4033^2 - 1\right)\right]$$
$$= 7660 \text{ psi}$$
$$\sigma_0 = 2500\left[2/\left(1.4033^2 - 1\right)\right]$$
$$= 5160 \text{ psi}$$

Both of these values are below the allowable stress of 7700 psi.

b) Creep Eq. (4.35)

Equation (4.35) can be solved once the value of n is known. The value of n is also needed in evaluating creep and fatigue analysis. Section VIII does not list any n values. However, an approximate value of n for a limited number of materials can be obtained from a stress-strain curve at a given temperature for a given number of hours listed in ASME II-D. Such curves are presented in isochronous charts. One such chart is shown in Figure 4.13. The value of n is determined by using Eq. (4.34) in conjunction with any of the curves in the chart. Hence, if Points A and B are chosen on the hot tensile curve, then the two unknowns, K' and n, in Eq. (4.34) can be calculated. The result is $n = 6.1$. Other curves in the chart could be used as well. For example, using Points C and D on the 300,000-h curve results in a calculated value of $n = 5.8$. This slight variation in the calculated values of n from various curves has no practical significance in calculating the stresses illustrated in Table 4.4. Using an average value of $n = 6.0$ in Eq. (4.35) results in the following stress at the inside and outside surfaces

$$S_{\text{inside}} = 5290 \text{ psi}$$
$$S_{\text{outside}} = 6940 \text{ psi}$$

The results indicate that there is about a 10% relaxation in the maximum stress values when the operating temperature increases from non-creep to the creep regime. This stress reduction is significant when evaluating fatigue and creep life.

The required thickness is obtained by using $n = 6.0$ in Eq. (4.35). This yields

$$7700 = 2500\left[\left(\frac{2-6}{6}\right)(1.0)^{2/6} + 1\right] / \left(\gamma^{2/6} - 1\right)$$

or

$$\gamma = 1.3611$$
$$R_0 = 24.50$$
$$t = 6.5 \text{ in.}$$

Accordingly, a thickness of 7.26 in. will result in a lower stress and a longer fatigue life than is implied by VIII-1.

c) Limit analysis Eq. (4.37)

$$7700 = 2500 / \ln(\gamma)$$

or

$$\gamma = 1.3836$$
$$R_0 = 24.90$$
$$t = 6.9 \text{ in.}$$

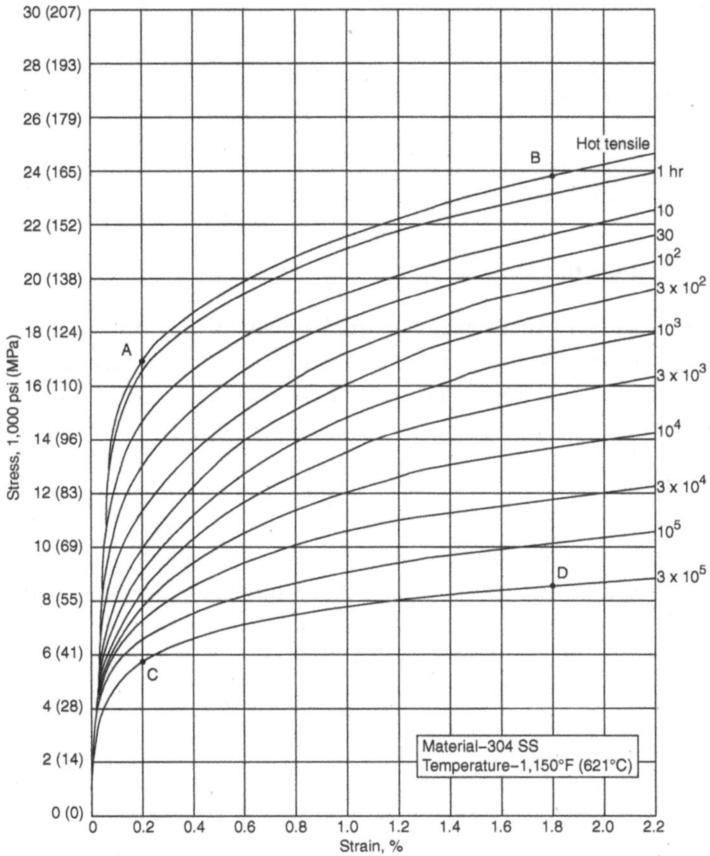

Figure 4.13 Average isochronous stress-strain curves for type 304 stainless steel at 1150°F (621°C) [ASME II-D].

d) Reference stress Eq. (4.38)

The procedure for the reference stress method is to first assume a thickness and then calculate the limit pressure in a component. Equation (4.38) is then used to obtain the reference stress. In this case, however, we used in Part (c) a closed-form solution to obtain a thickness of 6.9 in. The limit pressure P_u, based on this trial thickness, is calculated from Eq. (4.37)

$$\gamma = (18 + 6.9)/18 = 1.3833$$

$$\begin{aligned} P^{**} &= S_y[\ln(\gamma)] = S_y(\ln 1.3833). \\ &= 0.3245\,S_y \end{aligned}$$

From Eq. (4.38),

$$\begin{aligned} S_R &= (P / P_u) S_y \\ &= (2500 / 0.3245 S_y) S_y. \\ &= 7700 \text{ psi} \end{aligned}$$

Thus, the calculated thickness of 6.9 in. is adequate.

4.6.2 Spherical Shells

The behavior of spherical shells due to internal pressure is very similar to that of cylinders discussed in Section 4.6.1. Lame's equation for the stress in a thick spherical shell Den Hartog (1987) due to internal pressure is expressed as

$$S = P \frac{(1/2)(R_0 / R)^3 + 1}{(R_0 / R_i)^3 - 1}. \tag{4.39}$$

The governing equation for the circumferential stress in a spherical shell due to internal pressure in the creep range under a stationary stress condition Finnie and Heller (1959) is derived from Eq. (4.34) as

$$S = P \frac{\left(\frac{3 - 2n}{2n}\right)(R_0 / R)^{3/n} + 1}{(R_0 / R_i)^{3/n} - 1}. \tag{4.40}$$

The governing equation for the maximum circumferential stress in a spherical shell due to internal pressure using plastic theory (Hill (1950); Jones (2009)) is expressed as

$$S = P \left(\frac{1}{2 \ln(R_0 / R_i)} - 1 \right). \tag{4.41}$$

Equation (4.40) reverts to Eq. (4.39) when $n = 1.0$. The value of S in Eq. (4.40) approaches that given by Eq. (4.41) as n increases to infinity.

Example 4.7 A spherical head has an inside radius of 24 in., a thickness of 0.60 in., and is subjected to an internal pressure of 1000 psi. Calculate the following:

a) Maximum elastic stress from Eq. (4.39)
b) Maximum stress from Eq. (4.40) with n = 1, 6, 100
c) Maximum stress from Eq. (4.41)

Solution

a) From Eq. (4.39) with $R = 24$ in.,

$$S = (1000)\frac{(1/2)(24.6/24)^3 + 1}{(24.6/24)^3 - 1}$$
$$= 20{,}000 \text{ psi}$$

b) From Eq. (4.40) with $R = 24$ in.,

$$S = 20{,}000 \text{ psi for } n = 1$$
$$S = 19{,}370 \text{ psi for } n = 6$$
$$S = 19{,}260 \text{ psi for } n = 100$$

c) From Eq. (4.41),

$$S = (1000)\left(\frac{1}{2\ln(24.6/24)} - 1\right)$$
$$= 19{,}250 \text{ psi}$$

A comparison of the three methods shows that spherical shells behave essentially the same as cylindrical shells.

Problems

4.1 A pressure vessel is supported on legs. The membrane stresses, P_m, in the shell at the vicinity of the legs when the vessel is not pressurized are:
Condition 1

$$\sigma'_c = 0 \text{ psi} \quad \sigma'_\ell = -700 \text{ psi} \quad \tau'_{c\ell} = 350 \text{ psi} \quad \sigma'_r = 0 \text{ psi.}$$

The total membrane stresses in the shell when the vessel is pressurized are:
Condition 2

$$\sigma''_c = 1800 \text{ psi}, \quad \sigma''_\ell = 200 \text{ psi}, \quad \tau''_{c\ell} = 350 \text{ psi}, \quad \sigma''_r = -150 \text{ psi.}$$

Comment

The shear stresses τ_{cr} and τ_{1r} are equal to zero in this case. Moreover, the principal stresses in the $c\ell$ plane are obtained from the equation

$$\sigma_{1,2} = \frac{(\sigma_c + \sigma_\ell)}{2} \pm \left[\left(\frac{\sigma_c - \sigma_\ell}{2}\right)^2 + \tau_{c\ell}^2\right]^{1/2}. \tag{A}$$

Find the equivalent stress, P_m.

4.2 An internal tray has a diameter of 60 in. and is subjected to a 2 psi differential pressure. The design temperature is 1000°F and the material is 2¼ Cr-1Mo annealed SA 387-22 cl2.

Data:

- The tray is assumed to be simply supported
- $S = 7800$ psi at 1000°F from Table 1A of ASME II-D
- From Table 5A of ASME II-D for ASME VIII-2 values, $S_{av} = 14{,}000$ psi ($S = 20{,}000$ psi at room temperature and $S = 8000$ psi at 1000°F)
- From Table Y-1 of ASME II-D values, average $S_y = 26{,}850$ psi ($S_y = 30{,}000$ psi at room temperature and $S_y = 23{,}700$ psi at 1000°F)
- $S_m = 15{,}800$ psi at 1000°F from Figure E-100.4-4 of ASME II-D
- $S_{mt} = 5200$ psi at 1000°F and 300,000 h from Figure E-100.4-4 of ASME II-D
- $S_t = 5200$ psi at 1000°F and 300,000 h from Figure E-100.5-4 of ASME II-D

Perform the following:

a) Determine the thickness in accordance with VIII-1.
b) Evaluate the stresses in accordance with VIII-2.
c) Evaluate the stresses in accordance with the load-controlled criteria of ASME VIII-2 based on 300,000 h.

The elastic bending stress of a simply supported plate is given by

$$\sigma = 1.24PR^2 / t^2.$$

5

Analysis of Components

Strain and Deformation-Controlled Limits

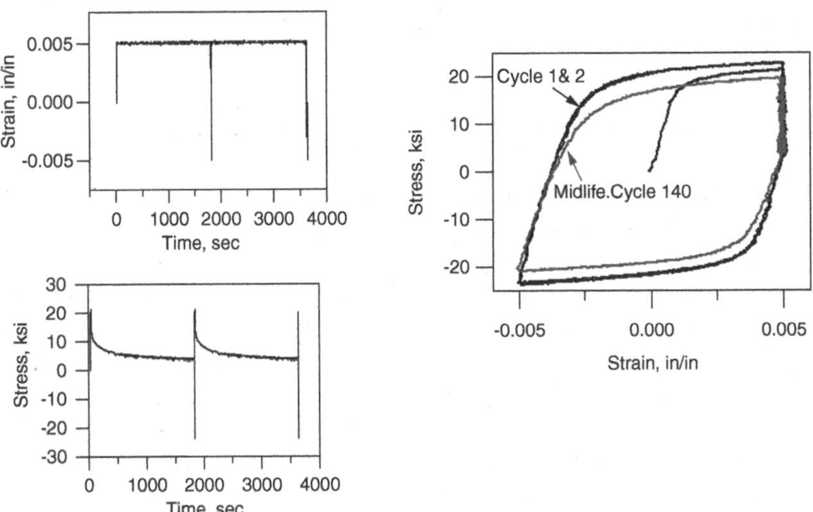

Type 316H stainless steel creep-fatigue plots [Yanli Wang, Oak Ridge National Laboratory].

5.1 Introduction

Stress analysis of a component in the creep regime requires both load- and strain-controlled evaluation. Load-controlled limits were discussed in Chapter 4 and the strain-controlled limits are described in this chapter. The evaluation for the strain-controlled limits is made in accordance with one of three criteria detailed in ASME VIII-2. The first criterion, based on elastic analysis, is the

Analysis of ASME Boiler, Pressure Vessel, and Nuclear Components in the Creep Range, Second Edition. Maan H. Jawad and Robert I. Jetter.
© 2022 John Wiley & Sons Ltd. Published 2022 by John Wiley & Sons Ltd.

easiest to use but is very conservative. When the stress calculations cannot satisfy the elastic limits then the component is redesigned or the analysis proceeds to the second criterion based on simplified inelastic analysis. This second criterion is costlier to perform than the elastic analysis and requires additional material data. However, the results obtained from the simplified inelastic analysis are more accurate than those obtained from an elastic analysis.

When the simplified inelastic stress analysis cannot satisfy the simplified inelastic stress limits, then the component is redesigned or the analysis proceeds to the third criterion based on inelastic analysis. To perform inelastic analysis, it is necessary to define the constitutive equations for the behavior of the material in the creep regime. These equations are not readily available for all materials. As a consequence, this third criterion is costly to perform but gives reasonably accurate results once performed.

5.2 Strain and Deformation-Controlled Limits

The procedure for strain and deformation limits is intended to prevent ratcheting. ASME VIII-2 gives the designer the option of using one of three methods of analysis. They are the elastic, simplified inelastic, and inelastic analyses. The object of all of these methods is to limit the strains in the operating condition to 1% for membrane, 2% for bending, and 5% for local stress. At welds, the allowable strain is one half of those values. It is of interest to note that some engineers believe that, in some instances, the limitation set on primary membrane strain of 1% may be too conservative because this criterion does not affect the overall failure of the component. Thus, a 1% strain in a flange may result in an unwanted leakage, whereas the same strain at the junction of a flat head to shell junction may be acceptable.

The elastic method, which is very conservative, is generally applicable when the primary plus secondary stresses are below the yield strength. The simplified inelastic analysis, which has less conservatism built into it compared to the elastic analysis, is based on bounding the accumulated membrane strain. The last option is to perform an inelastic analysis. The inelastic analysis yields accurate results but has the drawback of being expensive and time-consuming to perform. It requires a large amount of material property data that may not be readily available for the material under consideration.

One potential disadvantage of using the strain- and deformation-controlled limits of ASME VIII-2 is that it requires separate treatment of primary and secondary stress categories as shown in Figure 4.8. This condition is avoided in ASME VIII-2 Part 5 for analysis below the creep regime by combining the primary and secondary stress categories into one quantity. For simple structures, the separation of primary and secondary stress in ASME VIII-2 may not be a big problem. But for more complex structures with asymmetrical geometry and loading, it can be difficult, if not impossible, to sort out primary and secondary stress categories from a detailed finite element analysis. ASME VIII-2, however, does give the designer the option (called Test A-1) to use the combination of primary plus secondary stress limits similar to ASME VIII-2 Part 5 when the effects of creep are negligible. This creep modified shakedown limit avoids the potential problem of separating primary and secondary stresses and is used as an alternative to the standard ASME VIII-2 Supplement strain and deformation limit.

5.3 Elastic Analysis

The strain and deformation limits in the elastic stress analysis are considered to be satisfied if they meet the requirements of A-1, A-2, or A-3. A summary of the requirements for these three tests is shown in Figure 5.1. The membrane and primary bending stresses used in these tests are defined as

$$X = \left(P_L + P_b / K_t\right) / S_y \tag{5.1}$$

and the secondary stress is given by

$$Y = Q / S_y. \tag{5.2}$$

The value of K_t in Eq. (5.1) is approximated by ASME VIII-2 Supplement to a value of 1.25 for rectangular cross sections. The actual value of K_t for pipes and tubes is less than 1.25. Table 4.3 shows the K_t values for rectangular and circular cross sections.

Important definitions applicable to Tests A-2 and A-3
• X includes all primary membrane and bending stresses. • Y includes all secondary stresses. • S_y is the average of the yield stress at the maximum and minimum operating temperatures during the cycle.

5.3.1 Test A-1

This test, although applicable in all conditions, was originally intended for components that are in the creep range for only a portion of their expected design life. Compliance with this test indicates that creep is not an issue and the rules of ASME VIII-2 Part 5 may be used directly. The calculated stresses are satisfied when all of the following criteria are met:

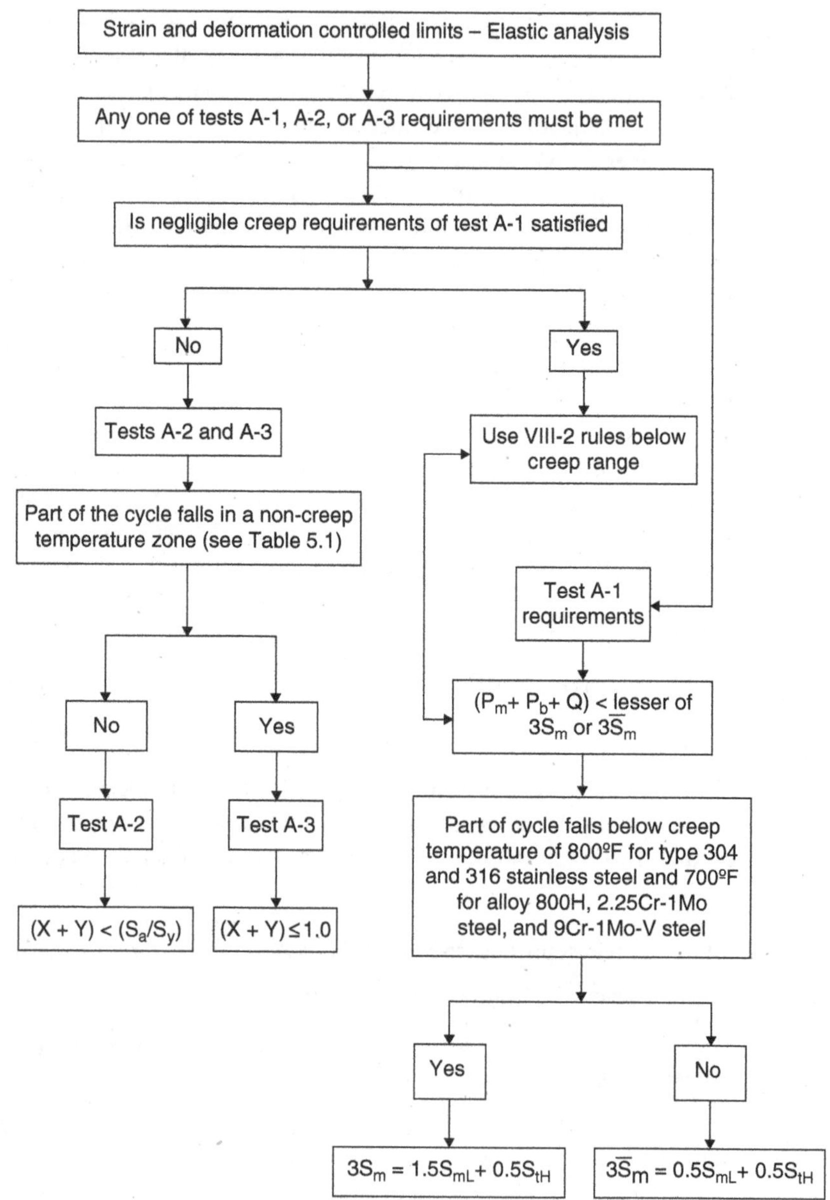

Figure 5.1 Strain-controlled limits – elastic criteria.

a) The combined primary and secondary stress are limited to the lesser of

$$P_L + P_b + Q < 3S_m \qquad (5.3)$$

(If Q is due to thermal transients, then S_m is the average of values taken at the hot and cold ends of the cycle. If pressure-induced loading is part of Q then S_m is the value at the hot end of the cycle.)

or

$$P_L + P_b + Q < 3\overline{S}_m \qquad (5.4)$$

The value of $\overline{S}_m$ is obtained from

1. When part of the cycle falls below the creep temperature

$$3\overline{S}_m = \left(1.5S_{mL} + 0.5S_{tH}\right).$$

2. When both temperature extremes of the cycle are above the creep temperature

$$3\overline{S}_m = \left(0.5S_{tL} + 0.5S_{tH}\right).$$

The extremes of the cycle are considered to be in the creep regime when the temperature is greater than 700°F(371°C) for 2.25Cr-1Mo and 9Cr-1Mo-V steels and greater than 800°F (427°C) for stainless and nickel alloys.

S_{mL} = the value of S_m determined at the maximum wall temperature at the low end of the cycle

S_{tH} = the value of S_t at the maximum wall temperature at the hot end of the cycle

S_{tL} = the value of S_t at the maximum wall temperature at the low end of the cycle

$0.5S_{tH}$ = an approximate conservative value for the relaxation stress, S_{rH}, at the hot end of the cycle.

$0.5S_{tL}$ = an approximate conservative value for the relaxation stress, S_{rH}, at the cold end of the cycle.

S_{rH} = hot relaxation stress. It is obtained by performing a pure uniaxial relaxation analysis starting with an initial stress of $1.5S_m$ and holding the initial strain throughout the time interval equal to the time of service in the creep regime. An acceptable, albeit conservative, alternate to this procedure is to use the elevated temperature creep dependent allowable stress level, $0.5S_t$, as a substitute for S_{rH}.

S_t = temperature- and time-dependent stress intensity limit.

b) Thermal stress ratcheting must be kept below a certain limit defined, for linear thermal distribution, by the equations

$$y' = 1/x \text{ for } 0 < x \leq 0.5 \tag{5.5}$$

$$y' = 4(1-x) \text{ for } 0.5 < x \leq 1.0. \tag{5.6}$$

For parabolic thermal distribution, the equations are

$$y' = 10.3855e^{-2.6785x} \text{ for } 0.3 < x \leq 0.615 \tag{5.7}$$

$$y' = 5.2(1-x) \text{ for } 0.615 < x \leq 1.0 \tag{5.8}$$

where

x = (general membrane stress, P_m)/S_y

y' = (range of thermal stress calculated elastically, ΔQ)/S_y.

c) Time duration limits are given by

$$\sum (t_i / t_{id}) < 0.1 \tag{5.9}$$

where

t_i = total time duration during the service lifetime of a component at the highest operating temperature

t_{id} = maximum allowable time as determined from Appendix E of ASME II-D by entering a stress-rupture chart at temperature T_i and a stress value of 1.5 times the yield stress at temperature T_i. The 1.5 factor is the product of 1.2 for strain hardening times 1.25 for increasing the rupture values from minimum to nominal.

d) Strain limit is given by

$$\sum \varepsilon_i < 0.2\% \tag{5.10}$$

where

ε_i = creep strain obtained from an isochronous stress-strain chart that would be expected from a stress level of $1.25S_{y|Ti}$ applied for the total duration of time during the service lifetime that the metal is at temperature T_i. Most of the isochronous stress-stain charts listed in Appendix E of ASME II-D are limited to a maximum strain of 2.2%. In cases where the actual ε_i corresponding to a given stress falls beyond the range of the isochronous stress-strain charts in ASME II-D, and an actual value is needed, other documents such as ASME FFS-1 may be used to obtain an approximate strain value.

5.3.2 Test A-2

This test applies for cycles where both extremes of the cycle are within the creep range of the material (see Test A-3 for definition of creep range for this application). The governing equation is

$$X + Y < S_a / S_y \tag{5.11}$$

where S_a is the lesser of:

a) $1.25S_t$ using the highest wall-averaged temperature during the cycle and time value of 10^4 h
b) the average of the two S_y values associated with the maximum and minimum wall-averaged temperatures during the cycle.

The requirement of item (a) is based on the concern that creep relaxation at both ends of the temperature cycle will exacerbate the potential for ratcheting. Experience has shown that using a value of S_t at 10^4 h as a stress criterion is a realistic assumption. The requirement of (b) is based on averaging the yield stress associated with the maximum and minimum temperatures as a good approximation.

5.3.3 Test A-3

This test is applicable for those cycles in which the average wall temperature at one of the stress extremes defining the maximum secondary stress range, Q, is below the temperature given in Table 5.1.

$$X + Y < 1.0. \tag{5.12}$$

Values shown in Table 5.1 are the approximate temperatures above which the material allowable stress values at 100,000 h are controlled by creep and rupture.

Note: Table 5.1 in Test A-3 has different temperatures than those defining T_i in Test A-1 for the onset of creep effects.

Table 5.1 Temperature limitation.

Material	Temperature, °F (°C)
304 stainless steel	948 (509)
316 stainless steel	1011 (544)
Alloy 800 H	1064 (573)
2.25Cr-1Mo steel	801 (427)
9Cr-1Mo-V	940 (504)

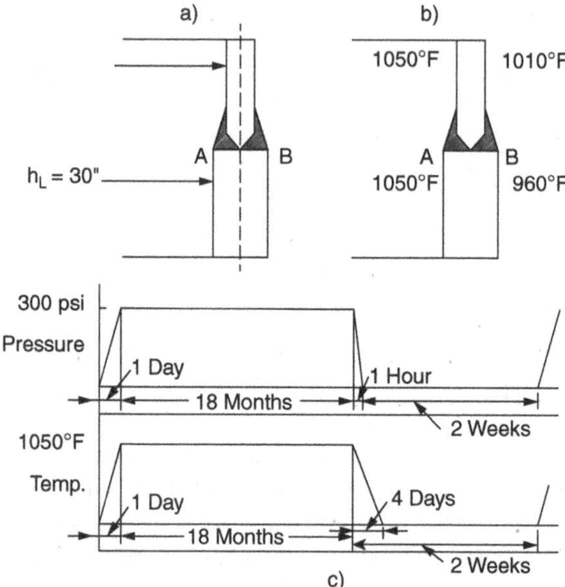

Figure 5.2 Vessel geometry and operating cycles.

Example 5.1 An VIII-1 vessel has a design temperature of 1100°F and a design pressure of 315 psi. The construction material is stainless steel grade 304. The operating temperature is 1050°F and the operating pressure is 300 psi. The required shell thickness due to design condition is 1.0 in. However, the thickness of part of the shell is increased to 2.0 in. to accommodate the reinforcement of some large nozzles in the shell. The junction between the 1.0-in. and 2.0-in. shells is shown in Figure 5.2a. Temperatures at the inside and outside surfaces of the shell during the steady-state operating condition are shown in Figure 5.2b. The operating cycle (Figure 5.2c), is as follows:

Pressure
The pressure increases from 0 psi to 300 psi in 24 h. It remains at 300 psi for about 18 months (13,000 h), and is then reduced to zero in 1 h. The same cycle is then repeated after a shutdown of two weeks for maintenance.

Temperature
The temperature increases from ambient to 1050°F in 24 h. It remains at 1050°F for about 18 months (13,000 h) and then reduces to ambient in four days. The same cycle is then repeated after the shutdown.

Evaluate the following:

1) Stress away from any discontinuities due to design condition.
2) Stress away from any discontinuities due to operating condition using elastic analysis.

Data:

$S = 9800$ psi at 1100°F (Appendix E of ASME II-D)
$E = 22{,}000$ ksi at 1100°F (Table TM-1 of ASME II-D)
$E = 22{,}400$ ksi at 1050°F (Table TM-1 of ASME II-D)
$S_m = 13{,}600$ psi at 1050°F (Appendix E of ASME II-D)
$S_m = 20{,}000$ psi at 100°F (Table 2A of ASME II-D)
$S_{mt} = 8500$ psi at 1050°F and at 130,000 h (Appendix E of ASME II-D)
$S_y = 30{,}000$ psi at 100°F (Table Y-1 of ASME II-D)
$S_y = 15{,}200$ psi at 1050°F (Appendix E of ASME II-D)
Average S_y during the cycle = 22,600 psi
$S_t = 12{,}200$ psi at 1050°F and at 10^4 h (Appendix E of ASME II-D)
$S_t = 8490$ psi at 1050°F and at 130,000 h (Appendix E of ASME II-D)
$\alpha = 10.4 \times 10^{-6}$ in./in./°F at 1050°F (Table TE-1 of ASME II-D)
$\mu = 0.3$.

Solution

Assumptions

- Calculations in this example are based on thin-shell equations to keep the calculations as simple as possible. This was done to demonstrate and highlight the method of creep analysis. For thick shells, Lame's equations must be used.
- The centerlines of the thin and thick shells are aligned.
- The following calculations are based on the criteria of Figure. 4.8.

Stress due to design condition

The design stress at the beginning of the cycle is zero. Membrane stress due to design pressure of 315 psi is shown in Table 5.2.

The maximum membrane equivalent stress factors, P_m, are obtained from Eq. (4.14) using the three principal stresses from Table 5.2

For 1-in. thick shell = 8457 psi
For 2-in. thick shell = 4229 psi.

From Figure 4.8, the value of P_m for both 1-in. and 2-in. shells is less than the design allowable stress of S_o (9800 psi). Hence, load-controlled design stress limits are adequate. The quantities P_L and P_b are not applicable in this case.

Table 5.2 Primary membrane stress due to design pressure of 315 psi.

Stress[1]	One-inch-thick shell[2]	Two-inch-thick shell[3]
$P_{m\theta}$	9608	4725
P_{mL}	4804	2363
P_{mr}	−158	−158

[1] $P_{m\theta} = PR_i/t$, $P_{mL} = PR_i/2t$, $P_{mr} = -P/2$.
[2] $R_i = 30.5$ in.
[3] $R_i = 30.0$ in.

Table 5.3 Primary membrane stress due to operating pressure of 300 psi.

Stress[1]	One-inch-thick shell[2]	Two-inch-thick shell[3]
$P_{m\theta}$	9150	4500
P_{mL}	4575	2250
P_{mr}	−150	−150

[1] $P_{m\theta} = PR_i/t$, $P_{mL} = PR_i/2t$, $P_{mr} = -P/2$.
[2] $R_i = 30.5$ in.
[3] $R_i = 30.0$ in.

Stress due to operating condition using elastic analysis
Both load- and strain-controlled limits must be satisfied. The load-controlled limits will be evaluated first in accordance with Figure 4.8.

(b.1) Load-controlled limits
The operating stress at the beginning of the cycle is zero. Stress due to an operating pressure of 300 psi is shown in Table 5.3.

The B maximum membrane equivalent stress factors, P_m, are obtained from Eq. (4.14) using the three principal stresses from Table 5.3

For 1-in. thick shell = 8053 psi

For 2-in. thick shell = 4027 psi.

From Figure 4.8, the values of P_m for the 1-in. and 2-in. thick shells are within the allowable stress of S_{mt} (8500 psi at 1050°F). P_L and P_b are not applicable in this case.

(b.2) Strain-controlled limits
The pressure as well as the temperature stresses must be included in the calculations. For thin shells, the governing equation for thermal stress is obtained from

$$\sigma_\theta = \sigma_l = \frac{E\alpha\Delta\tau_l}{2(1-\mu)}.$$

The stresses due to pressure and temperature of the 1.0-in. and 2.0-in. shells are given in Tables 5.4 and 5.5. Table 5.4 shows the inside surface stresses, and Table 5.5 shows the outside surface stresses.

Maximum equivalent stress values from Eq. (4.14)
Note that the calculated stresses shown below are based on the ASME VIII-2 definition of equivalent stress given by the Von Mises expression.

1.0-in. shell at inside surface (based on the three principal stresses from Table 5.4)

Membrane equivalent stress due to pressure, $P_{Lm} = 8053$ psi

Bending equivalent stress due to pressure, $Q = 150$ psi

Table 5.4 Stresses at inside surface due to pressure and temperature.

Stress	Stress due to pressure, psi	Stress due to temperature, psi
One-inch shell		
Membrane circumferential stress, $P_{m\theta}$	9150	0
Membrane axial stress, P_{mL}	4575	0
Membrane radial stress, P_{mr}	−150	0
Circumferential bending stress, $Q_{b\theta}$	0	−6655
Axial bending stress, Q_{bL}	0	−6655
Radial bending stress, Q_{br}	−150	0
Two-inch shell		
Membrane circumferential stress, $P_{m\theta}$	4500	0
Membrane axial stress, P_{mL}	2250	0
Membrane radial stress, P_{mr}	−150	0
Circumferential bending stress, $Q_{b\theta}$	0	−14,975
Axial bending stress, Q_{bL}	0	−14,975
Radial bending stress, Q_{br}	−150	0

Membrane equivalent stress due to temperature, $P_{Lm} = 0$ psi
Bending equivalent stress due to temperature, $Q = 6655$ psi.

- 1.0-in. shell at outside surface (based on the three principal stresses from Table 5.4)
  Membrane equivalent stress due to pressure, $P_{Lm} = 8053$ psi
  Bending equivalent stress due to pressure, $Q = 150$ psi
  Membrane equivalent stress due to temperature, $P_{Lm} = 0$ psi
  Bending equivalent stress due to temperature, $Q = 6655$ psi.
- 2.0-in. shell at inside surface (based on the three principal stresses from Table 5.5)
  Membrane equivalent stress due to pressure, $P_{Lm} = 4027$ psi
  Bending equivalent stress due to pressure, $Q = 150$ psi
  Membrane equivalent stress due to temperature, $P_{Lm} = 0$ psi
  Bending equivalent stress due to temperature, $Q = 14{,}975$ psi
- 2.0-in. shell at outside surface (based on the three principal stresses from Table 5.5)
  Membrane equivalent stress due to pressure, $P_{Lm} = 4027$ psi
  Bending equivalent stress due to pressure, $Q = 150$ psi
  Membrane equivalent stress due to temperature, $P_{Lm} = 0$ psi
  Bending equivalent stress due to temperature, $Q = 14{,}975$ psi

Table 5.5 Stresses at outside surface due to pressure and temperature.

Stress	Stress due to pressure, psi	Stress due to temperature, psi
One-inch shell		
Membrane circumferential stress, $P_{m\theta}$	9150	0
Membrane axial stress, P_{mL}	4575	0
Membrane radial stress, P_{mr}	−150	0
Circumferential bending stress, $Q_{b\theta}$	0	6655
Axial bending stress, Q_{bL}	0	6655
Radial bending stress, Q_{br}	−150	0
Two-inch shell		
Membrane circumferential stress, $P_{m\theta}$	4500	0
Membrane axial stress, P_{mL}	2250	0
Membrane radial stress, P_{mr}	−150	0
Circumferential bending stress, $Q_{b\theta}$	0	14,975
Axial bending stress, Q_{bL}	0	14,975
Radial bending stress, Q_{br}	150	0

Stress in 1.0-in. thick shell

From Eqs. (5.1) and (5.2),

$$X = (P_L + P_b / K) / S_y = (8053 + 0/1.25)/22{,}600 = 0.356$$
$$Y = Q / S_y = (6655 + 150)/22{,}600 = 0.301$$

All three tests (A-1, A-2, and A-3) will be checked in this example to demonstrate their applicability. In actual practice, the designer uses one of these tests to satisfy the limits. If the test limit is exceeded, then the other tests are evaluated or a new design is tried.

- Test A-1

The following two criteria will be checked first for this test. Use the lesser of

$$P_L + P_b + Q < 3S_m \tag{1}$$

or

$$P_L + P_b + Q < 3\overline{S}_m \tag{2}$$

from Eq. (1), keeping in mind that the stresses at the beginning of the cycle are zero.

Circumferential $P_L + P_b + Q$ stress at inside surface = 9150 – 6655 = 2495 psi
Longitudinal $P_L + P_b + Q$ stress at inside surface = 4575 – 6655 = −2080 psi
Radial $P_L + P_b + Q$ stress at inside surface = −150 – 150 = −300 psi

Effective $P_L + P_b + Q$ stress at inside surface = 0.707{[2495 − (−2080)]^2 + [(−2080) − (−300)]^2 + [(−300 − (2495)]^2}$^{0.5}$
Effective $P_L + P_b + Q$ stress at inside surface = 3993 psi
Circumferential $P_L + P_b + Q$ stress at outside surface = 9150 + 6655 = 15,805 psi
Longitudinal $P_L + P_b + Q$ stress at outside surface = 4575 + 6655 = 11,230 psi
Radial $P_L + P_b + Q$ stress at outside surface = −150 − 150 = −300 psi
Effective $P_L + P_b + Q$ stress at outside surface = 0.707{[15,805 − (11,230)]^2 + [(11,230) − (−300)]^2 + [(−300 − (15,805)]^2}$^{0.5}$
Effective $P_L + P_b + Q$ stress at outside surface = 14,372 psi

$$14{,}372 < 3(13{,}600)$$
$$14{,}372 < 40{,}800.$$

From Eq. (2),

$$14{,}372 \le \left[1.5(13{,}600) + 0.5(8490)\right]$$
$$14{,}372 < 24{,}600.$$

Hence, Eq. (2) controls.

It should be pointed out that the allowable stress of 13,600 psi is conservative. Allowable stress at actual temperature at the point may be used.

Next, the thermal ratchet will be checked. Assume linear thermal gradient in the shell.

From Eq. (5.5),
x = 8053/22,600 = 0.356,
then, y' = 1/x = 2.81
and the maximum permissible Q = (2.81)(22,600) = 63,483 psi

This value is larger than the actual secondary stresses in the shell. Thus, the stresses are satisfactory.

Next check Eq. (5.9) for time duration limits.
t_i = 130,000 h
1.5S_y at 1050°F = 1.5(15,200) = 22,800 psi.

From the rupture chart in ASME II-D for Type 304 stainless steel at 1050°F, and a stress value of 22,800 psi,
t_{id} = 2294 h.
Hence, Eq. (5.9) becomes
Σ(130,000/2294) = 56.7 > 0.1

Accordingly, Test A-1 fails and the analysis has to proceed based on creep criteria.

As a point of interest, check the strain limits of Eq. (5.10)
1.25S_y at 1050°F = 1.25(15,200) = 19,000 psi.

From Figure 5.6, the strain corresponding to a stress of 19,000 psi for 130,000 h is well over 2.2%. This value is well above the limit of 0.2% of Eq. (5.10) and this test fails as well. The designer needs to proceed to Test A-2 next.

- Test A-2
  S_a is the lesser of

$$1.25S_t = 1.25(12{,}200) = 15{,}250\,\text{psi}$$

  or

$$0.5\left(S_{yL} + S_{yH}\right) = 0.5(30{,}000 + 15{,}200) = 22{,}600\,\text{psi}$$

  Use $S_a = 15{,}250$ psi

$$\begin{aligned} 0.356 + 0.301 &\leq 15{,}250 / 22{,}600 \\ 0.657 &< 0.675 \end{aligned}$$

  Hence, stresses are satisfactory based on equivalent stress.
- Test A-3
  This test is applicable because the temperature at the cold side of the cycle is below that given in Table 5.1. From Eq. (5.12)

$$0.356 + 0.301 = 0.657 < 1.00.$$

  Thus, the stresses are satisfactory.

Stress in a 2.0-in.-thick shell

From Eqs. (5.1) and (5.2),

$$\begin{aligned} X &= \left(P_L + P_b / K\right) / S_y = (4025 + 0 / 1.25) / 22{,}600 = 0.178 \\ Y &= Q / S_y = 14{,}825 / 22{,}600 = 0.656 \end{aligned}$$

- Test A-1
  This test does not apply because the creep is not negligible.
- Test A-2

$$\begin{aligned} 0.178 + 0.656 &= 0.835 \\ 0.835 &> 0.675 \end{aligned}$$

  Thus, this test cannot be satisfied.
- Test A-3
  This test is applicable because the temperature at the cold side of the cycle is below that given in Table 5.5. From Eq. (5.12),

$$0.178 + 0.656 = 0.835 < 1.00.$$

  Thus, the stresses are satisfactory.

5.4 Simplified Inelastic Analysis

The simplified inelastic analysis is based on the concept of requiring membrane stress in the core of a cross section to remain elastic, whereas bending stress is allowed to extend in the plastic region. This concept is based on the Bree diagram Bree (1967, 1968), which assumes secondary stress to be mainly generated by thermal gradients. The diagram, Figure 5.3, is plotted with primary stress as the abscissa and secondary stress as the ordinate. The diagram is divided into various zones that define specific stress behavior of the shell. It assumes an axisymmetric thin shell with an axisymmetric loading. It also assumes the

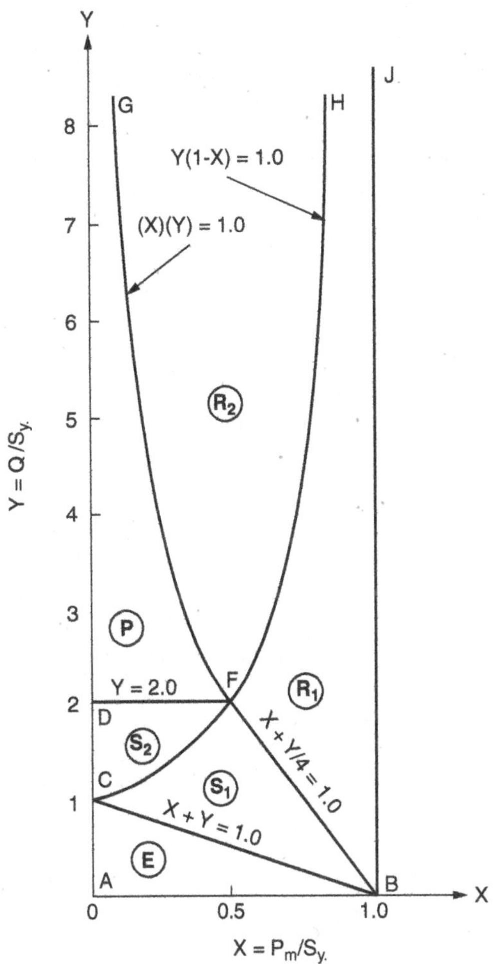

Figure 5.3 Bree diagram Bree (1967).

thermal stress to be linear across the thickness. The actual derivation and assumptions made in constructing the Bree diagram are detailed in Appendix D. The various limitations and zones of the diagram, Figure 5.3, are as follows: Limitations

- It is assumed that the material has an elastic perfectly plastic stress-strain diagram.
- Because mechanical stress is considered primary stress, it cannot exceed the yield stress value of the material. Thermal stress, on the other hand, is considered secondary stress and can thus exceed the yield stress of the material.
- Initial evaluation of the mechanical and thermal stresses in the elastic and plastic regions was made without any consideration to relaxation or creep.
- Final results were subsequently evaluated for relaxation and creep effect.
- It is assumed that stress due to pressure is held constant, while the thermal stress is cycled. Hence, pressure and temperature stress exist at the beginning of the first half of the cycle while pressure only exists at the end of the second half of the cycle.

Zone E

- This zone is bounded by axis lines *AB* and *AC* as well as line *BC*, which is defined by the equation $X + Y = 1.0$.
- Stress is elastic below the creep range.
- Ratcheting does not occur below the creep range.
- Stress redistributes to elastic value above the creep range.

Zone S_1

- This zone is bounded by axis line *CD* as well as line *BC* (defined by the equation $X + Y = 1.0$), line *DF* (defined by the equation $Y = 2$), and line *BF* (defined by the equation $X + Y/4 = 1.0$).
- Below the creep range, the stress is plastic on the outside surface of the shell during the first half of the first cycle. The stress shakes down to elastic in all subsequent cycles.
- Ratcheting does not occur below the creep range.
- Ratcheting occurs in the creep range.
- Shakedown is not possible at the creep range.

Zone S_2

- This zone is bounded by the axis line CD, line *DF* (expressed by the equation $Y = 2$), and line *FC* (defined by equation $Y(1 - X) = 1.0$).
- Zone S_2 is a subset of zone S_1.
- Below the creep range, stress is plastic on both surfaces of the shell during the first half of the first cycle. Stress shakes down to elastic in all subsequent cycles.

- Ratcheting does not occur below the creep range.
- Ratcheting occurs in the creep range.
- Shakedown is not possible at the creep range.

Zone P

- This zone is bounded by axis line *DY*, line *DF* (expressed by $Y = 2$), and line *FG* (expressed by the equation $XY = 1.0$).
- In this zone, alternating plasticity occurs in each cycle below the creep range.
- Shakedown is not possible below as well as in the creep range.
- Failure occurs due to low cycle fatigue below the creep range.
- Shakedown is not possible at the creep range.

Zone R_1

- This zone is bounded by line *BJ*, which is also the *Y* axis, line *BF* (expressed as $X + Y/4 = 1.0$), and line *FH* (expressed by the equation $Y(1 - X) = 1.0$).
- Ratcheting occurs below as well as in creep range.
- Shakedown is not possible below as well as in the creep range.

Zone R_2

- This zone is bounded by line *FG* (expressed as $XY = 1.0$), and line *FH* (expressed as $Y(1 - X) = 1.0$).
- Ratcheting occurs below as well as in creep range.
- Shakedown is not possible below as well as in the creep range.

The actual diagram used by ASME is shown in Figure 5.4. The figure includes Z lines of constant elastic core stress values O'Donnell and Porowski (1974). The key feature of the O'Donnell/Porowski technique is identifying an elastic core in a component subjected to primary loads and cyclic secondary loads. Once the magnitude of this elastic core has been established, the deformation of the component can be bounded by noting that the elastic core stress governs the net deformation of the section. Deformation in the ratcheting, R, regions of the Bree diagram can also be estimated by considering individual cyclic deformation. The ASME simplified inelastic analysis procedure for strain limits consists of satisfying test B-1, or test B-2. The following is a summary of these tests.

Important definitions applicable to tests B-1 and B-2
• *X* includes all membrane, primary bending, and secondary bending stresses due to pressure-induced, as well as thermal-induced, membrane stresses. • *Y* includes all thermal secondary stresses. • S_{yL} is the yield stress at the cold end of the cycle.

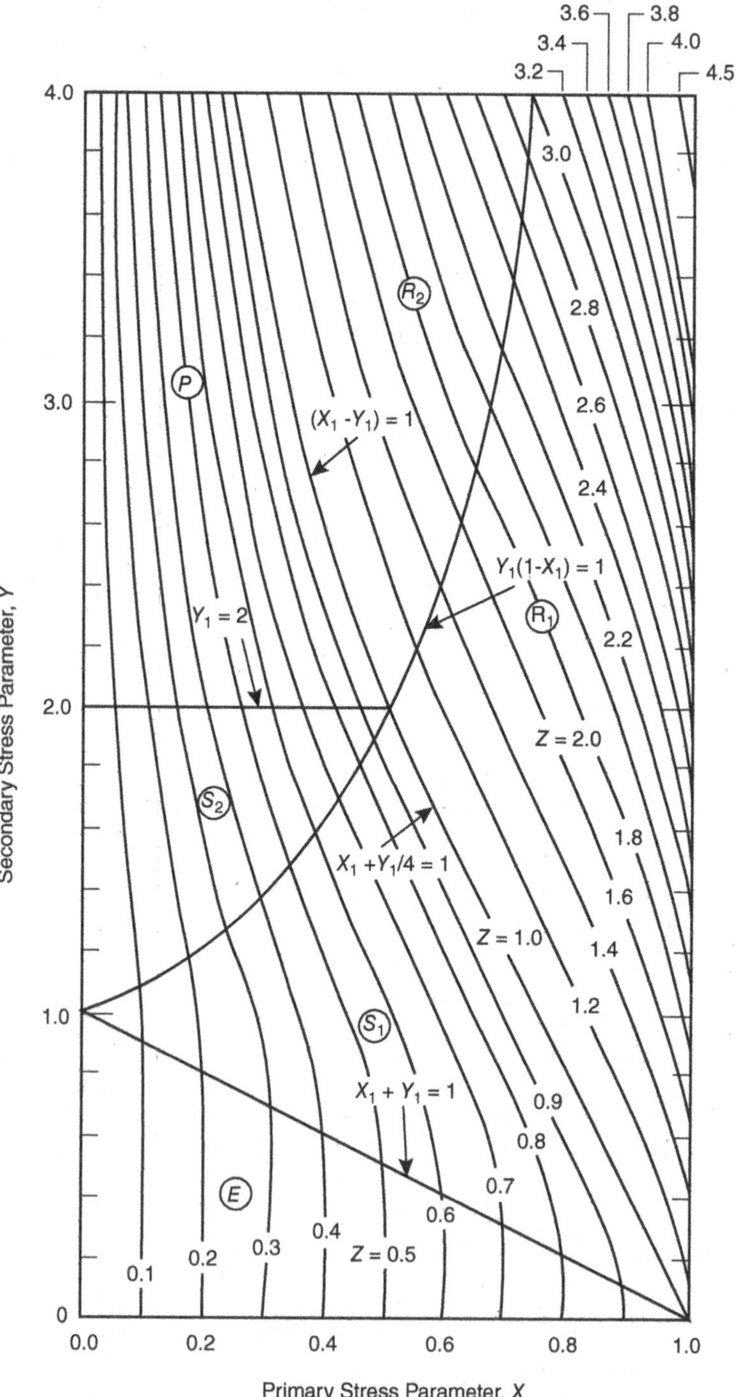

Figure 5.4 Effective creep stress parameter Z for simplified inelastic analysis using test numbers B-1 [ASME VIII-2].

5.4.1 Tests B-1 and B-2

The following conditions must be met to satisfy these tests.

1) The average wall temperature at one of the stress extremes defining each secondary equivalent stress range, Q, is below the applicable temperature in Table 5.1.
2) The individual cycle cannot be split into sub-cycles.
3) Pressure-induced membrane and bending stresses and thermal-induced membrane stresses are classified as primary stresses for the purposes of this evaluation.
4) Definitions of X and Y in Eqs. (5.1) and (5.2) apply for these two tests except that the value of S_y, as defined in these two equations, is replaced with S_{yL}, which is yield stress at the lower end of the cycle.
5) These tests are applicable only in regimes E, S_1, S_2, and P in Figure 5.4.

5.4.2 Test B-1

The following requirements apply to this test

1) The peak stress is negligible.
2) σ_c is less than the hot yield stress, S_{yH}.

The procedure for applying test B-1 consists of the following steps:

1) Determine Z values from either Figure 5.4 or the following values

$$\begin{aligned} &Z = X && \text{in region E} \\ &Z = Y + 1 - 2[(1 - X)Y]^{1/2} && \text{in region } S_1 \\ &Z = XY && \text{in region } S_2 \text{ and P.} \end{aligned} \tag{5.13}$$

2) Calculate the effective creep stress σ_c from

$$\sigma_c = Z(S_{yL}). \tag{5.14}$$

3) Calculate the quantity $1.25\sigma_c$.
4) Calculate the creep ratcheting strain from an isochronous curve using the quantity $1.25\sigma_c$ and the total hours for whole life.
5) The resulting strain should be less than 1% for the parent material and 1/2% for welded material.

5.4.3 Test B-2

This test is applicable to any structure and loading. However, Figure 5.5, which takes into consideration peak stresses, is applicable in lieu of Figure 5.4. The procedure for applying test B-2 consists of the following steps:

1) Determine Z values from Figure 5.5.
2) Calculate the effective creep stress σ_c from

$$\sigma_c = Z\left(S_{yL}\right) \tag{5.15}$$

3) Calculate the quantity 1.25 σ_c.
4) Calculate the creep ratcheting strain from and isochronous curve using the quantity $1.25\sigma_c$ and the total hours for whole life.
5) The resulting strain should be less than 1% for the parent material and 1/2% for weld material.

Example 5.2 Use the data from Example 5.1 to evaluate stress away from any discontinuities due to operating condition using simplified inelastic analysis. Use the isochronous curves shown in Figure 5.6.

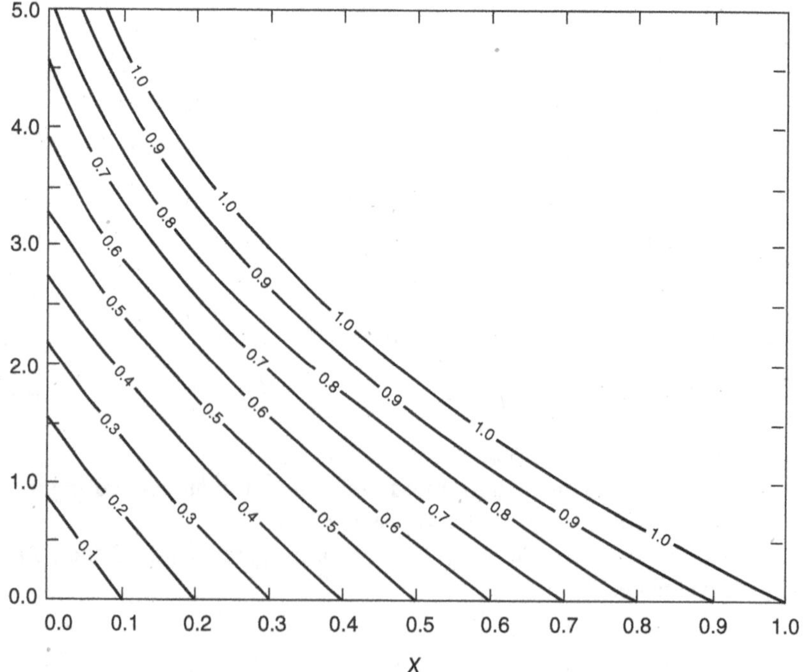

Figure 5.5 Effective creep stress parameter Z for simplified inelastic analysis using test number B-2 [ASME VIII-2].

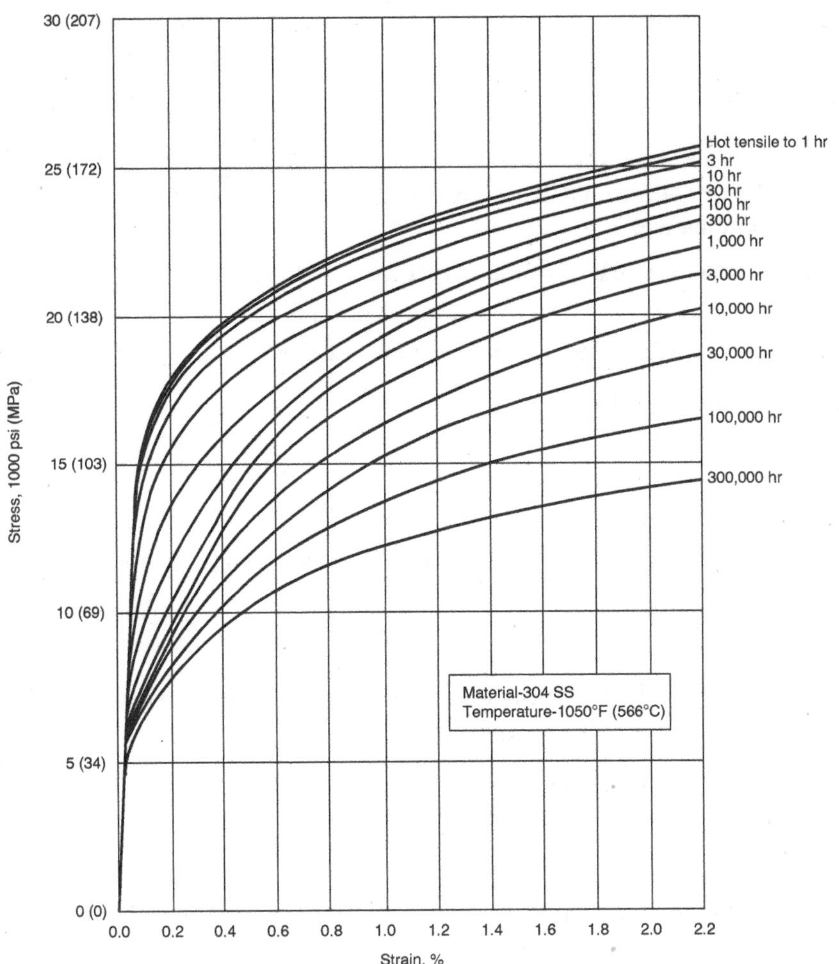

Figure 5.6 Isochronous stress-strain curves for Type 304 stainless steel [ASME II-D].

Solution

Normally, there is no need to use simplified inelastic analysis if the elastic analysis is satisfied. However, for this example, the simplified inelastic analysis is performed to compare the results of the two methods.

- Tests B-1 and B-2

  Calculations for X and Y are based on using the yield stress at ambient temperature from Eqs. (5.1) and (5.2).

One-inch shell

- Test B-1

$$X = P_m / S_y$$

= {membrane stress due to pressure + [(bending stress due to pressure)/1.25] + membrane stress due to temperature} /S_{yL}

From Table 5.5,

$$\text{circumferential } P_m = 9150 + (0)/1.25 + 0 = 9150 \text{ psi}$$
$$\text{axial } P_m = 4575 + (0)/1.25 + 0 = 4575 \text{ psi}$$
$$\text{radial } P_m = -150 + (-150)/1.25 + 0 = -270 \text{ psi}$$

and, from Eq. (4.14),

$P_m = 8160$ psi

$$X = 8160 / 30{,}000 = 0.272$$
$$Y = Q / S_y$$
$$= (\text{bending stress due to temperature}) / S_{yL} = 6655 / 30{,}000 = 0.272$$

From Figure 5.4,

$$Z = 0.272$$
$$S_{yL} = \text{ at the cold end of cycle is at room temperature} = 30{,}000 \text{ psi}$$
$$\sigma_c = 0.272(30{,}000) = 8160 \text{ psi}$$

From the isochronous curves (Figure 5.6), with stress = $1.25\sigma_c$ (10,200 psi) and expected life of 130,000 h, we obtain a strain of 0.42%. This value is acceptable because it is less than the permissible value of 1%.

- Test B-2

This test need not be performed because test B-1 is satisfied.

Two-inch shell

- Test B-1

$$X = P_m / S_y$$
$$= [\text{membrane stress due to pressure} + (\text{bending stress due to pressure})/1.25 + \text{membrane stress due to temperature}]/S_{yL}$$

From Table 5.5,

$$\text{circumferential } P_m = 4500 + (0)/1.25 + 0 = 4500 \text{ psi}$$
$$\text{axial } P_m = 2250 + (0)/1.25 + 0 \quad = 2250 \text{ psi}$$
$$\text{radial } P_m = -150 + (-150)/1.25 + 0 = -270 \text{ psi}$$

and, from Eq. (4.14),

$P_m = 4130$ psi

$$X = 4130 / 30{,}000 = 0.138$$
$$Y = Q / S_y$$
$$= (\text{bending stress due to temperature}) / S_{yL}$$
$$= 14{,}975 / 30{,}000 = 0.500$$

From Figure 5.4,

$$Z = 0.138$$
$$S_{yL} = \text{ at the cold end of cycle is at room temperature} = 30{,}000\,\text{psi}.$$
$$\sigma_c = 0.272(30{,}000) = 8160\,\text{psi}$$

From the isochronous curves (Figure 5.5), with stress = $1.25\sigma_c$ (5175 psi) and expected life of 130,000 h, we obtain a strain of 0.026%. This value is acceptable because it is less than the permissible value of 1%.

- Test B-2

  This test need not be performed because Test B-1 is satisfied.

  It should be pointed out at this time that the expected design life of 130,000 h for this shell is longer than the life of 100,000 h set for the allowable stress criterion for ASME I and VIII. Also, the allowable stress of 8500 psi in ASME VIII-2 Supplemental creep rules is conservatively based on a 67% criterion rather than the 80% set for ASME I and VIII. Additionally, the rules of ASME VIII-2 Supplemental creep rules for operating condition have to be complied with once the temperature is in the creep range. This criterion is not mandatory in ASME I or VIII below the creep range.

Example 5.3 Check the shell stress due to operating condition in accordance with ASME VIII-2 strain-controlled limits in Example 4.5 of Chapter 4. The following values are obtained from ASME II-D:

$S_y = 30$ ksi at room temperature
$S_y = 26.52$ ksi at 810°F
$S_t = 17{,}200$ psi for 100,000 h
$S_{mt} = 17{,}120$ psi

Solution

The equivalent stresses calculated in Example 4.5 are

Stress, psi	Shell	Head
P_L	7315	425
$P_L + P_b$	7315	910
$P_L + P_b/K_t$	7315	810
Q	27,540	225

Average $S_y = 28{,}260$ psi

Shell

- Elastic analysis

$$X = 7315/28{,}260 = 0.259,\ Y = 27{,}540/28{,}260 = 0.975$$

a) Test A-1
This test does not apply because the negligible creep criteria for Test A-1 are not satisfied.

b) Test A-2
This test, which is more conservative than Test A-3, does not apply because one temperature extreme of the stress cycle is below the creep threshold temperature for applicability of Test A-3.

c) Test A-3

$$X + Y = 1.233$$

This value is unacceptable because it exceeds 1.0.

- Simplified inelastic analysis
Test B-1

$$X = P_m / S_y = \left[P_L + \left(P_b + Q\right) / K_t\right] / S_{yL}$$

From Example 4.5,

$$\begin{aligned} \text{circumferential } P_m &= 0 + (0 + 9240)/1.25 = 7390 \text{ psi} \\ \text{axial } P_m &= 7200 + (0 + 30{,}800)/1.25 = 31{,}840 \text{ psi} \\ \text{radial } P_m &= -225 + (0 - 225)/1.25 + 0 = -405 \text{ psi} \end{aligned}$$

and, from Eq. (4.14), $P_m = 29{,}135$ psi

$$\begin{aligned} X &= 29{,}135/30{,}000 = 0.97 \\ Y &= Q/S_y \\ &= (\text{bending stress due to temperature})/S_{yL} = 0/30{,}000 = 0 \end{aligned}$$

From Figure 5.4,

$$\begin{aligned} Z &= 0.97 \\ S_{yL} &= \text{ at the cold end of the cycle is at room temperature} = 30{,}000 \text{ psi} \\ \sigma_c &= 0.97(30{,}000) = 29{,}135 \text{ psi} \end{aligned}$$

However, Test B-1 also requires that $\sigma_c \leq S_{yH}$. In this case, S_{yH} = 26,520 psi which is less than σ_c and Test B-1 is, thus, not applicable. Test B-2 is more conservative than Test B-1. Thus, the remaining alternatives are to consider a thicker section or resort to inelastic analysis.

Problems

5.1 A hydrotreater has an inside diameter of 12 ft and a length of 50 ft. The design pressure is 2400 psi and the design temperature is 975°F. The operating pressure is 2300 psi and the operating temperature is 950°F. The material of construction is 2.25Cr-1Mo steel. Expected life is 200,000 h.

a) Determine the required thickness in accordance with ASME VIII-1.
b) Evaluate the load-controlled limits.
c) Evaluate the strain- and deformation-controlled limits using
   i) Elastic tests A
   ii) Simplified inelastic tests B

Data:

$S = 9400$ psi at 975°F (Appendix E of ASME II-D)
$E_o = 1.0$ (weld joint efficiency)
$E = 24{,}930$ ksi at 975°F (Table TM-1 of ASME II-D)
$E = 25{,}150$ ksi at 950°F (Table TM-1 of ASME II-D)
$S_m = 17{,}600$ psi at 950°F (Appendix E of ASME II-D)
$S_m = 20{,}000$ psi at 100°F (Table 2A of ASME II-D)
$S_{mt} = 7850$ psi at 950°F and at 200,000 h (Appendix E of ASME II-D)
$S_y = 30{,}000$ psi at 100°F (Table Y-1 of ASME II-D)
$S_y = 24{,}800$ psi at 950°F (Appendix E of ASME II-D)
Average S_y during the cycle = 27,400 psi
$S_t = 11{,}300$ psi at 950°F and at 10^4 h (Appendix E of ASME II-D)
$S_t = 7850$ psi at 950°F and at 200,000 h (Appendix E of ASME II-D)

5.2 A boiler header has an outside diameter of 20 in. and a length of 20 ft. The design pressure is 2900 psi and the design temperature is 1000°F. The operating pressure is 2700 psi and the operating temperature is 975°F. The material of construction is 2.25Cr-1Mo steel. Expected life is 300,000 h.

a) Determine the required thickness in accordance with ASME-I.
b) Evaluate the load-controlled limits.
c) Evaluate the strain- and deformation-controlled limits using
   i) Elastic tests A
   ii) Simplified inelastic tests B

Data:

$S = 8000$ psi at 1000°F (Appendix E of ASME II-D)
$E_o = 0.65$ (ligament efficiency) for circumferential stress calculations

$E_o = 0.95$ (ligament efficiency) for longitudinal stress calculations
$E = 24{,}700$ ksi at 1000°F (Table TM-1 of ASME II-D)
$E = 24{,}930$ ksi at 975°F (Table TM-1 of ASME II-D)
$S_m = 16{,}000$ psi at 975°F (Appendix E of ASME II-D)
$S_m = 20{,}000$ psi at 100°F (Table 2A of ASME II-D)
$S_{mt} = 6250$ psi at 975°F and at 300,000 h (Appendix E of ASME II-D)
$S_y = 30{,}000$ psi at 100°F (Table Y-1 of ASME II-D)
$S_y = 24{,}250$ psi at 975°F (Appendix E of ASME II-D)
Average S_y during the cycle = 27,400 psi
$S_t = 10{,}000$ psi at 975°F and at 10^4 h (Appendix E of ASME II-D)
$S_t = 6250$ psi at 975°F and at 300,000 h (Appendix E of ASME II-D)

6

Creep-Fatigue Analysis

Failure of a steam tube in a boiler [Union Electric Company d/b/a Ameren Missouri]

6.1 Introduction

In this chapter, the strain methods detailed in ASME VIII-2 for designing pressure vessels under repetitive cyclic loading conditions in the creep range is discussed. Such cycles are generally encountered in power plants as well as petrochemical

Analysis of ASME Boiler, Pressure Vessel, and Nuclear Components in the Creep Range, Second Edition. Maan H. Jawad and Robert I. Jetter.
© 2022 John Wiley & Sons Ltd. Published 2022 by John Wiley & Sons Ltd.

plants under normal operating conditions. The assumption of repetitive cycles enables us to focus first on explaining the ASME procedure for designing components under cyclic loading in the creep range without having to deal with the complex problem of variable cyclic conditions. Variable cyclic loading due to upset, regeneration, or other emergency conditions are discussed later in this chapter.

Cyclic analysis is straightforward at temperatures below the creep range. The maximum calculated stresses are compared to fatigue curves obtained from experimental data with an appropriate factor of safety. The fatigue curves take into consideration such factors as average versus minimum stress values, effect of mean stress on fatigue life, and size effects.

In the creep range, cyclic life becomes more difficult to evaluate Jetter (2018). Stress relaxation at a given point affects the cyclic life of a component. The level of triaxiality and stress concentration factors play a significant role on creep-fatigue life at elevated temperatures and Poisson's ratio needs to be adjusted to account for inelastic stress levels. In addition, fatigue strength tends to decrease Frost et al. (1974) with an increase in temperature due to surface oxidation or chemical attack. These and other factors contribute to the tediousness and complexity of creep-fatigue evaluation.

The data required to evaluate the cyclic stress in a given material is extensive. Data needed for creep-rupture analysis at various temperatures include stress-strain diagrams, yield stress and tensile strength, creep and rupture data, modulus of elasticity, and isochronous curves. Large amounts of time and cost are involved in obtaining such data. As a consequence, data for creep analysis has been developed for only five materials in ASME VIII-2, They are 2.25Cr-1Mo and 9Cr steels, Type 304 and 316 stainless steels, and 800H nickel alloy. Additional materials will be added as data become available.

The ASME rules for cyclic loads in the creep range consist of determining points in a cycle time where the stresses at a given location in a vessel are at a maximum level. The stresses are then checked against limiting values for fatigue and creep. The analysis for cyclic loading consists of evaluating stresses based on load-controlled as well as strain-controlled limits similar to the procedure discussed in Chapters 4 and 5.

6.2 Creep-Fatigue Evaluation Using Elastic Analysis

The rules for creep-fatigue evaluation discussed in this section are applicable when

1. The rules of Section 5.4 for tests A-1 through A-3 are met and/or the rules of Section 5.5 for tests B-1 and B-2 with $Z < 1.0$ are met. However, the

contribution of stress due to radial thermal gradients to the secondary stress range may be excluded for this assessment of the applicability of elastic creep-fatigue rules, A-2 and A-3.

2. The $(P_L + P_b + Q) \le 3S'_m$ rule is met using for $3S'_m$ the lesser of $(3S_m)$ and $\left(3\bar{S}_m\right)$ as defined in Test A-1.
3. Pressure-induced membrane and bending stresses and thermal-induced membrane stresses are classified as primary (load-controlled) stresses.

The analysis procedure is performed by using the following seven steps.

Step 1
Determine the total amount of hours, t_H, expended at temperatures in the creep range.

Step 2
Define the hold temperature, T_{HT} , to be equal to the local metal temperature that occurs during sustained normal operation.

Step 3
Unless otherwise specified, for each cycle type j, define the average cycle time as

$$\bar{t}_j = t_H / \left(n_c\right)_j \tag{6.1}$$

where

$(n_c)_j$ = a specified number of applied repetitions of cycle type j

t_H = total number of hours at elevated temperatures for the entire service life as defined in Step 1

$\bar{t}_j$ = average cycle time for cycle type j.

Step 4
A modified strain, $\Delta\varepsilon_{mod}$, is calculated in this step. The procedure consists of calculating first a maximum strain, $\Delta\varepsilon_{max}$, and then modifying it to include the effect of stress concentration factors. The maximum elastic strain range during the cycle is calculated as

$$\Delta\varepsilon_{max} = 2S_{alt} / E \tag{6.2}$$

where

E = modulus of elasticity at the maximum metal temperature experienced during the cycle

$2S_{alt}$ = maximum stress range during the cycle excluding geometric stress concentrations = $P_L + P_b + Q$

$\Delta\varepsilon_{max}$ = maximum equivalent strain range.

The maximum elastic strain range is then used to calculate a modified strain $(\Delta\varepsilon_{mod})$ that includes the effect of local plasticity and creep, which can significantly increase the strain range at stress concentrations. ASME VIII-2 gives the designer the option of calculating this quantity by using any one of three different methods. These methods, of varying complexity and conservatism, are based on modifications of the Neuber equation Neuber (1961). Neuber's basic equation Bannantine et al. (1990) is of the form

$$K = \left(K_\sigma K_\varepsilon\right)^{1/2}. \tag{6.3}$$

This equation can be expressed as

$$K^2 Se = \sigma\varepsilon \tag{6.4}$$

where

e = strain away from concentration
K = theoretical stress concentration
K_ε = stress concentration due to strain
K_σ = stress concentration due to stress
S = stress away from concentration
ε = strain at concentration
σ = stress at concentration.

Equation (6.3) indicates that the total stress concentration at a point in the plastic or creep region is equal to the square root of the products K_σ and K_ε. The value of K_σ decreases, whereas that of K_ε increases with an increase in yield and creep levels. Equation (6.4), which is an alternate form of Eq. (6.3), shows that the total stress concentration is a function of the product of the actual strain and actual stress at a given point.

All three methods Severud and Winkel (1987); Severud (1991) of calculating modified strain, $\Delta\varepsilon_{mod}$, use a composite stress-strain curve, as shown by Figure 6.1. The composite stress-strain curve is constructed by adding the elastic stress-strain curve for the stress range S_{rH} to the appropriate time-independent (hot tensile) isochronous stress-strain curve for the material at a given temperature. The value of S_{rH} can be conservatively assumed as equal to $0.5S_t$. The conceptual basis for extending the elastic range is shown in Figure 1.21, illustrating how the elastic stress range is extended by an amount given by the hot relaxation strength, S_{rH}, when the stress range is limited to $\left(3\overline{S}_m\right)$.

- First method

  A conservative governing equation is

$$\Delta\varepsilon_{mod} = \left(S^* / \overline{S}\right) K_{sc}^2 \Delta\varepsilon_{max} \tag{6.5}$$

where

K_{sc} = either the equivalent stress concentration factor, as determined by test or analysis, or the maximum value of the theoretical elastic stress concentration factor in any direction for the local area under consideration. The equivalent stress concentration factor is defined as the effective (von Mises) primary plus secondary plus peak stress divided by the effective primary plus secondary stress. Note that fatigue strength reduction factors developed from low-temperature continuous cycling fatigue tests may not be acceptable for defining K_{sc} when creep effects are not negligible.

S^* = stress indicator determined by entering the composite stress-strain curve of Figure 6.1 at a strain range of ($\Delta\varepsilon_{max}$)

$\bar{S}$ = stress indicator determined by entering the composite stress-strain curve of Figure 6.1 at a strain range of ($K_{sc}\Delta\varepsilon_{max}$)

$\Delta\varepsilon_{mod}$ = modified maximum equivalent strain range that accounts for the effects of local plasticity and creep.

- Second method
  $\Delta\varepsilon_{mod}$ is calculated from a most conservative estimate by the equation

$$\Delta\varepsilon_{mod} = K_e K_{sc} \Delta\varepsilon_{max} \tag{6.6}$$

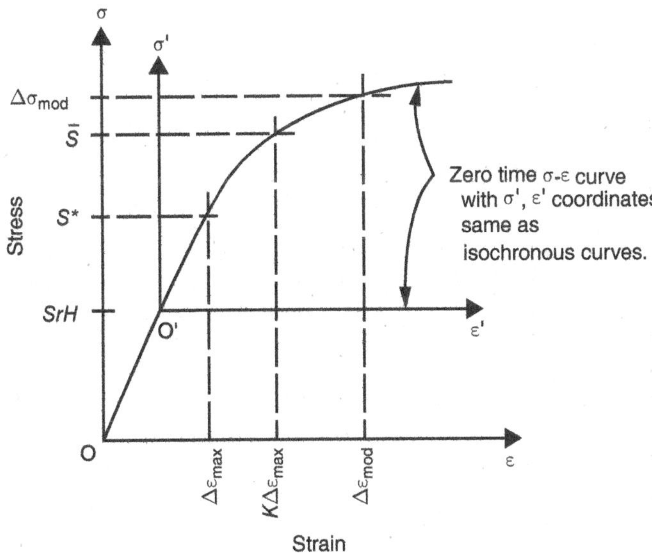

Figure 6.1 Stress-strain relationship [ASME VIII-2].

where

$$K_e = 1.0 \qquad \text{if } K_{sc}\Delta\varepsilon_{max} \leq \left(3\bar{S}_m\right)/E$$
$$K_e = K_{sc}\Delta\varepsilon_{max}E/\left(3\bar{S}_m\right) \quad \text{if } K_{sc}\Delta\varepsilon_{max} > \left(3\bar{S}_m\right)/E$$

- Third method

$\Delta\varepsilon_{mod}$ is given by the most accurate of the three methods by the equation

$$\Delta\varepsilon_{mod} = S^* K_{sc}^2 \Delta\varepsilon_{max} / \Delta\sigma_{mod} \tag{6.7}$$

where
S^* = stress obtained from an isochronous curve at a given value of $\Delta\varepsilon_{max}$
$\Delta\sigma_{mod}$ = range of effective stress that corresponds to the strain range $\Delta\sigma_{mod}$.
Both $\Delta\varepsilon_{mod}$ and $\Delta\sigma_{mod}$ in Eq. (6.7) are unknown, and they must be solved graphically by curve-fitting the appropriate composite stress-strain curve. For this reason, most designers opt to use either Eq. (6.5) or Eq. (6.6) because they are easier to solve.

Step 5
In this step, a stress, S_j, is obtained. It corresponds to a strain value that includes elastic, plastic, and creep considerations. Once the quantity $\Delta\varepsilon_{mod}$ is known, then the total strain range, ε_t, is obtained from the following equation

$$\varepsilon_t = K_v \Delta\varepsilon_{mod} + K_{sc}\Delta\varepsilon_c \tag{6.8}$$

where
ε_t = total strain range
$\Delta\varepsilon_c$ = creep strain increment
and

$$K_v = 1.0 + f\left(K_v' - 1.0\right), \text{ but not less than } 1.0 \tag{6.9}$$

where
f = triaxiality factor obtained from Figure 6.2.

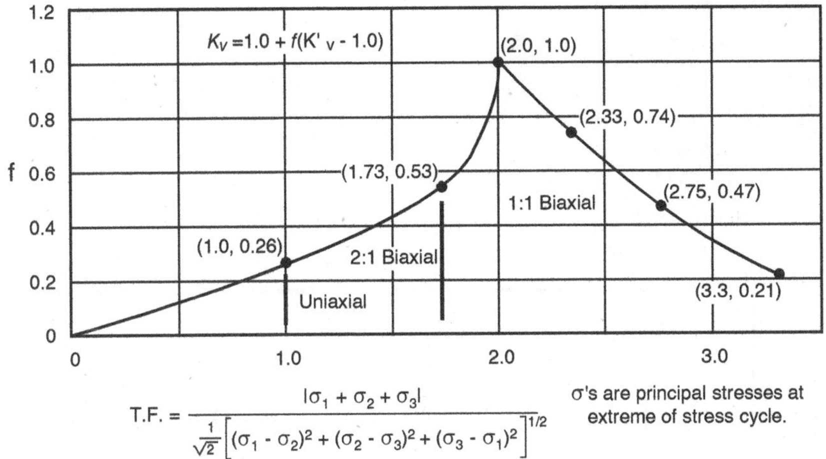

Figure 6.2 Inelastic multiaxial adjustments [ASME VIII-2].

This figure is based on experimental data by Severud (1991). The curves in Figure 6.2 can be approximated by

$$f = \frac{C_1 + C_2(T.F.) + C_3(T.F.)^2 + C_4(T.F.)^3}{1 + C_5(T.F.) + C_6(T.F.)^2 + C_7(T.F.)^3}. \tag{6.10}$$

C constants	0 < T.F. ≤ 1.73	1.73 < T.F. ≤ 2.0	2.0 < T.F. ≤ 3.3
C_1	0	0.367366	8.95
C_2	0.293	−0.53533476	−2.2882395
C_3	0	0.26042254	0
C_4	0	−0.042291728	0
C_5	0.283322	−1.4919987	1.6812905
C_6	−0.177575	0.74301006	0
C_7	0	−0.1234988	0

K_v' = plastic Poisson ratio adjustment factor obtained from Figure 6.3. This figure is based on the relationship between strain range and shakedown criteria. The curve in Figure 6.3 can be approximated by

$$K_v' = 1.0 \qquad \gamma \le 1.0$$

$$K_v' = \frac{C_8 + C_9\gamma}{1 + C_{10}\gamma + C_{11}\gamma^2} \qquad 1.0 < \gamma \le 50 \tag{6.11}$$

$$K_v' = 1.61 \qquad \gamma > 50$$

where

$$\gamma = K_e K \Delta\varepsilon_{max} E / 3\overline{S}_m$$

and

$C_1 = 1.88$
$C_2 = -2.5037475$
$C_3 = -1.6255583$
$C_4 = 0.0014771927.$

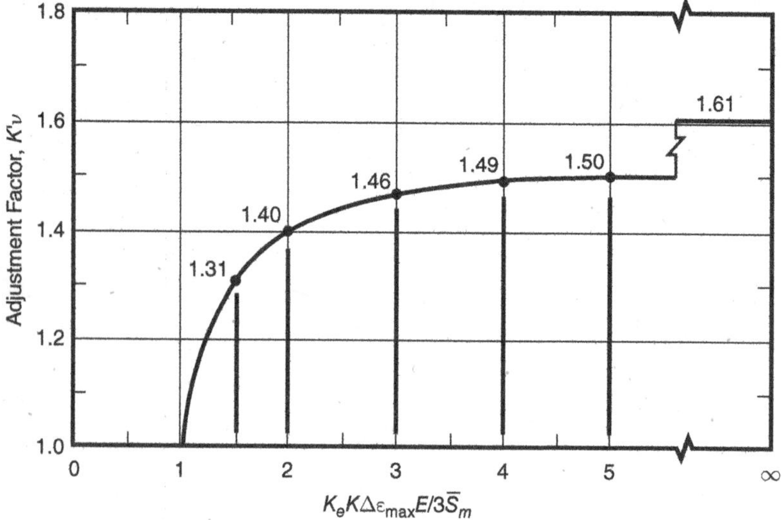

Figure 6.3 Adjustment for inelastic biaxial Poisson's ratio [ASME VIII-2].

The creep strain increment, ε_c, is obtained from an isochronous stress-strain curve similar to the one shown in Figure 5.6. The stress value for entering the figure is obtained from the quantity $1.25\sigma_c$, where σ_c is obtained from Section 5.5. The time used in the figure for determining ε_c is obtained by one of two methods as follows:

- Method (1): the time based on one cycle
- Method (2): the time based on the total number of hours during the life of the component and the resultant strain is then divided by the number of cycles

Method (1) is generally applicable to components with a small number of cycles and high membrane stress such as hydrotreaters.
Method (2) is generally applicable to components with repetitive cycles and small membrane stress such as headers in heat recovery steam generators.

Finally, a value of S_j is obtained from an appropriate isochronous chart using Eq. (6.8) for strain and the time-independent curve in the chart. S_j is defined as the initial stress level for a given cycle.

Step 6

The relaxation stress, $\bar{S}_r$, during a given cycle is evaluated in this step. Two methods are provided for this evaluation. The first method requires an analytical estimate of the uniaxial stress relaxation adjusted with correction factors to account for the retarding effects of multiaxiality and elastic follow-up. The adjusted relaxed stress level S'_r is thus a function of the initial stress determined from the analytically determined uniaxial relaxed stress, and a factor 0.8 G accounting for elastic follow-up and multiaxiality. The equation is expressed as

$$S'_r = S_j - 0.8G\left(S_j - \bar{S}_r\right) \tag{6.12}$$

where

G = the smallest value of the multiaxiality factor as determined for the stress state at each of the two extremes of the stress cycle.

The multiaxiality factor is defined as:

$$= \frac{\left[\sigma_1 - 0.5\left(\sigma_2 + \sigma_3\right)\right]}{\left[\sigma_1 - 0.3\left(\sigma_2 + \sigma_3\right)\right]} \quad \text{but not greater than 1.0.}$$

σ_1, σ_2, and σ_3 are principal stresses, exclusive of local geometric stress concentration factors, at the extremes of the stress cycle, and are defined by

$$|\sigma_1| \geq |\sigma_2| \geq |\sigma_3|$$

S_j = the initial stress level for cycle type j

S'_r = relaxed stress level at time t adjusted for the multiaxial stress state

$\bar{S}_r$ = relaxed stress level at time t based on a uniaxial relaxation model.

The second method for determining stress relaxation is based on the isochronous stress strain curves. Starting at the stress level determined from the time to relax to lower stress levels is determined by moving vertically down at a constant strain until intercepting the curve for the time of interest as shown in Figure 6.4. Because of the conservatism inherent in this approach Severud (1991), multiaxial and elastic follow-up corrections are not required.

The stress in either the “adjusted” analytical relaxation or that obtained from the isochronous curves is not allowed to relax below a factor of 1.25 times the elastic core stress, σ_c, as determined by the procedures for evaluation of the strain limits using simplified inelastic analysis. This lower stress value, S_{LB}, is illustrated in Figure 6.5.

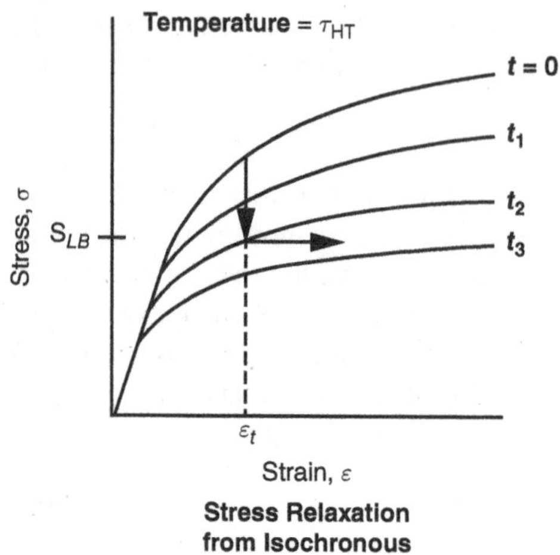

Figure 6.4 A method of determining relaxation [ASME VIII-2].

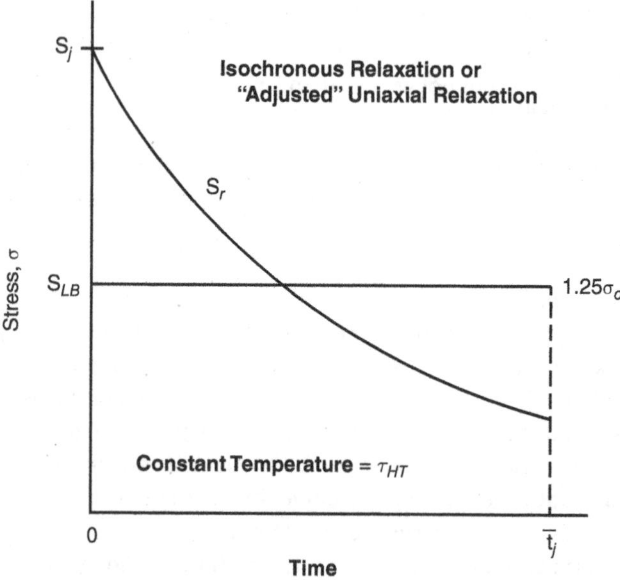

Figure 6.5 Stress-relaxation limits for creep damage [ASME VIII-2].

Step 7

The governing equation for creep fatigue Curran (1976) is given by

$$\left[\sum\left(n_c / N_d\right)_j + \sum(Dt / t_d)_k\right] < D_{cf} \tag{6.13}$$

where
D_{cf} = total creep-fatigue damage factor obtained from Figure 6.6
$K' = 0.90$
$(N_d)_j$ = number of design allowable cycles for cycle type, j, obtained from a design fatigue data using the maximum strain value during the cycle
$(n_c)_j$ = number of applied repetitions of cycle type, j
$(t_d)_k$ = allowable time duration determined from stress-to-rupture data for a given stress, (S_r'/K') for base material and $[S_r'/(K'R)]$ for weldments, and the maximum temperature at the point of interest and occurring during the time interval, k
$(\Delta t)_k$ = duration of the time interval, k
R = weld strength reduction factor from ASME II-D.

Figure 6.6 can be represented by the following equations

For 304 SS, 316 SS, and 9Cr-1Mo-V alloys

$$\sum(\Delta t/t_d) = (-2.3333)\left[(\sum(n/N_d)\right] + 1.0 \text{ for } 0.0 < \sum(n/N_d) \leq 0.3 \quad (6.14)$$

$$\sum(\Delta t/t_d) = (-0.429)\left[(\sum(n/N_d)\right] + 0.429 \text{ for } 0.3 < \sum(n/N_d) \leq 1.0 \quad (6.15)$$

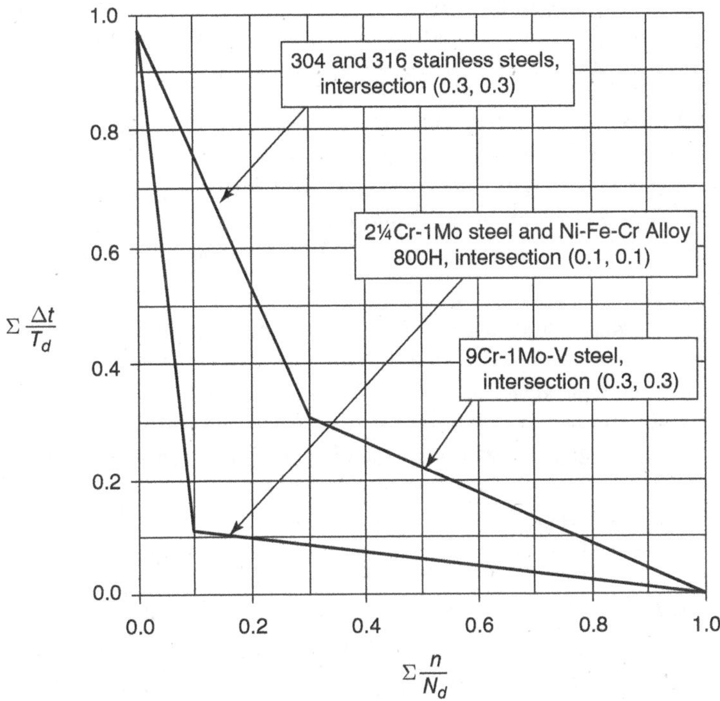

Figure 6.6 Creep-fatigue damage envelope [ASME VIII-2].

For 2.25Cr-1Mo annealed and 800H alloys

$$\sum(\Delta t/t_d)=(-9.0)[(\sum(n/N_d)]+1.0 \text{ for } 0.0<\sum(n/N_d)\leq 0.1 \quad (6.16)$$

$$\sum(\Delta t/t_d)=-(1/9)[(\sum(n/N_d)]+(1/9) \text{ for } 0.1<\sum(n/N_d)\leq 1.0 \quad (6.17)$$

The first term of Eq. (6.13) pertains to cyclic loading, whereas the second term pertains to creep duration. A substantial number of calculations is required before solving Eq. (6.13). The solution can be made using either elastic analysis or inelastic analysis. In this section, elastic analysis, which is the easiest to perform but the most conservative, is presented.

It should be noted that the quantity $\Sigma(\Delta t/t_d)_k$ in Eq. (6.13) is determined by one of two methods, as follows:

- Method (1): When the total strain range is $\varepsilon_t>(3\bar{S}_m)/E$, then the time interval k for stress relaxation is based on one cycle and the result is multiplied by the total number of cycles during the life of the component.
- Method (2): When the total strain range is $\varepsilon_t\leq(3\bar{S}_m)/E$, then shakedown occurs and the time interval k for stress relaxation can be defined as a single cycle for the entire design life. When applicable, Method (2) will yield a lower calculated creep damage than Method (1).

Example 6.1 Use the data from Examples 5.1 and 5.2 to evaluate the life cycle of the 2.0-in.-thick shell for 130,000 h (15 years) due to creep and fatigue conditions. Assume an arbitrary stress concentration factor, K_{sc}, of 1.10 for the longitudinal weld configuration. The design fatigue strain range values are shown in Table 6.1 and the stress-to-rupture values are given in Table 6.2.

Table 6.1 Design fatigue strain range for Type 304 stainless steel [ASME II-D].

	ε_t strain range (in./in.) at						
	100°F	800°F	900°F	1000°F	1100°F	1200°F	1300°F
Number of cycles, N_d	US customary units						
10	0.051	0.050	0.0465	0.0425	0.0382	0.0335	0.0297
20	0.036	0.0345	0.0315	0.0284	0.025	0.0217	0.0186
40	0.0263	0.0246	0.0222	0.0197	0.017	0.0146	0.0123
10^2	0.018	0.0164	0.0146	0.0128	0.011	0.0093	0.0077
2×10^2	0.0142	0.0125	0.011	0.0096	0.0082	0.0069	0.0057

(Continued)

Table 6.1 (Continued)

Number of cycles, N_d	ε_t strain range (in./in.) at						
	100°F	**800°F**	**900°F**	**1000°F**	**1100°F**	**1200°F**	**1300°F**
	US customary units						
4×10^2	0.0113	0.00965	0.00845	0.00735	0.0063	0.00525	0.00443
10^3	0.00845	0.00725	0.0063	0.0055	0.0047	0.00385	0.00333
2×10^3	0.0067	0.0059	0.0051	0.0045	0.0038	0.00315	0.00276
4×10^3	0.00545	0.00485	0.0042	0.00373	0.0032	0.00263	0.0023
10^4	0.0043	0.00385	0.00335	0.00298	0.0026	0.00215	0.00185
2×10^4	0.0037	0.0033	0.0029	0.00256	0.00226	0.00187	0.00158
4×10^4	0.0032	0.00287	0.00254	0.00224	0.00197	0.00162	0.00138
10^5	0.00272	0.00242	0.00213	0.00188	0.00164	0.00140	0.00117
2×10^5	0.0024	0.00215	0.0019	0.00167	0.00145	0.00123	0.00105
4×10^5	0.00215	0.00192	0.0017	0.0015	0.0013	0.0011	0.00094
10^6	0.0019	0.00169	0.00149	0.0013	0.00112	0.00098	0.00084

Table 6.2 Minimum stress-to-rupture values for Type 304 stainles steel [ASME II-D].

US customary units											
Temp. (°F)	**1 h**	**10 h**	**30 h**	**10^2h**	**3×10^2 h**	**10^3 h**	**3×10^3 h**	**10^4 h**	**3×10^4 h**	**10^5 h**	**3×10^5 h**
800	57	57	57	57	57	57	57	57	51	44.3	39
850	56.5	56.5	56.5	56.5	56.5	56.5	50.2	45.4	40	34.7	30.5
900	55.5	55.5	55.5	55.5	51.5	46.9	41.2	36.1	31.5	27.2	24
950	54.2	54.2	51	48.1	43	38.0	33.5	28.8	24.9	21.2	18.3
1000	52.5	50	44.5	39.8	35	30.9	26.5	22.9	19.7	16.6	14.9
1050	50	41.9	37	32.9	28.9	25.0	21.6	18.2	15.5	13.0	11.0
1100	45	35.2	31	27.2	23.9	20.3	17.3	14.5	12.3	10.2	8.6
1150	38	29.5	26	22.5	19.3	16.5	13.9	11.6	9.6	8.0	6.6
1200	32	24.7	21.5	18.6	15.9	13.4	11.1	9.2	7.6	6.2	5.0
1250	27	20.7	17.9	15.4	13	10.8	8.9	7.3	6.0	4.9	4.0
1300	23	17.4	15	12.7	10.5	8.8	7.2	5.8	4.8	3.8	3.1
1350	19.5	14.6	12.6	10.6	8.8	7.2	5.8	4.6	3.8	3.0	2.4
1400	16.5	12.1	10.3	8.8	7.2	5.8	4.7	3.7	3.0	2.3	1.9
1450	14.0	10.2	8.8	7.3	5.8	4.6	3.8	2.9	2.3	1.8	1.4
1500	12.0	8.6	7.2	6.0	4.9	3.8	3.0	2.4	1.8	1.4	1.1

Solution

Assumptions

- The calculations in this example are based on thin shell equations to keep the calculations as simple as possible. This is done to demonstrate and highlight the method of creep analysis. For thick shells, Lame's equations must be used.
- The following calculations are based on the criteria in Figure 4.8.
- $K_{sc} = 1.1$.

To apply this methodology, it is necessary to first verify that the prerequisites identified in Section 6.2, and repeated here, have been satisfied:

1. The rules of Section 5.4 for Tests A-1 through A-3 are met and/or the rules of Section 5.5 for Tests B-1 and B-2 with $Z < 1.0$ are met. However, the contribution of stress due to radial thermal gradients to the secondary stress range may be excluded for this assessment in Tests A-2 and A-3 for the applicability of elastic creep-fatigue rules.
2. The $(P_L + P_b + Q) \leq 3S'_m$ rule is met using for $3S'_m$ the lesser of $(3S_m)$ and $\left(\bar{S}_m\right)$ as defined in Test A-1.
3. Pressure-induced membrane and bending stresses and thermal-induced membrane stresses are classified as primary (load-controlled) stresses.

Requirement #1 is satisfied: Both A-3 and B-1 are satisfied.
Requirement #2 is satisfied.
Requirement #3 is satisfied.

Step 1

Total hours, $t_H = 130{,}000$ h (15 years)

Step 2

Hold temperature, $T_{HT} = 1050^\circ F$.

Step 3

$$\left(n_c\right)_j = \text{number of cycles } j = 10$$

$$\text{cycle time } \bar{t}_j = t_H / \left(n_c\right)_j$$

$$= 130{,}000/10 = 13{,}000 \text{ h} \sim 18 \text{ months}$$

a) Two-inch shell calculations

A review of Tables 5.4 and 5.5 shows the combination of pressure and thermal stresses in the 2.0-in. shell to be more severe than in the 1.0-in. shell. The tables also indicate that the membrane stress in the 1.0-in. shell is more severe. The following creep-fatigue calculations are performed on the 2.0-in. shell to illustrate the procedure. However, the designer must recognize that an analysis of the 1.0-in. shell may result in a more severe condition due to the higher core stress.

The values obtained for the outside stress of the 2.0-in. shell, given in Table 5.5, are summarized in Table 6.3.

The following values are obtained from Table 6.3:

Total circumferential stress = 4500 + 14,975 = 19,475 psi

Total longitudinal stress = 2250 + 14,975 = 17,225 psi

Total radial stress = − 150 + 150 = 0

The equivalent Von Mises stress is obtained from Eq. (4.14) as

$$S_e = 18{,}450 \text{ psi.}$$

Step 4

A modified strain, $\Delta\varepsilon_{mod}$, is calculated in this step. The procedure consists of calculating a maximum strain and then modifying it to include the effect of stress concentration factors. We start by calculating the maximum elastic strain range $\Delta\varepsilon_{max}$ in accordance with Eq. (6.5).

$$\begin{aligned}\Delta\varepsilon_{max} &= 2S_{alt} / E = S_e / E \\ &= 18{,}450 / 22{,}400{,}000 \\ &= 0.00082.\end{aligned}$$

Next, the value of modified strain $\Delta\varepsilon_{mod}$ is calculated. Step 4 gives the designer the choice of three different methods for calculating this quantity. The first two, given by Eqs. (6.4) and (6.5), will be used in order to compare the results.

- *First method* for calculating modified strain

The value of S^* is obtained from a composite stress-strain curve (Figure 6.1) with a strain value of $\Delta\varepsilon_{max}$; the value of $\overline{S}$ is obtained from the same curve

Table 6.3 Outside stress for 2.0 in. shell.

Two-inch shell	**Stress due to pressure, psi**	**Stress due to temperature, psi**
Membrane circumferential stress, $P_{m\theta}$	4500	0
Membrane axial stress, P_{mL}	2250	0
Membrane radial stress, P_{mr}	−150	0
Circumferential bending stress, $Q_{b\theta}$	0	14,975
Axial bending stress, Q_{bL}	0	14,975
Radial bending stress, Q_{br}	150	0

with a strain value of $(K_{sc})\Delta\varepsilon_{max}$. However, for this problem it is evident that the values of $\Delta\varepsilon_{max}$ and $(K_{sc})\Delta\varepsilon_{max}$ are within the proportional limit of Figure 6.1. Under these conditions, $\overline{S} = (K_{sc})S^*$ and the equation for $\Delta\varepsilon_{max}$ may be rewritten as

$$\begin{aligned}\Delta\varepsilon_{mod} &= \left(S^* / \overline{S}\right)K_{sc}^2\Delta\varepsilon_{max}\\ &= \left[S^* / \left(K_{sc}S^*\right)\right]K_{sc}^2\Delta\varepsilon_{max}\\ &= K_{sc}\Delta\varepsilon_{max}\\ &= 0.00090\end{aligned}$$

- *Second method* for calculating modified strain
  The value of K_e in Eq. (6.6) must first be calculated.

$$\begin{aligned}&3\overline{S}_m = 1.5S_m + S_t / 2 = 1.5(13{,}600) + (8490)/2 = 24{,}645\\ &\left(3\overline{S}_m\right)/E = 24{,}645 / 22{,}400{,}000 = 0.0011\\ &K_{sc}\Delta\varepsilon_{max} = (1.1)(0.00082) = 0.00090\end{aligned}$$

Because $(K_{sc}\Delta\varepsilon_{max})$ is less than $\left(3\overline{S}_m\right)/E$, the value of K_e is obtained from Eq. (6.6) as

$$K_e = 1.0.$$

The modified strain $\Delta\varepsilon_{mod}$ is calculated from Eq. (6.6) as

$$\begin{aligned}\Delta\varepsilon_{mod} &= K_e K_{sc}\Delta\varepsilon_{max}\\ &= (1.0)(1.1)(0.00082).\\ &= 0.00090\end{aligned}$$

This is identical to the result obtained with Method 1, which is the expected result within the proportional limit.

Step 5
The total strain is obtained from Eq. (6.8)

$$\varepsilon_t = K_v\Delta\varepsilon_{mod} + K_{sc}\Delta\varepsilon_c$$

where
$K_v = 1.0 + f(K_v' - 1.0)$, but not less than 1.0.
The quantity K'_v is obtained from Figure 6.3 with

$$\begin{aligned}&\left(K_e K_{sc}\Delta\varepsilon_{max}\right)\left[E / \left(3\overline{S}_m\right)\right] = (1.0)(1.1)(0.00082)(22{,}400{,}000 / 24{,}645) = 0.820\\ &K_v' = 1.0.\end{aligned}$$

The next step is to obtain the value of *f* from Figure 6.2 by using the triaxiality factor (TF). From Table 6.3,

$$\sigma_1 = 4500 + 14{,}976 = 19{,}476 \text{ psi}$$
$$\sigma_2 = -150 + 150 = 0 \text{ psi}$$
$$\sigma_3 = 2250 + 14{,}976 = 17{,}226 \text{ psi}$$
$$\text{TF} = \frac{(19{,}476 + 0 + 17{,}226)}{0.707\left[(19{,}476 - 0)^2 + (0 - 17{,}226)^2 + (17{,}226 - 19{,}476)\right]^{1/2}}$$
$$= 2.11 \text{ and from Fig. 6.2} \quad f = 0.907.$$

From Eq. (6.9),

$$K_v = 1.0 + (0.907)(1.0 - 1.0) = 1.0.$$

The elastic core stress, σ_c, is obtained from Example 5.2 and is equal to 4140 psi. From the isochronous curves (Figure 5.6), with stress of $1.25\sigma_c = 1.25(4140) = 5175$ psi, a strain value of 0.00026 is obtained. This strain value is on the elastic portion of the curve and is essentially independent of time duration, at least up to the total design life of 130,000 h. The ASME rules state that the creep strain increment may be determined for each cycle, 13,000 h in this example, or from the strain over the total life divided by the number of cycles. In the latter case, the applicable strain, $\Delta\varepsilon_c$, is equal to

$$\Delta\varepsilon_c = 0.00026 / 10 = 0.000026.$$

Table 6.4 Stress, S_j, versus time values.

Time (hr)	Stress (ksi)
S_j	15.0
30	13.7
100	12.0
300	10.5
1000	9.0
3000	8.0
10,000	7.7
30,000	7.2
100,000	6.8
300,000	6.6

From Eq. (6.8),

$$\varepsilon_t = (1.0)(0.00090) + (1.1)(0.000026) = 0.00093.$$

From Figure 5.6, the time-independent stress

$$S_j = 15{,}000\,\text{psi}.$$

Step 6

From Figure 5.6, stress-versus-time values are obtained, as shown in Table 6.4.

A plot of these values, with a limit of $1.25\sigma_c = (1.25)(4140) = 5175$ psi, would look similar to Figure 6.5.

Step 7

Equation (6.11) is solved in this step. The first part of the equation, $(n_c/N_d)_j$, is obtained as follows:

- First part of Eq. (6.11)

$$n_c = 10.$$

The value of N_d is obtained from Table 6.1. A cycle life of > 1,000,000 is obtained at a temperature of 1050°F and strain, $\varepsilon_t = 0.001$.

$$(n_c / N_d)_j = 10/1{,}000{,}000 \approx 0.0.$$

- Second part of Eq. (6.11)

The second part of Eq. (6.11), $(\Delta T/T_d)_k$, is determined numerically from Table 6.4 and from the differential equation $\int dT/T_d$. The quantity $(3\bar{S}_m)/E$, calculated above, is equal to 0.0011, whereas the value of ε_t is equal to 0.00093. Because $\varepsilon_t \le (3\bar{S}_m)/E$ shakedown occurs and the peak stress does not reset on each cycle. Under this scenario, peak stress can be assumed to relax throughout the design life and the cumulated creep damage is given by Table 6.4 at 130,000 h. Calculations are summarized in Table 6.5.

Column A in Table 6.5 gives the incremental locations for calculating stresses from Table 6.4. Column B shows the stresses obtained from Table 6.4 and Column H lists the incremental duration assumed for these stresses. Column C adjusts the stress values by the K factor and Column G lists the number of hours obtained from Table 6.2 using the stress values in Column C. Column I shows the ratio of Column H over Column G, which is the numerical value of the quantity ($\int dt/t_d$).

The 1050°F line in Table 6.2 terminates at the 11,000 psi stress level. Stress levels in Column C drop below the 11,000 psi level in Rows 6 through 10. The Larson-Miller parameter, P_{LM}, is used to obtain approximate rupture life by

Table 6.5 Numerical calculations of $(\Delta_t/t_d)_k$.

A	B	C	D	E	F	G	H	I
Location (hr)	S_r(ksi)	S_r/0.9 (ksi)	Equivalent temp. (°F)	P_{LM}	Time (hr)	Table 6.2	Table 6.4	H/G
S_j	15.0	16.67				21,300	30	0.0014
30	13.7	15.22				37,800	70	0.0019
100	12.0	13.33				90,800	200	0.0022
300	10.5	11.67				263,000	700	0.0027
1000	9.0	10.00	1071	39,006	673,700	673,700	2000	0.0030
3000	8.0	8.89	1094	39,591	1,657,600	1,657,600	7000	0.0042
10,000	7.7	8.56	1101	39,770	2,176,400	2,176,400	20,000	0.0092
30,000	7.2	8.00	1115	40,126	3,745,500	3,745,500	70,000	0.0187
100,000	6.8	7.56	1126	40,407	5,749,100	5,749,100	30,000	0.0052
Total								0.0485

using the higher temperature levels given in Table 6.2. These calculations are shown in Columns D, E, and F. Column D lists the temperatures from Table 6.2 corresponding to the stress levels in Column C. Column E calculates the corresponding P_{LM} from Eq. (1.3) for the temperature shown in Column D at 300,000 h. Column F recalculates the hours corresponding to a temperature of 1050°F from Eq. (1.3) with the corresponding P_{LM} factor.

From Table 6.5, the sum of $(\Delta t/t_d)_k$ for the 130,000-h cycle duration is 0.0485. Referring to Figure 6.6, with $(n_c/N_d)_j = 0.0$ and $(\Delta t/t_d)_k = 0.0485$, it is seen that the expected life of 130,000 h is well within the acceptable limits.

It is of interest to compare the above results with the evaluation of the creep damage using the hours for one cycle, 13,000, as a basis. In this case, the value of $(\Delta t/t_d)$ from Table 6.5 is equal to 0.0168. For 10 cycles the value is 0.168; over a factor of three greater, but still within acceptable limits.

Example 6.2 A 12-in.-diameter high-pressure superheater header in a heat recovery steam generator is constructed of modified 9Cr steel (SA 335-P91) and built in accordance with ASME I. The following design and operating data are given:

Data

- Design temperature = 1000°F; design pressure = 1300 psi.
- Longitudinal ligament efficiency (for circumferential stress): $E_o = 0.60$.

- Circumferential ligament efficiency (for longitudinal stress): $E_o = 0.85$, $y = 0.40$.
- Thickness, $t = 1.125$ in. (12-in. Sch-140 pipe).
- Maximum stress concentration factor due to holes in the shell is taken as $K_{sc} = 3.3$. This value is taken from the appendix on stress indices for nozzles in ASME VIII-2.
- Expected cycles = 10,000 (one full cycle per day for 25 years).
- Expected life = 200,000 h ≈ 25 years.
- The design longitudinal stress in the header due to tube weight, liquid, and bending is assumed as 7000 psi.
- Assume, for simplicity, that the design and operating conditions are the same.
- Assume that the temperature drops to 700°F before the start of a new cycle.
- Isochronous curves are given in Figure 6.7 for 1000°F.
- The design fatigue strain range for 9Cr-1M0-V steel is given in Table 6.6.
- The minimum stress-to-rupture values for 9Cr-1Mo-V steel are given in Table 6.7.

Determine whether the above conditions are adequate in accordance with the creep design rules in Chapters 4, 5, and 6.

Solution

The various allowable stress values are:

$S = 16{,}300$ psi at 1000°F (ASME II-D)

$E = 25{,}400$ ksi at 1000°F (Table TM-1 of ASME II-D)

$E = 27{,}500$ ksi at 700°F (Table TM-1 of ASME II-D)

$S_m = 19{,}000$ psi (Appendix E of ASME II-D)

$S_{mt} = 14{,}300$ psi at 200,000 h (Appendix E of ASME II-D)

$S_y = 53{,}200$ psi at 700°F (Table Y-1 of ASME II-D) which is the cold end of a cycle

$S_y = 40{,}200$ psi at 1000°F (Table Y-1 of ASME II-D)

$S_t = 14{,}300$ psi at 200,000 h (Appendix E of ASME II-D)

(a) Check design thickness in accordance with ASME-I

From Eq. (4.1)

$$t = \frac{(1300)(12.75)}{2(16{,}300)(0.60) + 2(0.4)(1300)} + 0.0$$
$$= 0.80 \text{ in.} < 1.125 \text{ in.}$$

Thus, 12-in. diameter schedule 140 pipe is adequate.

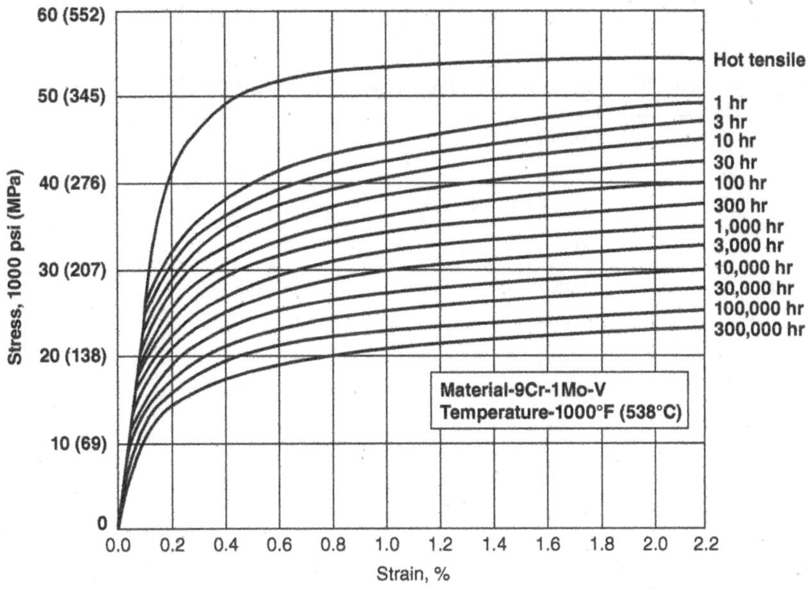

Figure 6.7 Isochronous stress-strain curves for 9Cr-1Mo-V steel at 1000°F [ASME II-D].

(b) Check load-controlled stress limits in accordance with ASME VIII-2 (Figure 4.8)

$$\gamma = R_o / R_i = 6.375 / 5.25 = 1.2143$$

From Example 4.4,

Circumferential stress at inside surface with a ligament efficiency of 0.60

membrane $= P_m = P/(0.6)(\gamma - 1) = 10{,}110$ psi

bending $= P_b = 0$

secondary $= Q = 645/0.6 = 1075$ psi

Longitudinal stress at inside surface with a ligament efficiency of 0.85

membrane $= P_m = P/(0.85)(\gamma^2 - 1) + 7000 = 10{,}225$ psi

bending $= P_b = 0$

secondary $= Q = 0$

Radial stress at inside surface

membrane $= P_m = -P/2 = -650$ psi

bending $= P_b = 0$

secondary $= Q = -650$ psi

Table 6.6 Design fatigue strain range for 9Cr-1Mo-V steel [ASME II-D].

Number of Cycles,[1] N_d	Strain range, ε_t [in./in. (m/m)] at 1000°F (540°C)
10	0.028
20	0.019
40	0.0138
10^2	0.0095
2×10^2	0.0075
4×10^2	0.0062
10^3	0.0050
2×10^3	0.0044
4×10^3	0.0039
10^4	0.0029
2×10^4	0.0024
4×10^4	0.0021
10^5	0.0019
2×10^5	0.00176
4×10^5	0.0017
10^6	0.00163
2×10^6	0.00155
4×10^6	0.00148
10^7	0.00140
2×10^7	0.00132
4×10^7	0.00125
10^8	0.00120

[1] Cycle strain rate: 4×10^{-3} in./ in./sec (m /m /sec).

The value of P_m is obtained from Eq. (4.14) as

$$P_m = 10{,}815 \quad \text{and} \quad P_L + P_b = 10{,}815.$$

Design limits

$$P_m < S_o \qquad 10{,}815 \text{ psi} < 16{,}300 \text{ psi}$$
$$P_L + P_b < 1.5S_o \quad 10{,}815 \text{ psi} < 24{,}450 \text{ psi}$$

Operating limits

$$P_m < S_{mt} \qquad 10{,}815 \text{ psi} < 14{,}300 \text{ psi}$$
$$P_L + P_b < 1.5S_m \quad 10{,}815 \text{ psi} < 28{,}500 \text{ psi}$$
$$P_L + P_b / k_t < S_t \quad 10{,}815 \text{ psi} < 14{,}300 \text{ psi}$$

Thus, the requirements of load-controlled limits are met.

(c) Check strain- and deformation-controlled limits in accordance with ASME VIII-2 (Figure 4.8)

Average yield stress = 46,700 psi

Calculate X and Y from Eqs. (5.1) and (5.2).

The value of Q is obtained from Eq. (4.14) as

$$Q = 1510 \text{psi}$$
$$X = 10{,}815 / 46{,}700 = 0.23 \quad Y = 1510 / 46{,}700 = 0.03$$

Check elastic analysis Test A-3,

$$X + Y = 0.26 < 1.0.$$

There is no need to check the simplified inelastic analysis Tests B because the requirement of Test A-3 is met.

Thus, the requirements of strain and deformation limits are met.

Table 6.7 Minimum stress-to-rupture values for 9Cr-1Mo-V steel [ASME II-D].

US customary units

Temp., °F	10 h	30 h	10^2 h	3×10^2 h	10^3 h	3×10^3 h	10^4 h	3×10^4 h	10^5 h	3×10^5 h
700	71.0	71.0	71.0	71.0	71.0	71.0	71.0	71.0	71.0	71.0
750	69.0	69.0	69.0	69.0	69.0	69.0	69.0	67.3	63.5	60.2
800	66.5	66.5	66.5	66.5	66.5	63.1	59.4	56.1	52.7	49.6
850	63.4	63.4	63.4	59.7	56.0	52.7	49.3	46.3	43.3	40.6
900	59.8	57.0	53.3	50.0	46.6	43.7	40.6	37.9	35.2	32.8
950	51.2	47.9	44.5	41.5	38.5	35.8	33.1	30.7	28.2	26.1
1000	42.8	39.9	36.8	34.1	31.4	29.0	26.6	24.5	22.3	20.5
1050	35.6	32.9	30.1	27.7	25.3	23.2	21.1	19.2	17.3	15.7
1100	29.2	26.8	24.4	22.3	20.1	18.3	16.4	14.8	13.1	11.7
1150	23.7	21.6	19.4	17.6	15.7	14.1	12.4	10.2	8.2	6.7
1200	19.0	17.1	15.2	13.6	11.9	10.5	8.0	6.5	4.9	3.7

(d) Check creep-fatigue requirements using elastic analysis in accordance with VIII-2 Eq. (6.11) Calculate the value of $3S'_m$, which is the smaller of

$$3S_m = (3)(19{,}000) = 57{,}000 \text{ psi}$$

or

$$3\bar{S}_m = 1.5S_m + 0.5S_t = 35{,}650 \text{ psi.}$$

Use $3S'_m = 35{,}650$ psi

$$3\bar{S}_m / E = 35{,}650 / 25{,}400{,}000 = 0.0014.$$

The requirements and steps outlined in Section 6.2 will be followed.

- The requirements of Test A-3 are met.
- The requirement of $(P_L + P_b + Q) \le 3S'_m$ must be met.

The value of $(P_L + P_b + Q)$ is obtained from Eq. (4.14) as

$$\left(P_L + P_b + Q\right) = 12{,}030 \text{ psi} < 35{,}650 \text{ psi.}$$

Step 1

Total number of hours, $t_H = 200{,}000$ h

Step 2

Hold temperature, $T_{HT} = 1000^\circ\text{F}$

Step 3

Average cycle time = 24 h/cycle

Step 4

The maximum strain during the cycle is given by

$$\Delta\varepsilon_{max} = 2S_{alt} / E = 12{,}030 / 25{,}400{,}000 = 0.000474 \text{ in./in.}$$

Because the magnitude of $\Delta\varepsilon_{max}$ is within the elastic limit, the value of $\Delta\varepsilon_{mod}$ can be written as

$$\Delta\varepsilon_{mod} = K_{sc}\Delta\varepsilon_{max} = (3.3)(0.000474) = 0.0016 \text{ in./in.}$$

Step 5

From Figure 6.2, let f be conservatively equal to 1.0.
From Figure 6.3, $K'_v = 1.0$.
Equation (6.9) gives $K_v = 1.0$.

From Eq. (5.11), with $Z = X$,

$$\sigma_c = (Z)(\text{yield stress at cold end of cycle})$$
$$= 0.23(53{,}200) = 12{,}240 \text{ psi}$$
$$1.25\sigma_c = 15{,}300 \text{ psi}$$

Enter the isochronous chart (Figure 6.7) with a stress of 15,300 psi and obtain the strain for either (a) 24 h (duration of one cycle from Step 3), which gives $\Delta\varepsilon_c = 0.00060$, or (b) 200,000 h, which gives $\Delta\varepsilon_c = 0.0025$ for the full life. The value for one cycle is then given by

$$\Delta\varepsilon_c = 0.0025 / 10{,}000 \approx 0.$$

The total strain is then obtained from Eq. (6.8) as

$$\varepsilon_t = (1.0)(0.0016) + (3.3)(0) = 0.0016.$$

From Figure 6.7, the time-independent stress, $S_j = 36.5$ ksi.

Step 6

In Figure 6.7, stress-versus-time values are obtained as shown in Table 6.8.

Step 7

Equation (6.11) is solved in this step. The first part of the equation, $(n_c/N_d)_j$, is first obtained as follows:

- First part of Eq. (6.11)

  $$n_c = 10{,}000 \text{ cycles}$$

  The value of N_d is obtained from Table 6.6, with $\varepsilon_t = 0.0016$. A cycle life of $\approx$ 1,000,000 is obtained at a temperature of 1000°F.

  $$(n_c / N_d)_j = 10{,}000 / 1{,}000{,}000 = 0.01$$

- Second part of Eq. (6.11)
  The second part of Eq. (6.11), $(\Delta t/t_d)_k$, is determined numerically from Table 6.8 and from the differential equation $\int dt/t_d$. Calculations for 200,000 h are summarized in Table 6.9.

Column A in Table 6.9 gives the incremental locations for calculating stresses from Table 6.8. Column B shows the stress values obtained from Table 6.8 and Column H lists the incremental duration assumed for these stresses. Column C adjusts the stress values by the K' factor and Column G lists the hours obtained from Table 6.7 using the stress values in Column C. Column I is the ratio of Column H over Column G, which is the numerical value of the quantity $(\int dt/t_d)$.

Table 6.8 Stress, S_j, versus time values for Example 6.2 (first iteration).

Time (hr)	Stress (ksi)
S_j	36.5
1	30.0
3	28.0
10	26.7
30	25.3
100	23.3
300	22.0
1000	21.0
3000	19.3
10,000	18.0
30,000	16.0 (use 16.3)
100,000	14.7 (use 16.3)
200,000	13.7 (use 16.3)

The 1000°F line in Table 6.7 terminates at the 20,500 psi stress level. Stress levels in Column C drop below the 20,500 psi level in Rows 1 through 12. The Larson-Miller parameter, P_{LM}, is used to obtain approximate rupture life by using higher temperature levels given in Table 6.7. These calculations are shown in columns D, E, and F. Column D lists the temperatures from Table 6.7 corresponding to the stress levels in Column C. Column E calculates the corresponding P_{LM} from Eq. (1.3) for the temperature shown in Column D at 300,000 h. Column F recalculates the hours corresponding to a temperature of 1000°F from Eq. (1.3) with the corresponding P_{LM} factor.

The quantity $\left(3\overline{S}_m\right)/E$, calculated earlier, is equal to 0.0014, whereas ε_t is equal to 0.0016. Because $\varepsilon_t > \left(3\overline{S}_m\right)/E$, shakedown does not occur and the peak stress does reset on each cycle. Under this scenario, peak stress can be assumed to relax through each cycle and the cumulated creep damage is given by Table 6.9.

In Table 6.9, the value of $(\Delta t/t_d)_k$ for 24 h is equal to 0.059. Note that most of the computed creep damage occurs in the first hour. This computed value can be more realistically evaluated by taking smaller time steps and computing the damage at each step. In this case, time steps of 0.25 h will reduce the damage during the first hour by more than a factor of two. However, even with the first hour damage reduction the total value of $(\Delta t/t_d)_k$ for 10,000 cycles is approximately 350, which is inadequate. Thus, the required thickness needs to be increased to accommodate 200,000 h.

Second iteration

A new trial thickness may be assumed by multiplying the original thickness by the ratio of strain ε_t over strain $\left(3\overline{S}_m\right)/E$. The result gives

$$t = (1.125)(0.0016/0.0014) = 1.286 \text{ in.}$$

Try a 12-in. Sch-160 pipe with OD = 12.75 in. and t = 1.3125 in.

(a) Check design thickness in accordance with ASME-I
By inspection, the newer thickness is adequate as per ASME-I.

(b) Check load-controlled stress limits in accordance with Figure 4.8

$$\gamma = R_o / R_i = 6.375/5.0625 = 1.2593$$

Table 6.9 Numerical calculations of $(\Delta_t/t_d)_k$ for example 6.2 (first iteration).

A	B	C	D	E	F	G	H	I
Location (h)	S_r (ksi)	S_r/0.9 (ksi)	Equivalent temp. (°F)	P_{LM}	Time (h)	Table 6.7 (h)	Table 6.8 (h)	H/G
S_j	36.5	40.5				23	1	0.043
1	30.0	33.3				507	2	0.004
3	28.0	31.1				1250	7	0.006
10	26.7	29.7				2417	20	0.008
30	25.3	28.1				5625	70	0.012
100	23.3	25.9				16,670	200	0.012
300	22.0	24.4				33,180	700	0.021
1000	21.0	23.3				68,180	2000	0.029
3000	19.3	21.4				200,000	7000	0.035
10,000	18.0	20.0	1005	37,324	366,750	366,750	20,000	0.055
30,000	16.5	18.3	1023	37,783	756,400	756,400	70,000	0.093
100,000	16.5	18.3	1023	37,783	756,400	756,400	100,000	0.132
200,000	16.5	18.3						
Total								0.450

From Example 4.4,
Circumferential stress at inside surface with a ligament efficiency of 0.60
membrane = $P_m = P/(0.6)(\gamma - 1) = 8355$ psi
bending = $P_b = 0$
secondary = $Q = 645/0.6 = 1075$ psi
Longitudinal stress at inside surface with a ligament efficiency of 0.85
membrane = $P_{m'} = P/(0.85)(\gamma^2 - 1) + 7000 = 9610$ psi
bending = $P_b = 0$
secondary = $Q = 0$
Radial stress at inside surface
membrane = $P_m = -P/2 = -650$ psi
bending = $P_b = 0$
secondary = $Q = -650$ psi
The value of P_m is obtained from Eq. (4.14) as

$$P_m = 9690\,\text{psi}, \quad \text{and } P_L + P_b = 9690$$

Design limits

$$\begin{aligned} &P_m < S_o && 9690\ \text{psi} < 16{,}300\ \text{psi} \\ &P_L + P_b < 1.5S_o && 9690\ \text{psi} < 24{,}450\ \text{psi} \end{aligned}$$

Operating limits

$$\begin{aligned} &P_m < S_{mt} && 9690\ \text{psi} < 14{,}300\ \text{psi} \\ &P_L + P_b < 1.5S_m && 9690\ \text{psi} < 28{,}500\ \text{psi} \\ &P_L + P_b / k_t < S_t && 9690\ \text{psi} < 14{,}300\ \text{psi} \end{aligned}$$

Thus, the requirements of load-controlled limits are met.

(c) Check strain- and deformation-controlled limits in accordance with Figure 4.8

Average yield stress = 46,700 psi
Calculate X and Y from Eqs. (5.1) and (5.2).
The value of Q is obtained from Eq. (4.14) as

$$Q = 1510\,\text{psi}$$
$$X = 10{,}815 / 46{,}700 = 0.23 \quad Y = 1510 / 46{,}700 = 0.03$$

Check elastic analysis Test A-3

$$X + Y = 0.24 < 1.0$$

There is no need to check the simplified inelastic analysis Tests B because Test A-3 is met.

Thus, the requirements of strain and deformation limits are met.

(d) Check creep-fatigue requirements using elastic analysis in accordance with ASME-2, Eq. (6.11*)* Calculate the value of $3S'_m$, which is the smaller of

$$3S_m = (3)(19{,}000) = 57{,}000 \text{ psi}$$

or

$$3\overline{S}_m = 1.5S_m + 0.5S_t = 35{,}650 \text{ psi}$$

Use $3S'_m$ = 35,650 psi.

$$3\overline{S}_m / E = 35{,}650 / 25{,}400{,}000 = 0.0014$$

The requirements and steps outlined in Section 6.2 will be followed.

- The requirements of Test A-3 are met.
- The requirement of $(P_L + P_b + Q) \leq 3S'_m$ must be met.

  The value of $(P_L + P_b + Q)$ is obtained from Eq. (4.14) as

$$\left(P_L + P_b + Q\right) = 10{,}820 \text{ psi}, \quad < 35{,}650 \text{ psi}$$

Step 1

$$\text{Total amount of hours, } T_H = 200{,}000 \text{ h}$$

Step 2

$$\text{Hold temperature, } T_{HT} = 1000^\circ F$$

Step 3

$$\text{Average cycle time} = 24 \text{ h/cycle}$$

Step 4
The maximum strain during the cycle is given by

$$\Delta\varepsilon_{max} = 2S_{alt} / E = 10{,}820 / 25{,}400{,}000 = 0.000426 \text{ in. /in.}$$

Because the magnitude of $\Delta\varepsilon_{max}$ is within the elastic limit, the value of $\Delta\varepsilon_{mod}$ can be written as

$$\Delta\varepsilon_{mod} = K_{sc}\Delta\varepsilon_{max} = (3.3)(0.000426) = 0.0014 \text{ in./in.}$$

Step 5

From Figure 6.2, let f be conservatively equal to 1.0.
From Figure 6.3, $K'_v = 1.0$.
Equation (6.9) gives $K_v = 1.0$.
From Eq. (5.11) with $Z = X$,

$$\sigma_c = (Z)(\text{yield stress at cold end of cycle})$$
$$= 0.21(53,200) = 11,200 \text{ psi}$$
$$1.25\sigma_c = 14,000 \text{ psi}$$

Enter the isochronous chart (Figure 6.7) with a stress of 14,300 psi, and obtain the strain for either (a) 24 h (duration of one cycle from Step 3), which gives $\Delta\varepsilon_c = 0.00056$, or (b) 200,000 h, which gives $\Delta\varepsilon_c = 0.0015$ for the full life. The value for one cycle is then given by $\Delta\varepsilon_c = 0.0015/10,000 \approx 0$
The total strain is then obtained from Eq. (6.8) as

$$\varepsilon_t = (1.0)(0.0014) + (3.3)(0) = 0.0014$$

From Figure 6.7, the time-independent stress

$$S_j = 34.0 \text{ksi}.$$

Table 6.10 Stress, S_j, versus time values for Example 6.2 (second iteration).

Time (hr)	Stress (ksi)
S_j	34.0
1	28.7
3	26.7
10	25.3
30	24.0
100	22.7
300	21.3
1000	20.0
3000	18.7
10,000	17.3
30,000	16.0 (use 16.3)
100,000	14.0 (use 16.3)
200,000	13.7 (use 16.3)

Step 6

In Figure 6.7, stress-versus-time values are obtained as shown in Table 6.10.

Step 7

Equation (6.11) is solved in this step. The first part of the equation, $(n_c/N_d)_j$, is obtained first as follows:

- First part of Eq. (6.11)

$$n_c = 10{,}000 \text{ cycles}$$

The value of N_d is obtained from Table 6.6 with $\varepsilon_t = 0.0014$. A cycle life of ≈ 10 000 000 is obtained at 1000°F.

$$\left(n_c / N_d\right)_j = 10{,}000 / 10{,}000{,}000 = 0.001 \approx 0$$

- Second part of Eq. (6.11)

  The second part of Eq. (6.11), $(\Delta t/t_d)_k$, is determined numerically from Table 6.10 and from the differential equation $\int dt/t_d$. Calculations for 200,000 h are summarized in Table 6.11.

  The quantity $\left(3\bar{S}_m\right)/E$, calculated earlier, is equal to 0.0014 and ε_t is also equal to 0.0014. Because $\varepsilon_t = \left(3\bar{S}_m\right)/E$, shakedown does occur and the peak stress does not reset on each cycle. Under this scenario, peak stress can be assumed to relax monotonically throughout the entire design life and the cumulated creep damage is given by Table 6.11.

  The total value of $(\Delta t/t_d)_k$ from Table 6.11 is 0.358. This value is acceptable from Figure 6.6 for an expected life of 200,000 h.

  The result indicates that the required original thickness of 1.125 in. is inadequate for a 10,000 cycle service with 200,000 h at 1000°F. The new thickness of 1.3125 in. is adequate for the intended service.

6.3 Welded Components

The procedure outlined in Section 6.2 is also applicable to welded components. Welded reduction factor, R_w, must be incorporated into stress calculations for creep rupture.

6.4 Variable Cyclic Loads

In Section 6.2 the analysis of components subjected to a repetitive cyclic load was presented. "Repetitive cyclic" in that context referred to a number of cycles of the same type and magnitude, such as startup followed by a period of

Table 6.11 Numerical calculations of $(\Delta_t/t_d)_k$ for Example 6.2 (second iteration).

A	B	C	D	E	F	G	H	I
Location (hr)	S_r (ksi)	$S_r/0.9$ (ksi)	Equivalent temp. (°F)	P_{LM}	Time (hr)	Table 6.7 (hr)	Table 6.8 (hr)	H/G
S_j	34.0	37.8				77	1	0.013
1	28.7	31.9				870	2	0.002
3	26.7	29.7				2417	7	0.003
10	25.3	28.1				5625	20	0.004
30	24.0	26.7				9708	70	0.007
100	22.7	25.2				23,330	200	0.009
300	21.3	23.7				55,460	700	0.013
1000	20.0	22.2				111,110	2000	0.018
3000	18.7	20.8				266,670	7000	0.026
10,000	17.3	19.2	1014	37,553	526,530	526,530	20,000	0.038
30,000	16.5	18.3	1023	37,783	756,400	756,400	70,000	0.093
100,000	16.5	18.3	1023	37,783	756,400	756,400	100,000	0.132
200,000	16.5	18.3						
Total								0.358

operation and then shutdown. The remainder of this chapter will address the situation where there are two or more types of cycles with varying magnitudes; for example, normal startup and shutdown cycles of a fixed number and magnitude, and a different number of thermal transient cycles of a different number and magnitude.

There are a number of methods developed to combine cyclic histories, many of which are described in some detail in Bannantine et al. (1990). The essential feature of these methods is to account for the additive effect of combining strain ranges. As discussed in the ASME criteria for Division 2 (1969), "When stress cycles of various frequencies are intermixed through the life of a vessel it is important to identify correctly the number and range of each type of cycle. It must be remembered that a small increase in stress range can produce a large decrease in fatigue life, and this relationship varies for different portions of the fatigue curve. Therefore, the effect of superposing two stress amplitudes cannot be evaluated by adding the usage factors obtained from each amplitude by itself. The stresses must be added before calculating the usage factors." Although the above is written in terms of stress, the discussion is equally applicable in terms of strain frequency and range.

Example 6.3 Consider the case of a thermal transient occurring in a pressurized vessel. At a point in the vessel, the peak stress due to pressure is 20,000 psi tension and the added stress from the thermal transient is 70,000 psi tension. The thermal stress occurs 10,000 times and the pressure stress occurs 1000 times. What is the combined effect?

Solution

Add the thermal stress range to the pressure stress range for 1000 cycles. Thus, the total usage factor is the sum of the usage factor for 1000 cycles with a stress range of 90,000 psi and the usage factor for 9000 cycles with a stress range of 70,000 psi.

6.5 Equivalent Stress Range Determination

When the design specification delineates a specific loading sequence, then such sequence should be used to determine the equivalent strain ranges. If the sequence is not specified, then the following procedures should be used.

6.5.1 Equivalent Strain Range Determination – Applicable to Rotating Principal Strains

Step 1

Calculate all strain components for each point, I, in time (ε_{xi}, ε_{yi}, ε_{zi}, γ_{xyi}, γ_{yzi}, γ_{zxi}) for the complete cycle. Note that when conducting elastic analysis, which is the basis of this discussion, peak strains from geometric discontinuities are not included because these effects are accounted for later in the procedure.

Step 2

Select a point when conditions are at an extreme for the cycle, either maximum or minimum, and refer to this Time Point by the subscript "o."

Step 3

Calculate the history of the change in strain components by subtracting the values at time o from the corresponding components at each point in time, i, during the cycle

$$\Delta\varepsilon_{xi} = \varepsilon_{xi} - \varepsilon_{xo}$$
$$\Delta\varepsilon_{yi} = \varepsilon_{yi} - \varepsilon_{yo}$$
$$\Delta\varepsilon_{zi} = \varepsilon_{zi} - \varepsilon_{z0}, \ldots$$

etc.

Step 4
Calculate the equivalent strain range for each point in time as

$$\Delta\varepsilon_{\text{equiv,i}} = \left[0.707/\left(1+\mu^*\right)\right]\left[\begin{array}{l}\left(\Delta\varepsilon_{xi}-\Delta\varepsilon_{yi}\right)^2 + \\ \left(\Delta\varepsilon_{yi}-\Delta\varepsilon_{zi}\right)^2 + \left(\Delta\varepsilon_{zi}-\Delta\varepsilon_{xi}\right)^2 \\ +1.5\left(\Delta\gamma_{xyi}^2+\Delta\gamma_{yzi}^2+\Delta\gamma_{zxi}^2\right)\end{array}\right]^{1/2} \tag{6.18}$$

where $\mu^* = 0.3$ when using elastic analysis.

6.5.2 Equivalent Strain Range Determination – Applicable When Principal Strains Do Not Rotate

Step 1
Determine the principal strains versus time for the cycle.

Step 2
At each time interval of Step 1, determine the strain differences $(\varepsilon_1 - \varepsilon_2)$, $(\varepsilon_2 - \varepsilon_3)$, $(\varepsilon_3 - \varepsilon_1)$.

Step 3
Select a point when conditions are at an extreme for the cycle, either maximum or minimum, and refer to this time by the subscript "o."

Step 4
Calculate the history of the change in strain differences by subtracting the values at time o from the corresponding components at each point in time, i, during the cycle. Designate these strain difference changes as

$$\begin{aligned}\Delta\left(\varepsilon_1-\varepsilon_2\right)_i &= \left(\varepsilon_1-\varepsilon_2\right)_i - \left(\varepsilon_1-\varepsilon_2\right)_0 \\ \Delta\left(\varepsilon_2-\varepsilon_3\right)_i &= \left(\varepsilon_2-\varepsilon_3\right)_i - \left(\varepsilon_2-\varepsilon_3\right)_0 \\ \Delta\left(\varepsilon_3-\varepsilon_1\right)_i &= \left(\varepsilon_3-\varepsilon_1\right)_i - \left(\varepsilon_3-\varepsilon_1\right)_0\end{aligned}$$

Step 5
For each point in time i, calculate the equivalent strain range as

$$\Delta\varepsilon_{\text{equiv, i}} = \left[0.707/\left(1+\mu^*\right)\right]\left\{\begin{array}{l}\left[\Delta\left(\varepsilon_1-\varepsilon_2\right)_{\text{i}}\right]^2 + \left[\Delta\left(\varepsilon_2-\varepsilon_3\right)_{\text{i}}\right]^2 \\ +\left[\Delta\left(\varepsilon_3-\varepsilon_1\right)_{\text{i}}\right]^2\end{array}\right\}^{1/2} \tag{6.19}$$

where $\mu^* = 0.3$ when using elastic analysis.

6.5.3 Equivalent Strain Range Determination – Acceptable Alternate When Performing Elastic Analysis

6.5.3.1 Constant Principal Stress Direction

Step 1

Determine the principal stresses versus time for the cycle.

Step 2

At each time interval of Step 2, determine the stress differences $(\sigma_1 - \sigma_2), (\sigma_2 - \sigma_3), (\sigma_3 - \sigma_1)$.

Step 3

Calculate the history of the change in stress differences by subtracting the values at a reference time from the corresponding components at each point in time, i, during the cycle. The maximum stress value of these differences is obtained from Eq. (4.14) and is equal to the value of $2S_{alt}$, which is the starting point of Step 4 under Section 6.2.

6.5.3.2 Rotating Principal Stress Direction

This procedure is basically the same as the procedure for rotating principal strain direction. The stress differences are first determined at the component level for all six components and then the principal stresses are determined from those component stress differences.

6.5.3.3 Variable Cycles

If there are two or more types of stress cycles which produce significant stresses, their cumulative effect shall be evaluated as shown:

Step 1

Designate the specified number of times each type of cycle (1, 2, 3, n) will be repeated during the life of the component as n_1, n_2, n_3, ..., n_n, respectively.
Note: In determining n_1, n_2, n_3, n_n, consideration shall be given to the superposition of cycles of various origins that produce a total stress difference range greater than the stress difference range of the individual cycles.

Step 2

For each type of stress cycle, determine the alternating stress intensity, S_{alt}, and designate these quantities $S_{alt\,1}$, $S_{alt\,2}$, $S_{alt\,3}$, ..., $S_{alt\,n}$.

Step 3

Proceed with evaluation as per Step 4 under Section 6.2

Example 6.4 Using the data previously developed for Examples 5.1, 5.2, and 6.1 to evaluate the life cycle of the 2.0-in.-thick shell for 130,000 h when subjected to the original maintenance shutdown cycle plus the unplanned process-induced shutdown as described below.

Pressure. The pressure remains constant.

Temperature. The temperature decreases such that the inner wall is 36°F colder than the outer wall, as shown in Figure 6.8. The process is then re-established and operation returned to normal conditions. This cycle, cool down and restart, takes approximately 8 h to complete. This cycle is expected to occur, on average, six times during each 18-month operational cycle.

Solution

The stress components for Example 5.1 for the outside surface of the 2.0-in. shell are tabulated in Table 6.12 with an additional column for the 36°F ΔT condition.

Step 1

Principal stresses for each time point in the cycle as identified in Figure 6.8 are computed from component stresses in Table 6.12 as shown below at Time 2.

$$\sigma_\theta = P_{m\theta} + Q_{b\theta} = 4500 + 14{,}975 = 19{,}475$$
$$\sigma_L = P_{mL} + Q_{bL} = 2250 + 14{,}975 = 17{,}225$$
$$\sigma_R = P_{mr} + Q_{br} = (-150) + 150 = 0$$

Step 2

Principal stress differences, or stress intensities, are computed from the principal stress as shown below for Time 2. The procedure is outlined in Section 4.4. The three resultant terms used in the von Mises Eq.(4.14) are

$$(\sigma_{\theta 2} - \sigma_{\theta 1}) - (\sigma_{L2} - \sigma_{L1})$$
$$(\sigma_{L2} - \sigma_{L1}) - (\sigma_{R2} - \sigma_{R1}).$$
$$(\sigma_{R2} - \sigma_{R1}) - (\sigma_{\theta 2} - \sigma_{\theta 1})$$

In cases where the principal stresses do not change directions during the cycle, the above equations can be rewritten as

$$(\sigma_\theta - \sigma_L)_2 - (\sigma_\theta - \sigma_L)_1$$
$$(\sigma_L - \sigma_R)_2 - (\sigma_L - \sigma_R)_1$$
$$(\sigma_R - \sigma_\theta)_2 - (\sigma_R - \sigma_\theta)_1$$

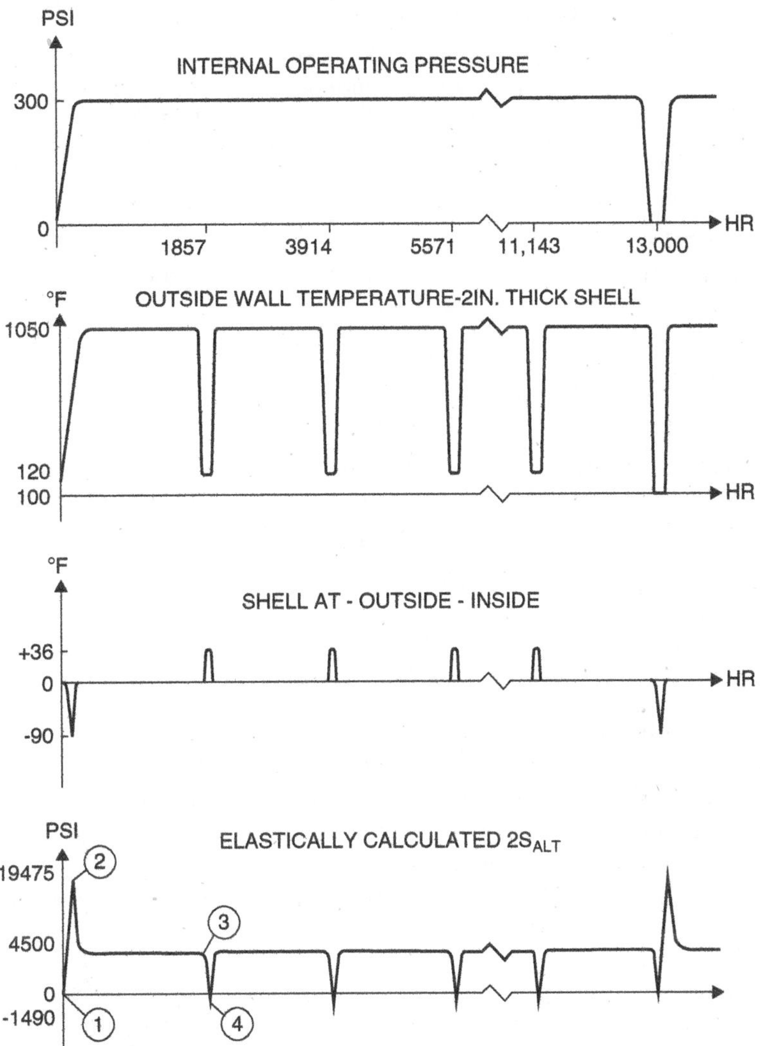

Figure 6.8 Pressure and temperature history.

Hence,

$$\sigma_\theta - \sigma_L = 19{,}475 - 17{,}225 = 2250$$
$$\sigma_L - \sigma_R = 17{,}225 - 0 = 17{,}220$$
$$\sigma_R - \sigma_\theta = 0 - 19{,}475 = -19{,}475$$

Step 3

Determine the range of principal stress differences between the reference points in time defining the cycles of interest. Shown below is the evaluation of

Table 6.12 Stress outside of the 2.0 in. shell.

Stress	Pressure stress	Thermal stress, ΔT = 90°F	Thermal stress, ΔT = −36°F
Membrane hoop stress, $P_{m\theta}$	4500	0	0
Membrane axial stress, P_{mL}	2250	0	0
Membrane radial stress, P_{mr}	−150	0	0
Circumferential bending stress, $Q_{b\theta}$	0	14,975	−5990
Axial bending stress, Q_{bL}	0	14,975	−5990
Radial bending stress, Q_{br}	150	0	0

the cycle range defined by Time Points 1 and 2. Note that this corresponds to the cycle range for Example 6.1. Because stress values at Time Point 1 are zero, the absolute value of the difference is straightforward.

$$\left(\sigma_\theta - \sigma_L\right)_2 - \left(\sigma_\theta - \sigma_L\right)_1 = 2250 - 0 = 2250$$
$$\left(\sigma_L - \sigma_R\right)_2 - \left(\sigma_L - \sigma_R\right)_1 = 17{,}225 - 0 = 17{,}225$$
$$\left(\sigma_R - \sigma_\theta\right)_2 - \left(\sigma_R - \sigma_\theta\right)_1 = -19{,}475 - 0 = -19{,}475$$

From Eq. (4.14), the effective stress is $2S_{alt\,1-2} = 18{,}450$ psi.

Stress differences (Table 6.13) are identified for the cycle defined by Time Points 1 and 2 in the fourth column ($S_{r1\text{-}2}$); for the cycle defined by Time Points 3 and 4 in Column 7 ($S_{r3\text{-}4}$); and for the composite cycle defined by Time Points 1, 2, 3, and 4 in the last column ($S_{r1\text{-}4}$).

From the above table, the maximum absolute values of the cycle range = $2S_{alt}$ are obtained from von Mises' equation (Eq. 4.14):

$$2S_{alt1-2} = 18{,}450 \text{ psi}$$

$$2S_{alt3-4} = 5990 \text{ psi}$$

$$2S_{alt1-4} = 20{,}965 \text{ psi.}$$

Cycles 1–4, with a cycle time of 13,000 h, and Cycles 3–4, with a negligible cycle time, will be selected to represent the combined loading history. The selection of combined cycles requires consideration of the cycle time and thus goes beyond the normal superposition considerations associated with non-time-dependent design criteria. This point will be discussed further when considering the evaluation of creep damage.

Table 6.13 Principal stresses and alternating stress intensity.

	Time 1	Time 2	S_{r1-2}	Time 3	Time 4	S_{r3-4}	S_{r1-4}
σ_θ	0	19,475	–	4500	−1490	–	–
σ_L	0	16,850	–	2250	−3740	–	–
σ_R	0	0	–	0	0	–	–
$\sigma_\theta - \sigma_L$	0	2250	2250	2250	−2250	4500	4500
$\sigma_L - \sigma_R$	0	16,825	16,825	2250	−3740	5990	20,565
$\sigma_R - \sigma_\theta$	0	−19,475	19,475	−4500	1490	5990	20,965

The following evaluation follows the same procedure used to evaluate creep-fatigue damage in Example 6.1.

Step 4
First, consider cycle defined by $2S_{\text{alt }1\text{–}4} = 20{,}965$ psi

$$\Delta\varepsilon_{max} = 2S_{alt1-4} / E = 20{,}965 / 22{,}400{,}000 = 0.000936.$$

Next, evaluate $\Delta\varepsilon_{\text{mod}}$ using Eq.(6.5) for Method 1.

$$\Delta\varepsilon_{\text{mod}} = \left(S^* / \bar{S}\right) K_{\text{sc}}^2 \Delta\varepsilon_{\text{max}}.$$

However, as before, $\Delta\varepsilon_{\text{max}}$ and $K_{\text{sc}}\Delta\varepsilon_{\text{max}}$ are within the proportional limit of the composite stress-strain curve, and $\Delta\varepsilon_{\text{mod}}$ reduces to

$$\begin{aligned}\Delta\varepsilon_{\text{mod}} &= K_{sc}\Delta\varepsilon_{\text{max}}\\ &= (1.1)(0.00936)\\ &= 0.00103\end{aligned}$$

Step 5
This step is essentially identical to the steps used in Example 6.1, and the assumptions therein are equally applicable. For $\Delta\varepsilon_{\text{mod}} = 0.00103$, the resulting expression for ε_t becomes:

$$\varepsilon_t = (1.0)(0.00103) + (1.1)(0.000026) = 0.000106.$$

Note that the creep strain increment per cycle, $\Delta\varepsilon_c$, remains the same because the pressure stress is the same and the value of Z in region E is equal to X.

From Figure 5.6, the time independent stress $S_j \approx 15{,}000$ psi.

Before proceeding with the creep damage assessment, it is necessary to assess the value of ε_t and S_j for the remaining cycle.

Step 3
From the preceding Step 3, the stress intensity range for Cycles 3–4 is given by

$$2S_{alt3-4} = 5990 \text{ psi.}$$

Step 4
As before,

$$\Delta\varepsilon_{max} = 2S_{alt3-4} / E = 5990 / 22{,}400{,}000 = 0.00027,$$

which is well within the elastic regime. And

$$\begin{aligned}\Delta\varepsilon_{\text{mod}} &= K_{sc}\Delta\varepsilon_{\text{max}} \\ &= (1.1)(0.00027) \\ &= 0.00030\end{aligned}$$

Step 5
As before, for $\Delta\varepsilon_{\text{mod}} = 0.00030$

$$\varepsilon_{\text{t}} = (1.0)(0.00030) + (1.1)(0.000026) = 0.00033.$$

Note that, per the presumed composite cycle definition, there is no significant cycle time and the incremental creep strain $\Delta\varepsilon_c$ will be equal to zero.

Conservatively based on $\varepsilon_{\text{t } 3\text{–}4} = 0.00033$

$$S_{\text{j}} = 7400 \text{psi.}$$

Step 6
It is clear that Cycles 3–4 satisfies the shakedown criteria and the assumed superposition is valid. Noting that $\varepsilon_{\text{t } 1\text{–}4} = 0.00106$ also satisfies the shakedown criteria, the resulting stress history for evaluating creep damage is quite closely approximated by the values in Table 6.5, and the overall creep damage fraction will be:

$$\Sigma(\Delta t / t_{\text{d}})_{\text{k}} = 0.0485.$$

In summary, the unplanned process-induced shutdowns have a very slight impact on the cyclic stress range that is wiped out within the first hour of sustained operation. The impact on vessel life is negligible. Physically, the loading for Example 6.1 resembles a hold time creep-fatigue test where there is no subsequent yielding after the initial startup sequence. The additional cycles in this example resemble additional minor, and more frequent, dips in the strain level in the creep-fatigue test with no apparent impact on cyclic life.

Problems

6.1 The hydrotreater given in Problem 5.1 is shut down after every three years (25,000 h) of operation for maintenance. The shutdown period is two weeks. Accordingly, the expected number of cycles is eight. Evaluate the life cycle of 200,000 h at 950°F.

6.2 The boiler header given in Problem 5.2 is shut down after every 2.5 years (20,000 h) of operation for maintenance. The shutdown period is six weeks. Accordingly, the expected number of cycles is 15. Evaluate the life cycle of 300,000 h at 975°F.

7

Creep-Fatigue Analysis Using the Remaining Life Method

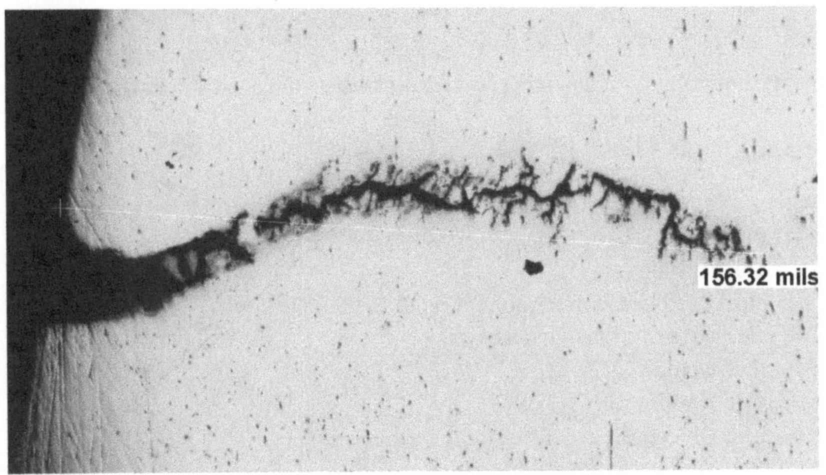

Creep-fatigue on the outside of a boiler tube [Union Electric Company d/b/a Ameren Missouri]

7.1 Basic Equations

The remaining life method, detailed in the ASME FFS-1 publication API 579/ASME FFS-1 (2021), is extensively used in the petroleum industry. The creep curve in this method disregards primary and secondary creep since they are assumed to be short in duration and considers the tertiary regime only as explained in Section 1.4.3 of Chapter 1 and shown in Figure 7.1.

Analysis of ASME Boiler, Pressure Vessel, and Nuclear Components in the Creep Range, Second Edition. Maan H. Jawad and Robert I. Jetter.
© 2022 John Wiley & Sons Ltd. Published 2022 by John Wiley & Sons Ltd.

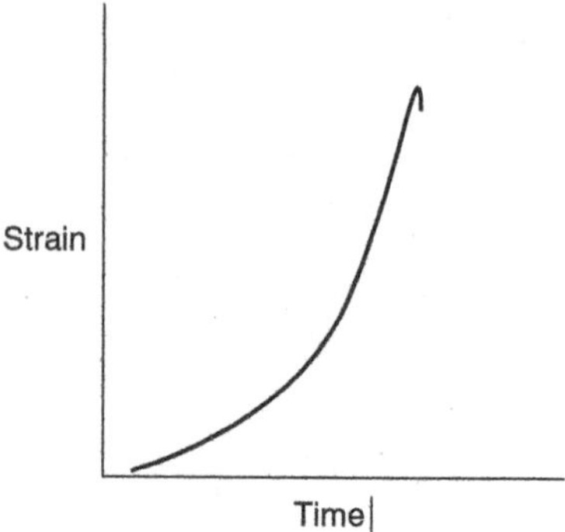

Figure 7.1 Tertiary creep versus time.

The equation for creep strain ε_c is expressed by the equation Prager (1995)

$$\dot{\varepsilon} = \dot{\varepsilon}_{co} \exp[(m + p + c)\varepsilon] \tag{7.1}$$

where
ε = strain
$\dot{\varepsilon} = d\varepsilon/dt$ = strain rate
m = modified Norton's component relating stress to strain
p = microstructural damage factors
c = other factors
$\dot{\varepsilon}_{co}$ = initial strain rate.
The term Omega, Ω, is defined by the expression

$$\frac{\partial \ln(\dot{\varepsilon})}{\partial \varepsilon} = \Omega = m + p + c. \tag{7.2}$$

Substituting this equation into Eq. (7.1) gives

$$\dot{\varepsilon} = \dot{\varepsilon}_{co} \exp(\Omega\, \varepsilon). \tag{7.3}$$

This equation can be rearranged as

$$\frac{d\varepsilon}{\exp(\Omega\, e)} = \dot{\varepsilon}_{co} dt. \tag{7.4}$$

Integrating this equation using the limits 0 to ε and 0 to t gives

$$t = \frac{1}{\dot{\varepsilon}_{co}\Omega}[1 - \exp(-\Omega\varepsilon)] \tag{7.5}$$

where

t = time.

Equation (7.5) can also be written in terms of creep strain as

$$\varepsilon_c = -(1/\Omega)\ \ln(1 - \dot{\varepsilon}_{co}\Omega t). \tag{7.6}$$

7.2 Equations for Creep-Fatigue Interaction

Equation (7.5) forms the basis for calculating the life of a component in the creep regime. It is simplified by assuming the strain to be large at failure and thus the term exp(− Ωε) approaches zero and can be deleted. The quantities $\dot{\varepsilon}_{co}$ and Ω are a function of material properties and experimental factors.

The procedure for calculating remaining life of a pressure vessel component using the Ω method is as follows:

Step 1

Calculate the stresses, in ksi, at any given point. If these stresses are "primary stresses" then they can be used directly in the analysis below. If they include "secondary" and "peak" stresses then they need to go through a relaxation analysis prior to using them in this analysis. This relaxation analysis may be performed by a finite element method using, as a material property, an isochronous stress-strain curve for the expected life at a given temperature.

Calculate the three principal stresses σ_1, σ_2, and σ_3, and then calculate the effective stress σ_e using the von Mises equation

$$\sigma_e = \frac{1}{\sqrt{2}}\sqrt{(\sigma_1 - \sigma_2)^2 + (\sigma_2 - \sigma_3)^2 + (\sigma_3 - \sigma_1)^2}. \tag{7.7}$$

Calculate the quantity

$$S_1 = \log_{10}(\sigma_e). \tag{7.8}$$

Step 2

Calculate the initial strain rate $\dot{\varepsilon}_{co}$ and Ω using the material property coefficients listed in Appendix E.

$$\mathrm{Log}_{10}(\dot{\varepsilon}_{co}) = -[(A_o + \Delta_\Omega^{sr}) + \left(A_1 + A_2S_1 + A_3S_1^2 + A_4S_1^3\right)/\left(460 + T\right)] \tag{7.9}$$

$$\mathrm{Log}_{10}(\Omega) = [(B_o + \Delta_\Omega^{cd}) + \left(B_1 + B_2S_1 + B_3S_1^2 + B_4S_1^3\right)/\left(460 + T\right)] \tag{7.10}$$

where

T = temperature, °F.

Constants A_i and B_i are based on actual data and are tabulated in the ASME FFS-1 code. Appendix E in this book lists some materials and their corresponding A_i and B_i constants.

The quantity Δ_Ω^{sr} is for material data scatter band and ranges between −0.5 and +0.5 with −0.5 indicating the bottom of the scatter band and +0.5 indicating the top of the scatter band. A conservative value for life evaluation may be taken as −0.5 (the bottom of the scatter band) unless more data is available to the designer. Δ_Ω^{cd} is for material ductility factor and ranges between −0.3 and +0.3, with −0.3 indicating ductile behavior and +0.3 indicating brittle behavior. A conservative value for life evaluation is normally taken as 0.0 unless more information is available regarding the material.

Step 3

Calculate the triaxial adjustment factor δ_Ω and the Bailey Norton Coefficient, n_{BN}

$$\delta_\Omega = \frac{\sigma_1 + \sigma_2 + \sigma_3}{3\sigma_e} - 1/3 \tag{7.11}$$

$$n_{BN} = \frac{A_2 + 2A_3S_1 + 3A_4S_1^2}{460 + T}. \tag{7.12}$$

Step 4

Calculate the adjusted omega factors

$$n_{BN} = \frac{A_2 + 2A_3S_1 + 3A_4S_1^2}{460 + T} \tag{7.13}$$

$$\lambda = \delta_\Omega + 1$$

$$\Omega_m = \Omega_n^\lambda + (\alpha\Omega)(n_{BN}) \tag{7.14}$$

where

α_Ω = 3.0 for spherical heads
= 2.0 for cylindrical and conical shells
= 1.0 for other components.

Step 5

The expected life of the component due to creep is calculated from Eq. (7.5) as

$$L = \frac{1}{(\dot{\varepsilon}_{co})(\Omega_m)} \tag{7.15}$$

where
L = life in hours.

Step 6
Calculate the creep and fatigue fractions from the equations

$$D_c = t / L \tag{7.16}$$

and

$$D_f = n / N \tag{7.17}$$

where
D_c = the ratio of required operating hours to permissible hours
D_f = the ratio of required operating cycles to permissible cycles
L = permissible hours obtained from Eq. (7.15)
n = required operating cycles
N = permissible operating cycles obtained from fatigue curves
t = required operating hours.

Step 7
Use Figure 7.2 to determine creep-fatigue interaction acceptance
The envelope in Figure 7.2 can be expressed by the following equations

$$D_c = \frac{(D_{cm} - 1)D_f}{D_{fm}} + 1 \qquad 0 < D_f \le D_{fm} \tag{7.18}$$

$$D_c = \frac{D_{cm}(1 - D_f)}{1 - D_{fm}} \qquad D_{fm} < D_f \le 1.0. \tag{7.19}$$

The following example illustrates the applicability of the above equations to a cylindrical shell.

Example 7.1 The principal stresses on the inside surface of a cylindrical shell subjected to internal pressure are $\sigma_1 = 8.0$ ksi, $\sigma_2 = 4.0$ ksi, $\sigma_3 = -1.0$ ksi. Material of construction is 2.25Cr-1Mo annealed steel. Design temperature is 1000°F. Let $n = 1000$ cycles and $N = 90{,}000$ cycles. Determine if an expected life of the shell of 100,000 h is adequate for the following conditions:

a) $\Delta_\Omega^{sr} = -0.5$ (bottom of material scatter band) and $\Delta_\Omega^{cd} = 0.0$ (average between brittle and ductile material)
b) $\Delta_\Omega^{sr} = 0.0$ (average of material scatter band) and $\Delta_\Omega^{cd} = -0.3$ (ductile material)

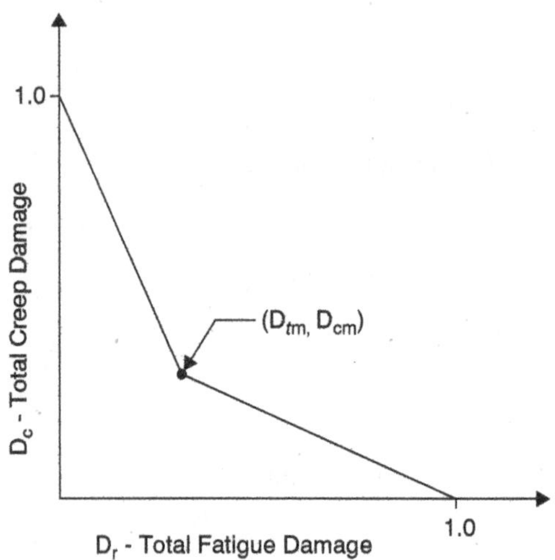

Figure 7.2 Creep-fatigue envelope [ASME, FFS-1].

Solution
Condition (a).

Step 1
From Eqs. (7.7) and (7.8),

$$\sigma_e = (0.707)\left[(8.0-4.0)^2 + (4.0-(-1.0))^2 + ((-1.0)-8.0)^2\right]^{0.5} = 7.809$$

$$S_1 = \log_{10}(7.809) = 0.893.$$

Step 2
The values of A_i and B_i for this material are listed in Appendix E.
From Eq. (7.9),

$$\begin{aligned}\text{Log}_{10}(\dot{\varepsilon}_{co}) &= -\{(-21.86+(-0.5))+[51{,}669+(-7597.4)(0.893)\\ &\quad +(-2131)(0.893)^2+(-199.3)(0.893)^3]/(460+1000)\}\\ &= -7.12493\\ \dot{\varepsilon}_{co} &= 10^{-7.12493}\\ \dot{\varepsilon}_{co} &= 7.50\times 10^{-8}\end{aligned}$$

From Eq. (7.10),

$$\begin{aligned}\text{Log}_{10}(\Omega) &= \{(-1.85+(0.0)) + [7195+(-2358)(0.893)\\&\quad +(-127.4)(0.893)^2+(56.2)(0.893)^3]/(460+1000)\}\\&= 1.5943\\\Omega &= 10^{1.5943}\\\Omega &= 39.29388\end{aligned}$$

Step 3

From Eqs. (7.11) and (7.12),

$$\begin{aligned}\delta_\Omega &= \frac{8.0+4.0-1.0}{3(7.809)} - 1/3\\&= 0.13607\\n_{BN} &= \frac{-7597.4+2(-2131)(0.893)+3(-199.3)(0.893)^2}{460+1000}\\&= 8.1356.\end{aligned}$$

Step 4

From Eqs. (7.13) and (7.14),

$$\begin{aligned}\Omega_n &= \max([(39.29388-8.1356),3.0])\\&= 31.1582\\\lambda &= 0.13607+1\\&= 1.13607\\\Omega_m &= 31.2411^{1.13607}+(2.0)(8.1356)\\&= 66.0224\end{aligned}$$

Step 5

The remaining life based on creep analysis is obtained from Eq. (7.15) as

$$\begin{aligned}L &= \frac{1}{(7.50\times10^{-8})(66.0224)}\\&= 201{,}950 \text{ h}.\end{aligned}$$

Step 6

Calculate the creep and fatigue fractions from Eqs. (7.16) and (7.17) as

$$\begin{aligned} D_c &= 100{,}000/201{,}950 \\ &= 0.50 \\ D_f &= 1000/90{,}000 \\ &= 0.011. \end{aligned}$$

Step 7

From Figure 7.2, $D_{fm} = 0.15$ and $D_{cm} = 0.15$ for this material.

From Eq. (7.18),

$$\begin{aligned} D_c &= \frac{(0.15-1)(0.011)}{0.15} + 1 \\ &= 0.94 > 0.50. \end{aligned}$$

Hence, the expected life of 100,000 h is acceptable.

Condition (b).

Step 1

From Eqs. (7.7) and (7.8),

$$\sigma_e = (0.707)\left[(8.0-4.0)^2 + (4.0-(-1.0))^2 + ((-1.0)-8.0)^2\right]^{0.5} = 7.809$$

$$S_1 = \log_{10}(7.809) = 0.893.$$

Step 2

The values of A_i and B_i for this material are listed in Appendix E.

From Eq. (7.9),

$$\begin{aligned} \text{Log}_{10}(\dot{\varepsilon}_{co}) &= -\{(-21.86+0) + [51{,}669 + (-7597.4)(0.893) + (-2131)(0.893)^2 \\ &\quad + (-199.3)(0.893)^3]/(460+1000)\} \\ &= -7.625. \end{aligned}$$

$$\dot{\varepsilon}_{co} = 10^{-7.625}$$

$$\dot{\varepsilon}_{co} = 2.372 \times 10^{-8}$$

From Eq. (7.10),

$$
\begin{aligned}
\text{Log}_{10}(\Omega) &= \{(-1.85+(-0.3)) + [7195+(-2358)(0.893)+(-127.4)(0.893)^2 \\
&\quad +(56.2)(0.893)^3]/(460+1000)\} \\
&= 1.29433 \\
\Omega &= 10^{1.29433} \\
\Omega &= 19.6936.
\end{aligned}
$$

Step 3

From Eqs. (7.11) and (7.12),

$$
\begin{aligned}
\delta_\Omega &= \frac{8.0+4.0-1.0}{3(7.809)} - 1/3 \\
&= 0.13607 \\
n_{BN} &= -\frac{-7597.4+2(-2131)(0.893)+3(-199.3)(0.893)^2}{460+1000} \\
&= 8.1356.
\end{aligned}
$$

Step 4

From Eqs. (7.13) and (7.14),

$$
\begin{aligned}
\Omega_n &= \max([(19.6936-8.1356),3.0] \\
&= 11.5579 \\
\lambda &= 0.13607+1 \\
&= 1.13607 \\
\Omega_m &= 11.5579^{1.13607} + (2.0)(8.1356) \\
&= 32.3965.
\end{aligned}
$$

Step 5

The remaining life based on creep analysis is obtained from Eq. (7.15) as

$$
\begin{aligned}
L &= \frac{1}{(2.372x10^{-8})(32.3965)} \\
&= 1{,}301{,}500\,\text{h}.
\end{aligned}
$$

Step 6

Calculate the creep and fatigue fractions from Eqs. (7.16) and (7.17) as

$$D_c = 100{,}000 / 1{,}301{,}500$$
$$= 0.08$$
$$D_f = 1000 / 90{,}000$$
$$= 0.011.$$

Step 7

From Figure 7.2, $D_{fm} = 0.15$ and $D_{cm} = 0.15$ for this material

From Eq. (7.18),

$$D_c = \frac{(0.15 - 1)(0.011)}{0.15} + 1$$
$$= 0.94 > 0.080.$$

Hence, the expected life of 100,000 h is acceptable.

Notice the substantial increase in expected life in Condition (b) compared to Condition (a) due to change in ductility and material scatter band. The designer's choice of the appropriate values of $\Delta_\Omega{}^{sr}$, and $\Delta_\Omega{}^{cd}$ becomes extremely important in evaluating creep life using the remaining life method.

7.3 Equations for Constructing Ishochronous Stress-Strain Curves

Isochronous stress-strain curves for any temperature and time for the materials listed in Appendix E can be constructed in the remaining life method. The isochronous stress-strain curves are a composite of elastic, plastic, and creep strains. The pertinent equations are detailed in this section.

The total strain is given by

$$\varepsilon_t = \varepsilon_e + \varepsilon_p + \varepsilon_c \tag{7.20}$$

where

ε_c = creep strain

ε_e = elastic strain

ε_p = plastic strain

ε_t = total strain.
The equation for elastic strain ε_e is defined by

$$\varepsilon_e = \sigma / E \tag{7.21}$$

where
E = modulus of elasticity
σ = effective stress.
The equation for plastic strain ε_p is expressed as

$$\varepsilon_p = \gamma_1 + \gamma_2. \tag{7.22}$$

The expressions for γ_1 and γ_2 are fairly lengthy and involve many constants as a function of yield stress and tensile strength given in FFS-1. Substituting the various constants in the expression of γ_1 gives

$$\gamma_1 = 0.5\left(\sigma_t / \alpha_1\right)^{(1/m_1)}\left\{1 - \tanh\left[2\left(\sigma_t - \alpha_2\right)/\alpha_3\right]\right\} \tag{7.23}$$

where

$$R = \sigma_{ys} / \sigma_{ult}$$

$$m_1 = \frac{[\ln(R) + (\varepsilon'_p - \varepsilon_{ys})]}{\ln\left[\frac{\ln(1 + \varepsilon'_p)}{\ln(1 + \varepsilon_{ys})}\right]}$$

$$\alpha_1 = [\sigma_{ys}(1 + \varepsilon_{ys})] / [\ln(1 + \varepsilon_{ys})]^{m_1}$$

$$K = 1.5R^{1.5} - 0.5R^{2.5} - R^{3.5}$$

$$\alpha_2 = \alpha_{ys} + K(\alpha_{ult} - \alpha_{ys})$$

$$\alpha_3 = K(\alpha_{ult} - \alpha_{ys}).$$

Similarly, the expression for γ_2 is

$$\gamma_2 = 0.5\left(\sigma_t / \alpha_4\right)^{(1/m_2)}\left\{1 + \tanh[2(\sigma_t - \alpha_2)/\alpha_3]\right\} \tag{7.24}$$

where

$$\alpha_4 = [\sigma_{uts}\exp(m_2)] / [m_2^{m_2}]$$

m_2 = material parameter given in Appendix E.

ε'_p = material parameter given in Appendix E.

There is a temperature limitation on ε'_p and m_2. For example, the values for ε'_p and m_2 for low and high-alloy steels are limited to 900°F.

The creep strain, ε_c, is obtained from Eq. (7.6) where the quantity Δ_Ω^{sr} is for material data scatter and ranges between −0.5 and + 0.5. The value normally used for constructing average isochronous stress strain curves is +0.5. Δ_Ω^{cd} is for material ductility factor and ranges between −0.3 and + 0.3. A conservative value used for constructing average isochronous stress strain curves is normally taken as 0.0 for new materials.

Hence, isochronous curves may be drawn from Eq. (7.20) where ε_e is obtained from Eq. (7.21), ε_p is obtained from Eqs. (7.22) – (7.24), and ε_c is obtained from Eqs. (7.6) and (7.8) through (7.10).

Example 7.2 Construct an isochronous stress-strain curve for Type 304 stainless steel material at 1200°F and 100,000 h.

Solution

The total strain is given by Eq. (7.20).

The elastic strain, ε_e, is given by Eq. (7.21), with the elastic modulus obtained From ASME II-D as E = 21,200 ksi.

The plastic strain, ε_p, is given by Eq. (7.22). However, it is set to zero in this example since the material constants m_2 and ε'_p are limited to 900°F, which is less than the given temperature of 1200°F for this problem.

The creep strain, ε_c, is given by Eq. (7.6).

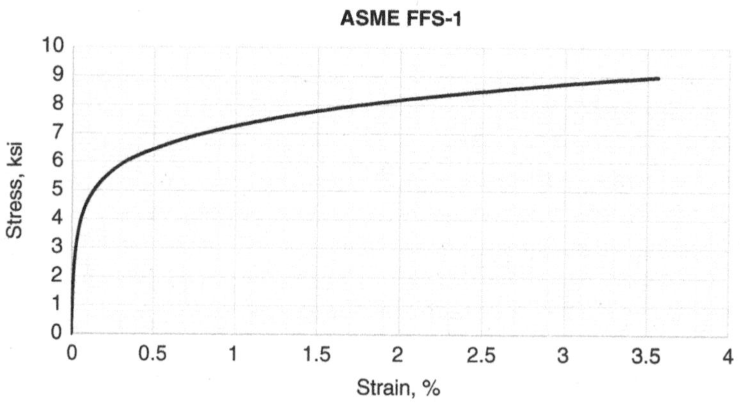

Figure 7.3 Isochronous stress-strain curve for Type 304 stainless steel at 1200°F and 100,000 h.

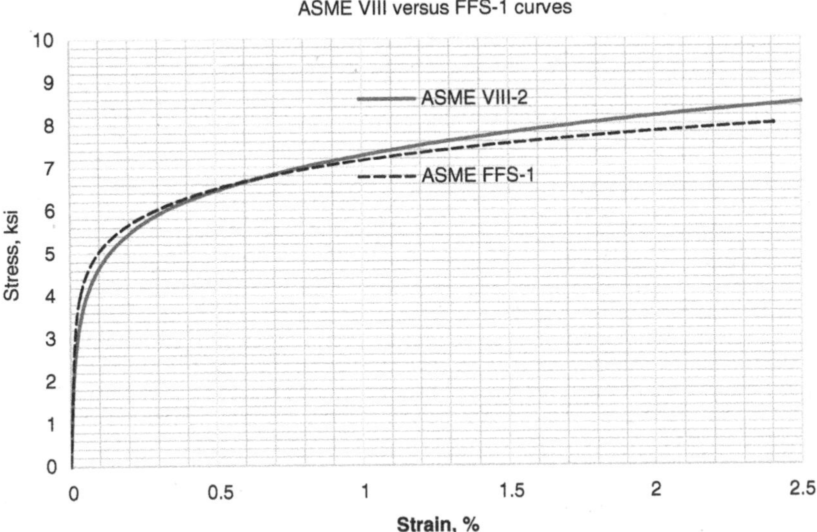

Figure 7.4 Comparison of isochronous stress-strain curves for Type 304 stainless steel at 1200°F and 100,000 h obtained from ASME VIII-2 and FFS-1.

The sum of the elastic and creep strains are then obtained for various stress values and an average isochronous stress-strain diagram is then drawn as shown in Figure 7.3.

A comparison of the isochronous curve obtained from FFS-1 and that obtained from ASME VIII-2 is shown in Figure 7.4.

Given the scatter in the data and the different methodologies used to generate these curves, this is considered to be good agreement.

8

Nuclear Components Operating in the Creep Regime

Experimental Breeder Nuclear Reactor [Argonne National Laboratory]

8.1 Introduction

Construction rules for high temperature reactors (HTRs) are published in ASME III-5, addressing both high temperature gas-cooled reactors (HTGRs) and high temperature liquid-cooled reactors (HTLRs). The rules address both low temperature operating conditions, typically 700°F (370°C) or less, for carbon and martensitic steels and 800°F (425°C) or less for austenitic or high

Analysis of ASME Boiler, Pressure Vessel, and Nuclear Components in the Creep Range, Second Edition. Maan H. Jawad and Robert I. Jetter.
© 2022 John Wiley & Sons Ltd. Published 2022 by John Wiley & Sons Ltd.

nickel alloys as well as elevated temperatures conditions and both metallic and nonmetallic materials. Effective from the 2015 Edition, the ASME III-NH was discontinued and the subject matter transferred to ASME III-5, Subsection HB.

The scope of the rules for nuclear components in ASME III is significantly broader than that of ASME I and VIII. ASME III Division 1 has separate subsection books for different component classes. These classes are classified as follows Byk (2018):

> *Subsection NB addresses items which are intended to conform to the requirements for Class 1 construction – Components that are part of the primary core cooling system.*
>
> *Subsection NC addresses items which are intended to conform to the requirements for Class 2 construction – Components that are part of various important-to-safety emergency core cooling system.*
>
> *Subsection ND addresses items which are intended to conform to the requirements for Class 3 construction – Components that are part of the various systems needed for plant operation.*
>
> *Subsection NE addresses items which are intended to conform to the requirements for Class MC or metal containment construction.* They are normally a gas-tight shell or other enclosure around a nuclear reactor to confine fission products that otherwise might be released to the atmosphere in the event of an accident. Such enclosures are usually dome-shaped and made of steel or steel-reinforced concrete.

The component classification – for example Class 1, Class 2, Class 3, etc. – is intended to recognize the different levels of importance associated with each class with regard to the safe operation of the nuclear facility. It is the responsibility of the owner to apply system safety criteria to classify equipment, which is required in the component's Design Specification. The range of component categories is also much broader. Thus, at temperatures below the creep range, ASME III-NB (the highest safety classification) categorizes components as NB-3300 Vessels, NB-3400 Pumps, NB-3500 Valves, and NB-3600 Piping. ASME III-NCD cover similar components at a lower safety classification. (ASME III-C and D have been combined into a single subsection covering both C and D due to their significant similarities.) ASME III-NG covers core support structures, ASME III-NF covers component supports and ASME III-NH (prior to 2015) covered Elevated Temperature Class 1 components; that is, vessels, pumps, valves and piping. There were additional code cases covering elevated temperature Class 2 and 3 and Core Support Structures.

With the exception of the rules for graphite components, ASME III-5 was written basically as an editorial compilation of existing subsections and code cases addressing components that operate at both conventional (below the creep regime) and elevated temperatures. Thus ASME III-NB, NCD, NF, and NG are referenced completely or in part; the older nuclear code cases for elevated temperature components were incorporated into ASME III-5 (and then the code cases themselves were annulled); and, the aforementioned graphite rules were included in the main text of Division 5. Therefore, for metallic components, what may have appeared to be new division rules were actually just the existing rules reformatted, and adjusted to meet the Division 5 structure and terminology where appropriate. For example, to differentiate the rules for high temperature reactor types, components identified as Class 1 in ASME III-NB are identified as Class A under the provisions of ASME III-5. Similarly, components identified as Class 2 in ASME III-NCD are identified as Class B in ASME III-5. There is no equivalent of Class 3 in ASME III-5 – the rules for Class B would be applicable.

Any alterations to the existing ASME III-1 rules are specifically identified in the corresponding ASME III-5 subsections, via the renumbered paragraphs and subparagraphs. These ASME III-5 Subsections HA, HB, HC, HF, and HG, Subparts A (applying below the creep regime) are very short (six pages or less). When incorporating the existing rules of the older code cases (some dating back to the early 1980s), new ASME III-5 paragraphs and subparagraphs were created, but the rules are essentially the existing rules. ASME III-5 Subsections HB, HC, and HG, Subparts B (applying at elevated temperature) have more pages since the existing code case rules were incorporated directly into ASME III-5.

The purpose of this chapter is to focus, in a summary fashion, on the rules for design, and to a lesser extent, materials, in ASME III-5. Basically, Code Case 2843, which is applicable to ASME VIII-2, is a "cleaned-up," and somewhat simplified, version of the rules from ASME III-NH. As discussed above, the NH rules are now contained in ASME III-5, so the preceding chapters discussing Code Case 2843 are generally applicable to ASME III-5, Subsection HB, Subpart B (hereinafter referred to as HBB).

8.2 High Temperature Reactor Characteristics

In the last several years there has been worldwide renewed interest in HTRs; in particular, the HTGR, the liquid metal reactor (LMR), and more recently, the molten salt reactor (MSR). The HTGR is characterized by high gas outlet temperatures, 1380°F to 1740°F (750°C to 950°C), and moderately high pressure, about 1000 psi (7MPa). However, most of the code pressure boundary components are shielded from these very high gas temperatures by circulating the cooler core inlet flow next to the code pressure boundary. The exception is the core support structure and the heat transfer interface between the primary and secondary heat transport systems. Because of the massive core heat capacitance, thermal transients are generally quite slow. The LMR operating temperature is generally in the range of 930°F to 1020°F (500°C to 550°C) and much lower pressure, about 150 psi to 300 psi (1 MPa to 2 MPa). Because of the relatively low heat capacitance of the core, and high heat capacitance of the liquid metal coolant, thermal transients can be quite rapid, particularly for Service Level B and C events. The MSR operating temperature is generally hotter than the range of the LMR, 1100°F–1550°F (600°C to 850°C), but the operating pressure and thermal transients are similar. For metallic components, although there are major differences in configuration and operating conditions between the HTGR, LMR, and MSR, the underlying design technology, materials, and fabrication and examination requirements have much in common such that the most of the code rules will apply to both systems. The following figures are representations of the various reactor types. Figure 8.1 shows a conceptual design of a sodium cooled reactor. Figure 8.2 shows a sketch of a molten salt reactor and Figure 8.3 shows an outline of a gas cooled reactor.

Figure 8.1 GE Hitachi PRISM – sodium cooled nuclear reactor [GE Power].

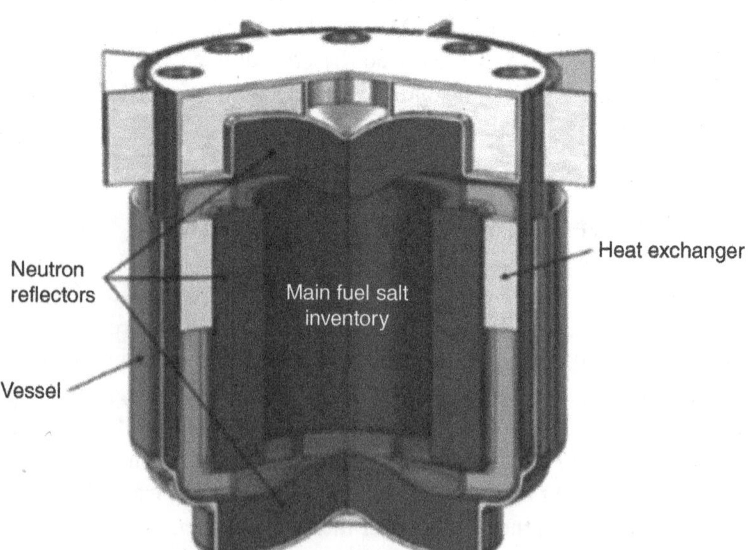

Figure 8.2 Terra Power MCFR – molten salt nuclear reactor [Terra Power].

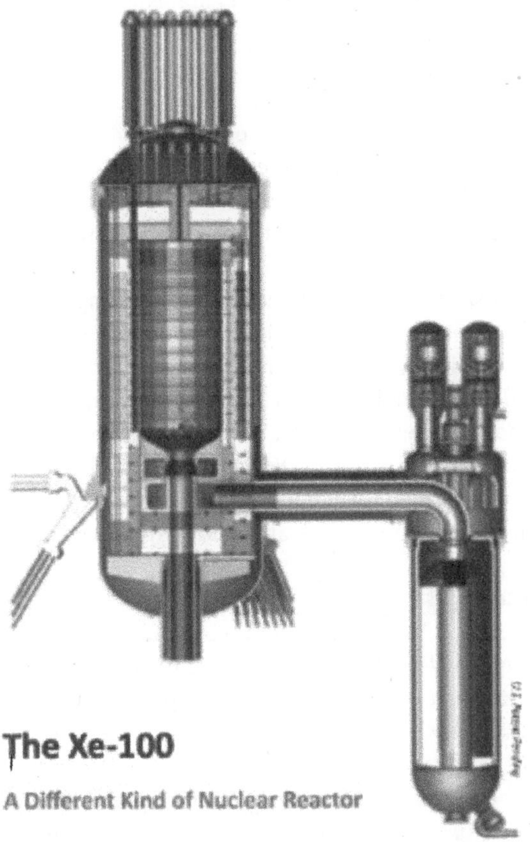

Figure 8.3 X-Energy Xe 100 – gas cooled nuclear reactor [X-Energy].

8.3 Materials and Design of Class A Components

8.3.1 Materials

There are currently six base materials and three bolting materials approved for ASME III Class A construction at elevated temperature. The base materials are Types 304 and 316 stainless steel, Alloy 800H, 2.25Cr-1Mo (Gr22) annealed and 9Cr-1Mo-V (Gr91) steel, and recently through nuclear Code Case N-898, Alloy 617. The bolting materials are Types 304 and 316 stainless steel, and Alloy 718. The relatively small number of allowable materials is due to the extensive testing required to qualify a material for Class A construction. Guidelines for data needs are provided in Appendix HBB-Y of ASME III-5. Unlike Code Case 2843 (shown in Appendix A of this book), except for data in the standard tables in

ASME II-D, all the allowable stress values and material parameters for implementation of the Class A rules are provided in HBB of ASME III-5.

8.3.1.1 Thermal Aging Effects

One of the features of both HBB of ASME III-5 and Code Case 2843 of ASME VIII-2 is the use of time-dependent allowable stress criteria for operating loads. This avoids the dilemma of using allowable stresses based on long-term creep properties for loading conditions where there is little creep; for example, a short duration over temperature, or no creep, such as a seismic event. However, evaluation of short-term events requires accounting for the potential softening effect of long-term thermal aging on yield and ultimate tensile strength. This is accomplished in Code Case 2843 in Paragraph 4.1 and in HBB-2160 of ASME III-5. The reduction factors in Code Case 2843 of ASME VIII-2 are a function of temperature only, similar to the factors for Types 304 and 316 stainless steel and Alloy 800H in HBB of ASME III-5. However, the factors for 2.25Cr-1Mo annealed and 9Cr-1Mo-V steel in HBB of ASME III-5 are a function of both temperature and cumulative service life.

8.3.1.2 Creep-Fatigue Acceptance Test

Another difference between ASME VIII-2 and III-5 is that there is a creep-fatigue acceptance test for Types 304 and 316 stainless steel in HBB of ASME III-5. This requirement was based on very poor creep-fatigue performance of a particular heat of 304 stainless steel.

8.3.1.3 Restricted Material Specifications to Improve Performance

There is also a non-mandatory Appendix HBB-U in ASME III-5 Guidelines for Restricted Material Specifications to Improve Performance in Certain Service Applications. The incentive to optimize specifications for Types 304 and 316 stainless steel arose from observations that the mechanical properties on these steels vary considerably from heat to heat, particularly within the creep range. The issue that brought this concern into focus was a debate on the proper choice of flow and strength characteristics to be used when performing inelastic analyses to satisfy the strain and creep-fatigue damage limits of Appendix HBB-T in ASME III-5. The choice is complex. For example, a low assumed yield strength can maximize predicted ratcheting and strain accumulation, but it will tend to minimize residual stresses and computed creep-rupture damage. It is also not necessarily true that the material with the highest flow resistance will have the highest damage resistance. The effects can also be shown to depend on the magnitude of the applied loading. It was decided that the best approach was for the inelastic analyses to be performed using the most probable (nominal) behavior and to have the various safety and design factors account for scatter. Thus, one way to improve design confidence is to reduce scatter in the flow properties and optimize the strength characteristics.

The intent of Appendix HBB-U of ASME III-5 is to recommend more restrictive composition, melting, and fabrication practices within existing specifications for austenitic stainless steels to reduce data scatter, particularly in high-temperature creep-related properties. These changes were not intended to significantly modify the strengths. In considering most of the mechanical properties of interest to the designer, grain size variations have been shown to be a particularly significant cause of heat-to-heat variability. Generally, a small grain size is most desirable to optimize behavior; however, it may be impractical in product forms such as forgings to specify a grain size range. The most cost-effective measures that can be taken have specifications that involve "good practice" and that reduce the range of allowable chemical composition variables on those elements with the greatest influence, particularly on creep-related properties.

8.3.2 Design by Analysis

The design requirements in Code Case 2843 of ASME VIII-2 are generally the same as the requirements in HBB of ASME III-5 for HBB-3200 *Design by Analysis*, and in Appendix HBB-T *Rules for Strain, Deformation, and Fatigue at Elevated Temperature*. However, the HBB rules are more comprehensive. For example, the load-controlled limits in the creep regime in ASME VIII-2 are given in Figure 4.8, which is a simplified version of the one given in HBB-3200 of ASME III-5 and shown in Figure 8.4. Some additional differences are discussed subsequently.

8.3.2.1 Equivalent Stress Definition

One important difference between ASME VIII-2 and III-5 is the definition of equivalent stress. In VIII-2, by reference to VIII-2 Part 5, the equivalent stress is given by the von Mises formula:

$$\sigma_{eff} = (1/\sqrt{2}).[(\sigma_1 - \sigma_2)^2 + (\sigma_2 - \sigma_3)^2 + (\sigma_3 - \sigma_1)^2]^{1/2} \tag{8.1}$$

where σ_1, σ_2 and σ_3 are principal stresses.

Conversely, for ASME III-5 the equivalent stress is given by the Tresca definition, the largest absolute value of the maximum of the principal differences: $(\sigma_1 - \sigma_2)$, $(\sigma_2 - \sigma_3)$ and $(\sigma_3 - \sigma_1)$.

For both ASME III-5 and VIII-2 the required thickness of a pressurized circular cylinder remote from discontinuities is governed by:

$$P_m \leq S_{mt} \tag{8.2}$$

where P_m is the general primary membrane intensity and S_{mt} is the associated allowable stress for the applicable temperature and time.

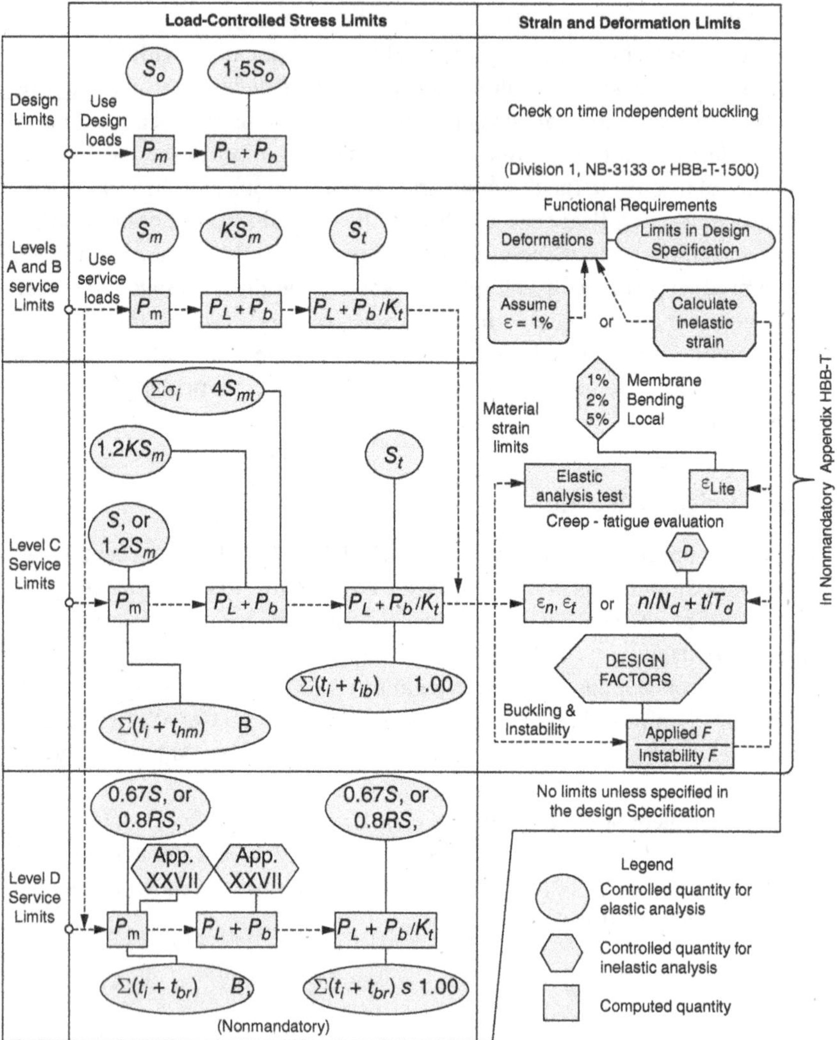

Figure 8.4 Flow diagram for elevated temperature design (ASME III-5).

For a pressurized cylinder remote from discontinuities, the principal membrane stresses in the tangential, longitudinal, and radial directions are given by:

$$\sigma_1 = \sigma_t = P/(\gamma - 1) \tag{8.3}$$

$$\sigma_2 = \sigma_l = P/(\gamma^2 - 1) \tag{8.4}$$

$$\sigma_3 = -P/2 \tag{8.5}$$

where P = pressure, $\gamma = R_o/R_i$, and R_o and R_i are the outer and inner radii of the cylinder respectively.

For the ASME III-5 Tresca criteria the governing difference in principal stresses is given by

$$(\sigma_3 - \sigma_1) = P(\gamma + 1)/[2(\gamma - 1)]. \tag{8.6}$$

This can also be expressed in terms of the inner radius R_i as:

$$(\sigma_3 - \sigma_1) = PR_i/t + P/2. \tag{8.7}$$

This can be further simplified as:

$$(\sigma_3 - \sigma_1) = PR_m/t \tag{8.8}$$

where $R_m = (R_o + R_i)/2.$

Equations (8.6), (8.7), and (8.8) are equal and represent the value of the primary general membrane stress in a pressurized cylinder in terms of γ, R_i, and R_m, respectively, using the Tresca equivalent stress.

For a thin shell, $\gamma = 1.1$ and, substituting into Eq (8.1) and (8.6), gives the following ratio of calculated general primary membrane stress, P_m, based on von Mises used in ASME VIII-2 as compared to the Tresca used in ASME III-5:

$$(P_m)_{\text{VIII-2}}/(P_m)_{\text{III-5}} = 0.866.$$

Which means that the calculated stress using von Mises equivalent stress is 13% lower than the Tresca equivalent stress for a ratio of outer radius to inner radius of 1.1.

The above example is based on the general primary membrane stress, P_m. The principal stress components that constitute P_m are the same for both ASME VIII-2 and III-5. The difference is the formula for equivalent stress.

In ASME VIII-1 and I there are explicit formulas provided for calculating required wall thickness. That design procedure is referred to as Design-By-Rule (DBR) and is discussed more thoroughly in Chapter 4.

8.3.2.2 Rules for Bolting

Another difference between ASME VIII-1 and III-5 is that HBB of III-5 provides specific rules for bolts, HBB-3230 *Stress Limits for Load-Controlled Stresses in Bolts,* including allowable stress values. These rules address the allowable stress due to pressure loading as well as those produced by a combination of preload, pressure, and thermal expansion. The allowable pressure-induced stress is half that for the comparable base material and the allowance for the combined load is twice the value for pressure only.

8.3.2.3 Weldment Strength Reduction Factors

Weldments are another area in which there are differences between Code Case 2843 of ASME VIII-2 and HBB of ASME III-5. In both rules, the weld is modeled as base metal in determining the stress distribution in the component. Also, in both cases there is consideration of the potential for reduced creep rupture strength in the deposited weld metal. Both rules provide tables with allowable weld materials and their associated ratios of weld creep-rupture strength to the base metal creep rupture strength. However, there are additional weldment considerations in HBB-T-1700 of ASME III-5. In the vicinity of the weld the allowable number of design cycles is reduced to one-half of the value permitted for base material and the allowable strain limits are also reduced by one half.

8.3.2.4 Constitutive Models for Inelastic Analysis

In HBB of ASME III-5, the use of inelastic analysis is specifically addressed for the evaluation of strain limits and creep-fatigue damage evaluation, HBB-T-1310 and HBB-T-1420, respectively. General guidance is provided for the characteristics of constitutive models for inelastic analysis and specific guidance and models for some materials are provided in ASME III-5 Appendix Z.

8.3.2.5 A-1, A-2, and A-3 Test Order

At a more detailed level, another difference between ASME VIII-2 and III-5 is the order in which the various procedures for evaluation of strain limits based on elastic analysis are provided. In III-5 the first two procedures, Test A-1 and Test A-2, are applicable to the creep regime and Test A-3 is applicable to the regime where creep effects are negligible, based on the criteria provided therein. However, in VIII-2, Test A-1, technically virtually the same as Test A-3 in III-5, comes first followed by Tests A-2 and A-3, which are virtually the same as Tests A-1 and A-2 in III-5. The purpose of the reordering was to put the simpler-to-execute rules applicable to negligible creep first among the options.

8.3.2.6 Determination of Relaxation Stress, S_r

There is also a technical simplification applicable to the A-1 negligible creep regime in ASME VIII-2. The evaluation of primary plus secondary stress limits $(P + Q)$ requires evaluation of a creep relaxation modified elastic shakedown limit, $3\overline{S}_m$, instead of the traditional $3S_m$ limit below the creep regime. In ASME III-5 this requires determination of the relaxation strength, S_r, the remaining value of stress due to relaxation at the end of a design life. Procedures are given for determination of S_r in ASME III-5 but tabulated values are not provided. In

ASME VIII-2 the value of $0.5S_t$, the time-dependent allowable stress for the design life, is provided as a conservative and convenient approximation of S_r.

8.3.2.7 Buckling and Instability

Buckling and instability rules are not addressed in Code Case 2843 of ASME VIII-2. They are addressed in Code Case 2964 of ASME VIII-1 and in Chapter 9 of this book. The rules in HBB of ASME III-5 consider both time-independent, negligible creep, and time-dependent buckling where creep is significant. An example is shown in Figure 8.5 for axial compression of cylindrical shells. The figure shows the time below which the rules of ASME III-NB below creep are applicable and above which creep-buckling in accordance with ASME III-5 is applicable for various materials at given temperatures. The time-dependent rules for buckling are based on inelastic analysis and load factors are specified for various loading conditions.

The approach to evaluation of buckling and instability is different in III-5. The rules for the creep regime are based on inelastic analysis using load factors for the various conditions under consideration. In addition, time-temperature charts are provided that permit the use of the time-independent buckling design limits in ASME III-1 NB-3313, when creep is not significant.

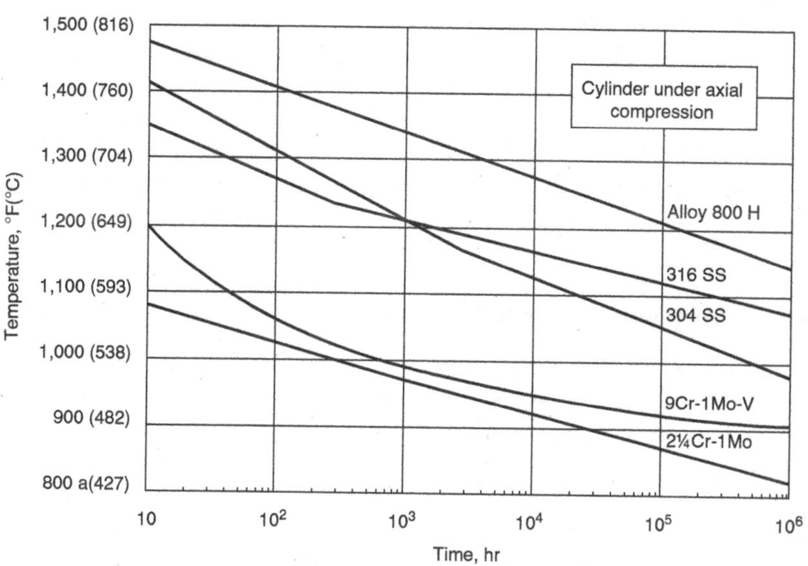

Figure 8.5 Time-temperature limits for application of creep buckling analysis (ASME III-5).

8.3.2.8 D Diagram Differences

The creep-fatigue damage interaction diagrams in ASME III-5, VIII-2, and FFS-1 also have differences. For ASME III-5 and VIII-2 the difference is in the curve for 9Cr-1M0-V (Gr91). In ASME III-5 this curve was based on experimental data from the United States and Japan. Because of the strong strain-softening effect in Gr91 it is difficult to assess a method for the separate effects of fatigue and creep damage that is consistent with the various creep-fatigue damage evaluation methods, so the ASME III-5 approach was to reflect the most conservative approach – intending to update as needed. A step in that direction was taken in ASME Code Case N-812-1, where it was shown that the approach to creep-damage calculation based on determination of the stress relaxation based on isochronous stress-strain curves, the most common approach, could conservatively be correlated with an 0.3, 0.3 damage intercept vs. the 0.01, 0.1 reference intercept in ASME III-5. Further, the corresponding design code in Japan uses an intercept of 0.3, 0.3 for Gr91. On that basis, the intercept selected for use in ASME VIII-2 was taken to be 0.3, 0.3. The creep-fatigue intercepts in ASME FFS-1 were developed independently and contain data from component service exposed material, as well as virgin material used for ASME III-5 and VIII-2, which accounts in part for the differences with the ASME FFS-1 isochronous stress-strain curves.

8.3.2.9 Isochronous Stress-Strain Curve Differences

In the current ASME III-5 (2021 edition) the most enduring isochronous stress-strain curves for 2.25Cr-1Mo annealed steel are for 300,000 h. However, for earlier editions (2017 and earlier) the longest time curves were for 500,000 h. The current isochronous curves for ASME VIII-2 were based on the earlier versions of ASME III-5 and have the 500,000 curves for 2.25Cr-1Mo annealed steel. The change to 300,000 h was made to bring the curves for 2.25Cr-1Mo annealed steel in agreement with the other materials shown in ASME III-5.

8.3.3 Component Design Rules

In addition to the design by analysis rules, HBB of ASME III-5 has specific component design rules, as described below.

Vessel design is covered in HBB-3300. With minor modification, the ASME III-NB rules for reinforcement of openings apply to HBB. For deformation-controlled limits in HBB, the stress indices from NB-3338 may be used to determine stress components. Note that HBB-3353 provides requirements for defining the surface geometry of welds to be used in meeting HBB deformation-controlled limits.

Pump and valve designs are covered in HBB-3400 and HBB-3500 of ASME III-5. The analysis methods in NB-3400 and NB-3500 of ASME III-1 apply to elastic analysis for load-controlled limits. Though limited in scope, the NB analysis methods may be used for deformation-controlled limits only if creep effects are negligible.

Piping design is covered in HBB-3600. For piping, the primary plus secondary stress indices (B and C) and associated stress equations may be used for load-controlled stress limits and deformation-controlled strain limits. The stress indices for detailed analysis from NB-3684 and NB-3685 may also be used for strain limits but only these detailed indices may be used for creep-fatigue evaluation. The restriction on the applicability of the B and C indices is due to the fact that they do not represent a specific stress component and location, as is required for the creep-fatigue evaluation. In the case of their applicability to strain limits, it was judged that the indices were sufficiently conservative for use.

8.4 Class B Components

The rules for construction of Class B elevated temperature components were previously contained in a series of code cases. These code cases have been consolidated in ASME III-5 Subsection HC, "Class B Metallic Pressure Boundary Components," Subpart B, "Elevated Temperature Service," including three supporting appendices. Articles HCB-2000, -3000 and -4000 – for materials, design, and fabrication and installation – and their associated appendices are covered in more detail below, as are HCB-6000 and -7000 for testing and overpressure protection. Revised or supplemental requirements are with respect to the baseline requirements in Subsection NC.

8.4.1 Materials

The currently allowable materials for ASME III-1 Class B components at elevated temperature are provided in Appendix HCB-II of ASME III-5. The materials in Appendix HCB-II correspond generally to those used at lower temperatures in ASME III-NCD. In Appendix HCB-II, two sets of allowable stresses are provided: one set, HCB-II-2000, is for the case where, due to the short time at the temperature defined in Appendix HCB-III, creep effects are not significant; and the other, HCB-II-3000, is for the general case where creep is significant. The allowable stresses in Appendix HCB-II are based on the same criteria as those used to develop the allowable stresses for pressure vessels constructed in accordance with the rules for ASME VIII-1, except that reduction factors are also provided to account for the reduction in creep rupture strength of weld metal and weldments. Similar to ASME VIII-1, the allowable stresses in HCB-3000 of ASME III-5 are not a function of load duration. Because of the complexity of when certain factors apply and to which allowable stress value, a flow chart showing their applicability is included in Appendix HCB-II (concurrent with the 2015 edition, the allowable stresses were expanded to include Service Level D).

8.4.2 Design

The design approach in HCB-3300, -3400, and -3500 of ASME III-5 is based on extending the design-by-formula rules in NCD-3300, NCD-3400, and NCD-3500 of ASME III-1 to elevated temperatures using the allowable stresses in Appendix HCB-II of ASME III-5. Note that these rules do not explicitly address fatigue damage resulting from cyclic service.

The approach to piping structural design criteria, HCB-3600 of ASME III-5, is somewhat different than for the other components since a primary loading mechanism in piping systems, particularly at elevated temperature, is restrained thermal expansion. Two approaches are provided. The first applies when creep effects are negligible in accordance with Appendix HCB-III of ASME III-5. In this case, the design rules of NCD–3600 of ASME III-1 apply using the appropriate allowable stresses from Appendix HCB-II of ASME III-5. If creep effects are not negligible, the design-by-analysis procedure of HCB-3630 of ASME III-5 applies. The piping design-by-analysis procedures are also an extension of the rules in NCD–3600 of ASME III-1 using the allowable stresses from Appendix HCB-II of ASME III-5; however, in this case they are modified to account for creep effects. The most significant modification for creep effects in HCB–3630 of ASME III-5 is the definition of the stress reduction factor "*f*," as shown in Appendix HCB-I of ASME III-5. Conceptually, this is an extension of the definition of the stress reduction factor in NCD-3611.2 of ASME III-1. However, for elevated temperatures, the factor, r_1, has been modified to include a term to account for the higher of the peak stresses due to either the through-the-wall temperature gradients or the axial temperature difference. The second modification is in the stress reduction factor Table HCB-I-2000-1 and Table HCB-I-2000-2 of ASME III-5 for N_1. Both of these tables have been modified to take into account the effects of creep on cyclic life. Depending upon the material and service temperature the effects can be quite significant.

The rules for buckling and instability are given in HCB-3130 of ASME III-5 for all components. As usual, there are two approaches. If creep is not significant per Appendix HCB-III of ASME III-5, then the rules of NCD-3133 of ASME III-1 may be used. However, when creep is significant the rules are similar to those in Subsection HB, Subpart B.

Effective from the 2017 edition, the Subsection HB Class A rules of ASME III-5 may be used for Class B construction and still stamped as a Class B component. This permits the evaluation of cyclic service in accordance with the more detailed requirements of Class A. However, it is important to note that it is necessary to comply with all the construction requirements of Class A to take advantage of this option.

8.5 Core Support Structures

The rules for construction of elevated temperature core support structures are provided in ASME III-5 Subsection HG, Subpart B *Class SM Metallic Core Support Structures*. These rules were formerly provided in Code Case N-201-1. More specifically, general rules for elevated temperature are provided in Articles HGB-1000 through 8000. However, for materials for which creep effects are negligible in accordance with the time-temperature limits of Mandatory Appendix HGB-IV, an alternative set of rules are provided in Mandatory Appendix HGB-II. The rules of HGB-1000 to 8000 apply when the time-temperature limits of Appendix HGB-IV are not satisfied and the effects of creep must be accounted for.

The materials used in HGB-2000 and 3000 are essentially the same as those in Subsection HB, Subpart B, except that Alloy 718 is not included as an acceptable material for bolting. Generally, the design rules and allowable stress values follow the rules and allowable stresses in Subsection HB, Subpart B. The deformation-controlled loading requirements in Mandatory Appendix HGB-I, "Rules for Strain, Deformation, and Fatigue Limits at Elevated Temperature" are the same as the rules in Nonmandatory Appendix HBB-T. One notable exception is the design rules for bolting, which are referred to as threaded structural fasteners. The core support rules permit significantly higher design and operating stresses, provided that all loading conditions are identified and the strain and deformation limits and creep-fatigue damage are evaluated, including the effects of ratcheting, creep, and eventual retightening. The fabrication and installation rules, and the requirements for examination, closely follow the requirements in Subsection NG as supplemented by the rules of HBB-4000 and 5000, respectively; however, the requirement for double volumetric examination (radiographic as well as ultrasonic) only applies to Category A welds, as opposed to the more general requirements in HBB-5000. Also, as is the case for Subsection NG, the rules for testing and overpressure protection are not applicable to core support structures.

As noted above, Appendix HGB-II provides rules when creep effects are negligible, in accordance with Appendix HGB-IV. The allowable stresses for the materials permitted in Appendix HGB- IV are based on the criteria for S_m used in Subsections NB and Subsection HB, Subpart B. Delta ferrite restrictions are added for weld rod material. With a few exceptions, the design rules in Appendix HGB-II are the same as those in Subsection NG. Appendix HGB- II supplements the rules for external pressure design in Subsection NG with more general rules in Appendix HBG-III that are applicable to geometry and loading conditions not covered by Subsection NG. Tables are provided in Appendix HBG-II that extend the fatigue design curves in Subsection NG for the condition where creep effects are negligible, as per Appendix HGB-IV.

9

Members in Compression

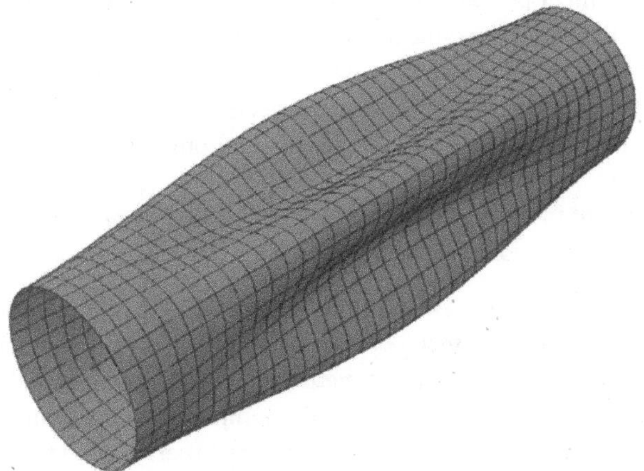

Shell buckled in lobe mode [Chithranjan Nadarajah]

9.1 Introduction

Creep-buckling phenomena can be thought of in terms of the shell or column creeping under a sustained compressive load up to a deformation level where instantaneous buckling takes place. Accordingly, a time factor must be included in the buckling equations. The analysis of shells and columns operating in the creep regime Odqvist (1966) is extremely complicated due to numerous factors, some of which are not necessarily well known. These include eccentricity, the effect of primary and secondary creep on buckling, the interaction between

Analysis of ASME Boiler, Pressure Vessel, and Nuclear Components in the Creep Range, First Edition. Maan H. Jawad and Robert I. Jetter.
© 2022 John Wiley & Sons Ltd. Published 2022 by John Wiley & Sons Ltd.

elastic and creep buckling, and the properties of material over time at elevated temperatures. These factors require numerous assumptions on the part of the designer in order to develop design criteria for compression in the creep range. Some of these factors and assumptions that are of particular interest to the designer are:

- Creep buckling in shells and columns with an eccentricity will occur at any axial compressive load Kraus (1980) no matter how small it is when the shell or column are subjected to temperatures in the creep range over a certain period.
- Creep buckling does not occur instantaneously Boyle and Spence (1983) but rather after a certain time lapse, when the imperfection grows and then results in instantaneous buckling. Thus, the design premise aims to determine a *critical* period that is longer than the intended service of a component, rather than the *critical* buckling force.
- The elastic buckling equation does not necessarily apply in the creep range due to the nonlinear relationship between stress and strain.
- Creep buckling occurs at a specified time only for nonlinear stress-strain behavior.
- The tangent modulus, E_t, obtained from the isochronous curves, gives a simplified equation for designing in the creep regime.

Since buckling analysis in the creep regime is an extremely complicated subject, different simplified methods of analysis have been proposed and are available to the designer. Each of these methods has its advantages and disadvantages regarding the accuracy of results and the complexity of the solution. The simplified methods presented in this chapter are easy to perform but fairly conservative. They are based on the methodology used in the ASME VIII-1 code. They consist of constructing External Pressure Charts from Isochronous stress-strain curves and using them to determine an allowable compressive stress in the creep regime.

9.2 Construction of External Pressure Charts (EPC) Using Isochronous Stress-Strain Curves

External Pressure Charts used in the creep regime are developed from the isochronous stress-strain curves as follows:

1) For any given average isochronous stress strain curve determine the proportional limit, P_L, modulus of elasticity, E, and the x–y coordinates of the nonlinear portion of the curve. The modulus of elasticity obtained from the curve may differ slightly from that published in the tabular values of ASME II-D.
2) The tangent modulus, E_t, is obtained for the nonlinear portion of the curve Jawad and Griffin (2011) and Jawad et al. (2016). This is needed for

calculating allowable compressive stress. Some commonly used methods for obtaining E_t are

a) Plot the nonlinear curve and manually draw a tangent line at various locations along the curve. The tangent modulus, E_t, is the slope of the tangent line at that location. This method tends to be laborious and only approximate.
b) Use a finite difference approach to determine the tangent modulus ($E_t = d\sigma/d\varepsilon$) in small increments of stress, σ, and strain, ε, at various locations along the nonlinear curve. This method lends itself to computerization. ASME uses a computer program algorithm Chen et al. (2021) for generating external pressure charts that utilize the finite difference method to obtain E_t. The algorithm uses the Moving Average statistical procedure Spence (1991) to smooth out the resultant curve.
c) If the isochronous curves were originally generated from a set of stress-strain equations, then E_t is obtained by taking the derivative of these equations.

3) Determine the values of Factor A, defined as the ratio of stress over tangent modulus ($A = S_c/E_t$) for various stress levels along the curve
4) Plot the external pressure chart using various values of S_c versus corresponding values of Factor A obtained from Step 3. The maximum value of S_c in the external pressure chart must not exceed the allowable creep rupture stress, S_r or 0.8 times the yield stress, whichever is less.
5) The nonlinear portion of the A versus S_c curve in the external pressure chart in Step 4 is based on average isochronous stress-strain values. Accordingly, it must be reduced by 20% to obtain minimum values in accordance with the ASME criterion listed in Appendix 3 of II-D. The strain corresponding to each reduced stress point is adjusted by the ratio ($0.2S_c/E$).
6) The reduced nonlinear curve of the external pressure chart, EPC, is merged with the elastic curve near the proportional limit point, P_L. To have a smooth transition length, a splice curve is normally used to link the two curves. One type of splice curve is given by the equation $\ln(\sigma) = \alpha + \beta \ln(\varepsilon)$ where Constants α and β are obtained from the boundary conditions of the splice curve.

The following example illustrates this procedure.

Example 9.1 Determine the external pressure chart for the 100,000 h isochronous stress-strain curve in Figure 9.1 for Type 304 stainless steel at 1200°F by the following methods

a) Finite difference
b) Derivative of the stress-strain equations.

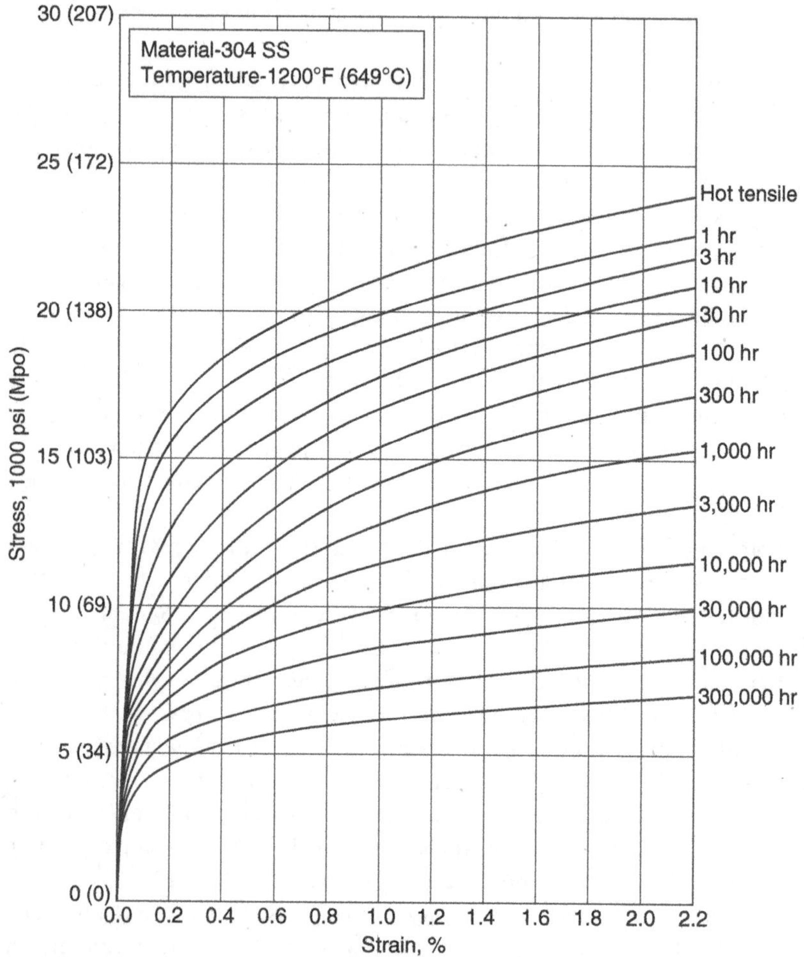

Figure 9.1 Average Isochronous Stress-Strain Curves for Type 304 Material at 1200°F (649°C) [ASME II-D].

The modulus of elasticity, E, is measured as 20,930 ksi. The proportional limit, P_L, for the 100,000 h curve is measured as 1.0 ksi. The stress to rupture is $S_r = 6.2$ ksi for 100,000 h obtained from ASME and $0.8S_y = 11.2$ ksi.

Solution

a) Finite Difference Method

1) Obtain x–y coordinates for the 100,000-h curve in Figure 9.1. These coordinates are determined from the stress-strain equations in Appendix B,

Section B.1.1 for stress increments of 0.02 ksi. A total of about 450 sets of points are used.

2) A finite difference procedure is used to obtain E_t values for the nonlinear portion of the curve.
3) Factor A is then obtained from the relationship $A = S_c/E$ for the elastic portion and $A = S_c/E_t$ for the nonlinear portion.
4) An external pressure curve of A versus average S_c is then obtained. The maximum stress is limited to $S_r = 6.2$ ksi.
5) An external pressure curve of A versus minimum S_c is obtained by multiplying the nonlinear curve between $A = 0.0001$ and 0.01 by a factor of 0.8. A splice curve between $A = 0.00005$ and 0.0001 of the form $\ln(S_c) = (10.911246 + 0.402251 \ln(A))$ is used to connect the nonlinear curve with the elastic curve. A straight line is also used between $A = 0.01$ and $A = 0.1$. The resultant minimum curve is shown in Figure 9.2.

b) Derivative of the Stress-Strain Equations Method

1) The 100,000 h curve in Figure 9.1 is obtained from the average stress-strain equations shown in Appendix B. The total strain consists of the three terms elastic, plastic, and creep.
2) The derivative of each of the three terms of the average stress-strain equation is obtained as shown in Appendix C. The derivatives of the elastic and

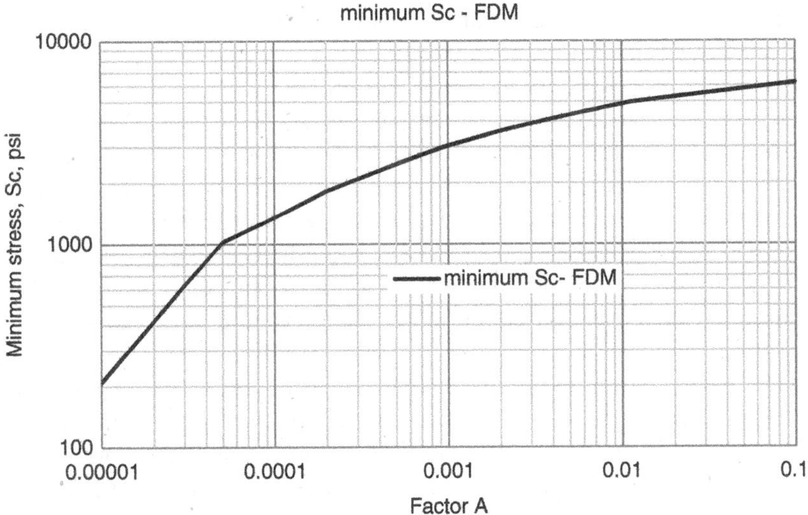

Figure 9.2 Minimum External Pressure line for Type 304 Stainless Steel at 100,000 h and 1200°F Obtained from Finite Difference Evaluation.

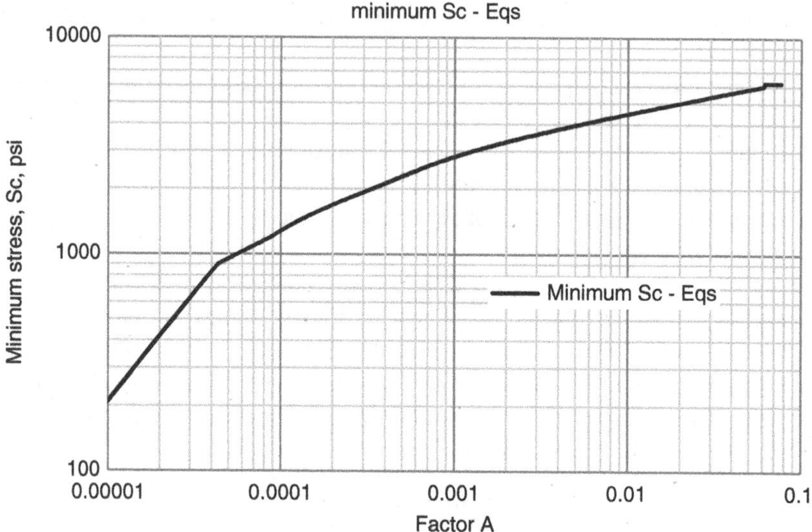

Figure 9.3 Minimum External Pressure line for Type 304 Stainless Steel at 100,000 h and 1200°F Obtained from Differentiating the Stress-Strain Equation.

plastic components are straightforward. The derivative of the creep component is fairly complicated and requires the use of a commercial Symbolic Math program.

3) The value of E_t at various points is obtained from Eq. (C.3) of Appendix C.
4) Factor A is then obtained from the relationship $A = S_c/E$ for the elastic portion and $A = S_c/E_t$ for the nonlinear portion.
5) An external pressure curve of A versus average S_c is then obtained. The maximus stress is limited to the creep rupture stress, $S_r = 6.2$ ksi.
6) An external pressure curve of A versus minimum S_c is obtained by multiplying the nonlinear curve by a factor of 0.8. A splice curve between $A = 0.00005$ and 0.0001 is used with value $\ln(S_c) = 10.9332 + 0.4116 \ln(A)$ to connect the nonlinear curve to the elastic curve. A line is also used between $A = 0.01$ and $A = 0.1$. The maximum stress is capped at 6.2 ksi.
7) The resultant minimum curve is shown in Figure 9.3. A comparison between the minimum external pressure curve using the finite difference method and the minimum curve using the derivative method is shown in Figure 9.4. The two methods of analysis give essentially the same result.

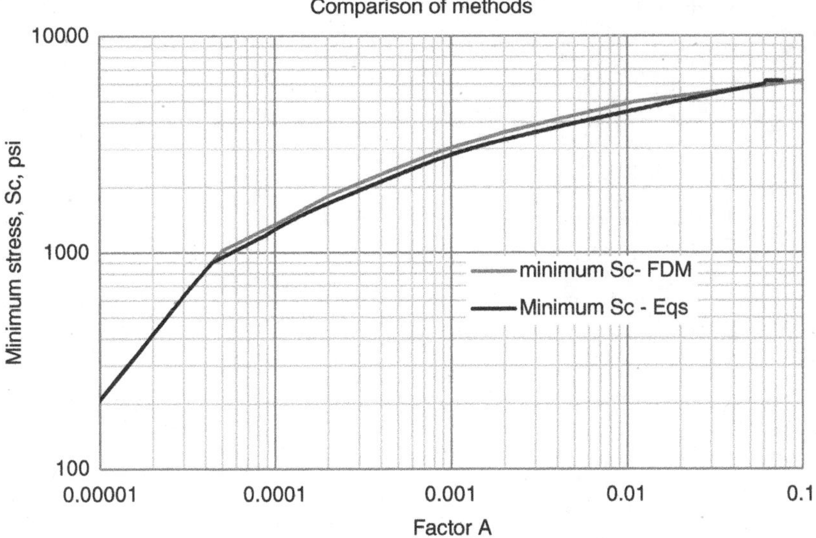

Figure 9.4 Minimum External Pressure lines for Type 304 Stainless Steel at 100,000 h and 1200°F Obtained from Finite Difference Evaluation and Equation Derivatives.

9.3 Cylindrical Shells Under Axial Compression

Buckling under axial compression may occur in one of three modes. The first is the classical Euler buckling, as discussed in Section 9.7, where the shell deforms as a beam. The second mode is local buckling in a lobe pattern, as shown in the introductory figure to this chapter. The third mode is local deformation in a ring pattern, as shown in Figure 9.5. The methodology presented in this section is based on the ASME Section VIII-1 code and assumes failure in a second or third mode. Other international codes use similar methods. The critical elastic axial buckling strain in a long cylinder, for $\mu = 0.3$, is given by the equation Timoshenko and Gere (2009)

$$\varepsilon_c = \frac{0.61}{\left(R_0 / t\right)} \tag{9.1}$$

where

ε_c = critical strain

R_o = outside radius

t = thickness of cylinder.

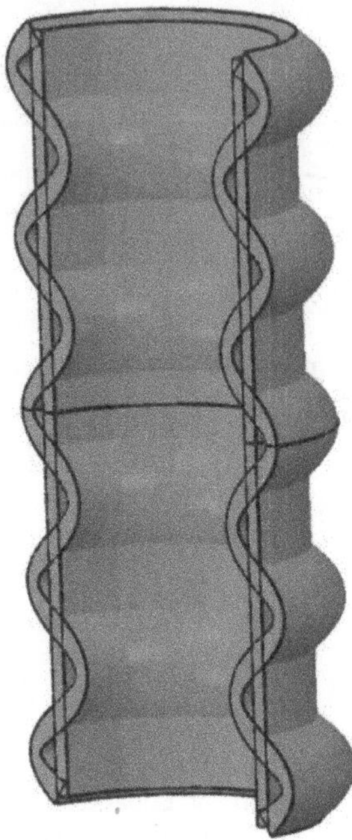

Figure 9.5 Shell buckled in local ring mode [Peter Carter].

For large R_o/t ratios, experiments have shown that the theoretical axial critical strain could be off by a large margin. This is due to such factors as non-homogeneity of the material physical properties, inaccuracy of determining modulus of elasticity and variation in thickness. Accordingly, the ASME code applies a knock-down factor, *KD*, to Eq. (9.1)

$$\varepsilon = \frac{0.61}{(KD)(R_0/t)}. \tag{9.2}$$

The relationship between elastic stress and strain is given by

$$S_c = \varepsilon_c E \tag{9.3}$$

where E = elastic modulus.

ASME also applies a design safety factor, *FS*, to Eq. (9.3) to take into consideration such items as local out-of-roundness and inaccuracy in the magnitude of applied loads. Equation (9.3) becomes

$$S = \frac{\varepsilon_c E}{FS}. \tag{9.4}$$

For axial compression ASME uses a knock-down factor, *KD*, of 5.0. Equations (9.2) and (9.4) are then written as

$$A = \frac{0.125}{(R_0 / t)} \tag{9.5}$$

and

$$S = \frac{AE}{FS} \tag{9.6}$$

where

A = factor

S = allowable compressive stress.

In the creep regime, Eqs. (9.5) and (9.6) are also applicable for cylindrical shells Jawad et al. (2016). The knock-down factor $KD = 5.0$ is applicable but the design factor, *FS*, which is set to a value of 2.0 in the non-creep regime by ASME, needs modification. The modification is required because in the creep regime the design factor needs to be a function of time rather than load. Griffin Griffin (1981) explained this by referring to the Norton equation

$$\varepsilon = KS^n \tag{9.7}$$

The value of n is normally between five and six for most Cr-Mo steels as well as 304 and 316 type stainless steels. A time factor of 10 in time, consistent with the scatter of data, is used in the creep regime. Hence, using a time factor of 10 and $n = 5.5$ in Eq. (9.7) results in a load factor of about 1.50. Accordingly, ASME VIII-1 uses $FS = 1.5$ in the creep regime. It indexes the value to 100,000 h and gradually increases it to a value of $FS = 2.0$ at time $\tau = 0$ to match the design factor used by ASME in the non-creep regime. The transition criterion is

$$FS = 2.0 \qquad \tau \le 1\text{ h} \tag{9.8a}$$

$$FS = \frac{2.0}{1 + 0.0288\ln(\tau)} \qquad 1 < \tau \le 100{,}000\text{ h} \tag{9.8b}$$

$$FS = 1.5 \qquad \tau > 100{,}000\text{ h} \tag{9.8c}$$

where τ is time in hours.

The above equations are applicable in the elastic region. It has been shown Gerard (1962) that, in the nonlinear portion of the stress-strain curve, the critical buckling stress equation is in terms of a reduced modulus that is a function of both the tangent and secant moduli, with the secant modulus being larger than the tangent modulus. Accordingly, the reduced modulus can be approximated to the first order by the tangent modulus only Jawad and Griffin (2012).

The resulting equation is the same as that given by Eq. (9.6), with the exception of substituting the tangent modulus, E_t, in lieu of E. Accordingly, an external pressure chart similar to the one shown in Figure 9.2 is used which includes elastic and nonlinear parts.

The design procedure in the creep regime is as follows:

1) Construct an external pressure chart as a function of A and S_c, as described in Section 9.2.
2) Calculate the value of A for given shell dimensions from Eq. (9.5).
3) Calculate the factor of safety FS from Eq. (9.9) for two conditions. The first is at $\tau < 1.0$ h and the second at the operating condition.
4) Calculate the allowable compressive stress, S, from the smaller of the following two equations for the two conditions mentioned in Step 3 above

$$S = AE / FS \tag{9.9}$$

or,

$$S = S_c / FS. \tag{9.10}$$

Example 9.2 Determine the allowable compressive stress in a cylinder with $R_o = 24$ in. and $t = 0.625$ in., which is made of Type 304 stainless steel, at 1200°F and 100,000 h. The modulus of elasticity $E = 20{,}930$ ksi. The external pressure chart for this condition is shown in Figure 9.2.

Solution

The value of A is obtained from Eq. (9.5) as

$$A = \frac{0.125}{(24 / 0.625)} = 0.0033$$

From Figure 9.2, $S_c = 3800$ psi
From Eq. (9.8b),

$$FS = \frac{2.0}{1 + 0.0288 \ln(100{,}000)} = 1.50.$$

From Eq. (9.9),

$$S = (0.0033)(20{,}930{,}000) / 1.5 = 46{,}046 \text{ psi}.$$

From Eq. (9.10)

$$S = 3800 / 1.5 = 2530 \text{ psi}.$$

Hence, the allowable compressive stress for 100,000 h is 2530 psi.

It is of interest to note that the allowable compressive stress at 1200°F for Type 304 stainless steel, calculated from an external pressure chart in ASME II-D and using the rules of ASME VIII-1 without consideration of creep, is 5200 psi.

9.4 Cylindrical Shells Under External Pressure

The methodology for calculating the allowable external pressure in cylindrical shells in the creep range is based on Sturm's paper Sturm (1941). Sturm's equation for critical pressure in a cylindrical shell is

$$P_{cr} = \frac{\kappa E}{\left(D_0 / t\right)^3} \tag{9.11}$$

where

D_o = outside diameter

E = elastic modulus of elasticity

κ = factor given in Figure 9.6. It is obtained from the solution of the differential equations for buckling and is a function of length, diameter, thickness, and Poisson's ratio.

P_{cr} = critical external buckling pressure

t = thickness.

Equation (9.11) can be written in terms of critical stress, S_c, as

$$S_c = \left[\frac{\kappa}{2\left(D_0 / t\right)^2}\right] E \tag{9.12}$$

where the bracketed term in Eq. (9.12) is the strain.

Define Factor A as,

$$A = \frac{\kappa}{2\left(D_0 / t\right)^2} \tag{9.13}$$

Equation (9.12) then becomes

$$S_c = AE. \tag{9.14}$$

Figure 9.6 can be replotted in terms of Factor A rather than κ by using Eq. 9.12. Performing such an operation using L/D_o rather than L/R, and smoothing out the resultant curves for design purposes, results in the ASME Geometric Chart shown in Figure 9.7. Factor A is obtained from the chart by entering the values

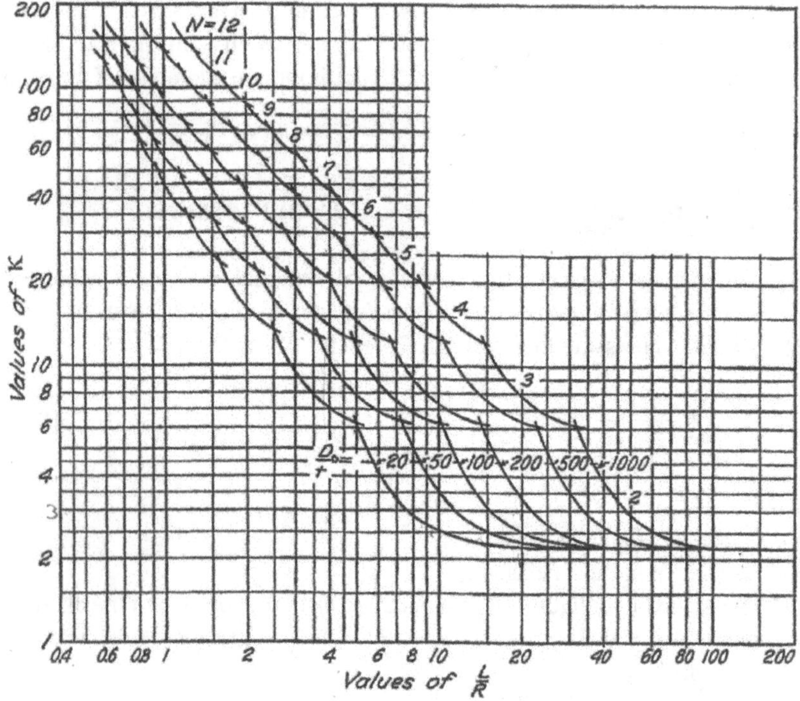

Figure 9.6 Collapse coefficients of cylinders with pressure on sides and ends, edges simply supported, μ = 0.3 Sturm (1941). *Source*: Sturm, R. G., 1941. A Study of the Collapsing Pressure of Thin-Walled Cylinders. University of Illinois Engineering Experiment Station Bulletin 329, Urbana, Illinois.

of L/D_0 and D_0/t. Equation (9.14) can then be solved to obtain the critical elastic buckling stress in a cylinder due to external pressure.

In the nonlinear regime, the value of tangent modulus, E_t, is used rather than E, as explained in Section 9.3. Accordingly, the design procedure is

1) For a cylinder with given D_0 and L, assume a value of t.
2) Calculate D_0/t and L/D_0.
3) Enter Figure 9.7 and determine Factor A.
4) Use an external pressure chart similar to Figure 9.2 to obtain a value of S_c.
5) Determine the allowable external pressure from the smaller of the following two equations

$$P = \frac{2AE}{(FS)(D_0/t)} \tag{9.15}$$

$$P = \frac{2S_c}{(FS)(D_0/t)} \tag{9.16}$$

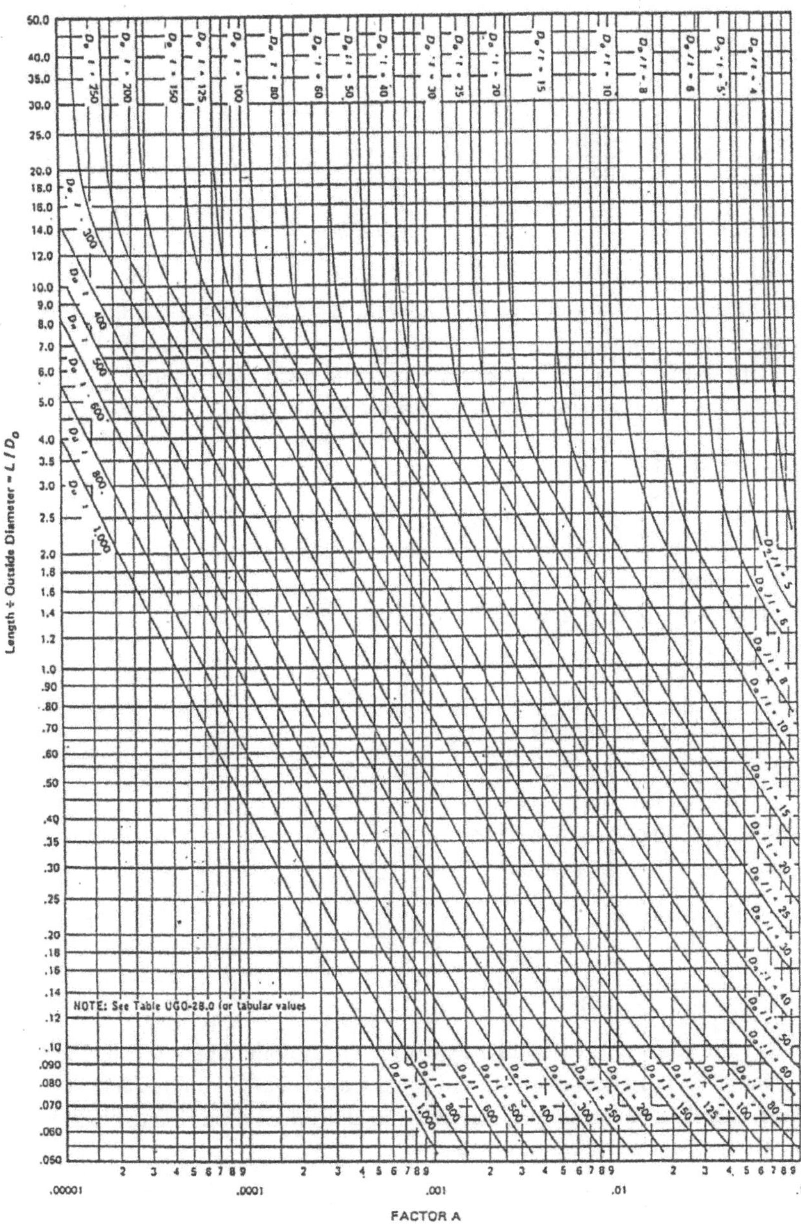

Figure 9.7 Geometric chart [ASME II-D].

where

$$FS = 3.0 \qquad \tau \leq 1\,\text{h} \tag{9.17a}$$

$$FS = \frac{3.0}{1 + 0.0434\ln(\tau)} \qquad 1 < \tau \leq 100{,}000\,\text{h} \tag{9.17b}$$

$$FS = 2.0 \qquad \tau > 100{,}000\,\text{h} \tag{9.17c}$$

Example 9.2 A cylinder has $D_o = 48.0$ in., $L = 120$ in., $t = 0.625$ in., and operates at 1200°F. Material of construction is Type 304 stainless steel. The modulus of elasticity E = 20,930 ksi. What is the allowable external pressure at 100,000 h? Use Figure 9.2 for an external pressure chart.

Solution

$$L / D_0 = 2.50, \quad D_0 / t = 76.8$$

From Figure 9.6, $A = 0.0008$
From Figure 9.2, $S_c = 2800$ psi
From Eq. (9.17),

$$FS = \frac{3.0}{1 + 0.0434\ln(100{,}000)} = 2.0$$

From Eq. (9.15),

$$P = \frac{2(0.0008)(20{,}930{,}000)}{(2.0)(76.8)} = 218\ \text{psi}$$

From Eq. (9.16),

$$P = \frac{2(2800)}{(2.0)(76.8)} = 36.5\ \text{psi}$$

Hence, the allowable external pressure for this cylinder at 100,000 h is 36.5 psi.

Note that the allowable external pressure at 1200°F for Type 304 stainless steel, calculated from an external pressure chart in Section II-D using the rules of VIII-1 without consideration of creep, is 67.5 psi.

9.5 Spherical Shells Under External Pressure

The classical equation for the buckling external pressure of a spherical shell Timoshenko and Gere (2009) is given by

$$S_c = \frac{E}{(R_0/t)[3(1-\mu^2)]^{0.5}} \tag{9.18}$$

$$S_c = \frac{0.60E}{(R_0/t)} \text{ for } \mu = 0.3. \tag{9.19}$$

A buckling equation for spherical shells was derived by von Karman Von Karman and Tsien (1960) that takes local imperfections into consideration. The minimum value of the equation is

$$S_c = \frac{0.18E}{(R_0/t)}. \tag{9.20}$$

This equation, while substantially more accurate than Eq. (9.19), still requires a knock-down factor to match experimental data. ASME uses a knock-down factor of 1.5 to obtain

$$S_c \approx \frac{0.125E}{(R_0/t)}. \tag{9.21}$$

Substituting

$$S_c = \frac{PR_0}{2t} \tag{9.22}$$

into Eq. (9.21) and applying a factor of safety gives

$$P = \frac{0.25E}{(FS)(R_0/t)^2}, \tag{9.23}$$

which is the allowable external pressure on a spherical shell in the elastic region.

Define

$$A = \frac{0.125}{(R_0/t)}. \tag{9.24}$$

Substituting Eq. (9.24) into Eq. (9.21) gives

$$S_c = AE. \tag{9.25}$$

Equation (9.25) is used in the inelastic region by substituting E_t for E. Hence,

$$S_c = AE_t. \tag{9.26}$$

In the inelastic region the value of S_c is obtained from an external pressure chart as defined by Eq. (9.26). The allowable external pressure in the inelastic region is then obtained from Eq. (9.22) as

$$P = \frac{2S_c}{(FS)(R_0/t)}. \tag{9.27}$$

The rules for calculating allowable external pressure on spherical shells in the creep regime are the same as those in the time-independent regime, with the exception of using isochronous curves to construct external pressure charts and using a variable factor of safety, as explained in Section 9.3. The procedure is as follows:

1) Calculate R_o/t.
2) Determine Factor A from Eq. (9.24).
3) From an external pressure chart obtained from an isochronous stress-strain diagram, calculate the critical compressive stress S_c for time equal zero and operating time.
4) Calculate the Factor of Safety from the following equations

$$FS = 4.0 \qquad \tau \leq 1\,\text{h} \tag{9.28a}$$

$$FS = \frac{4.0}{1 + 0.0869\,\ln(\tau)} \qquad 1 < \tau \leq 100{,}000\,\text{h} \tag{9.28b}$$

$$FS = 2.0 \qquad \tau > 100{,}000\,\text{h}. \tag{9.28c}$$

5) Calculate the allowable external pressure from the smaller of Eqs. (9.23) and (9.27).

Example 9.3 A hemispherical head has D_o = 48.0 in., t = 0.3125 in., and operates at 1200°F. The material of construction is Type 304 stainless steel. The modulus of elasticity E = 20,930 ksi. What is the allowable external pressure at 100,000 h? Use Figure 9.2 for an external pressure chart.

Solution

$$R_0/t = 76.8$$

From Eq. (9.24)

$$A = 0.0016$$

From Figure 9.2,

$$S_c = 3100\ \text{psi}$$

From Eq. (9.28),

$$FS = 2.0 \text{ for } 100{,}000 \text{ h}$$

From Eqs. (9.23) and (9.27),

$$P = \frac{(0.25)(20{,}930{,}000)}{(2.0)(76.8)^2} = 443.6 \text{ psi}$$

$$P = \frac{2(3100)}{(2.0)(76.8)} = 40.4 \text{ psi} \qquad \text{governs at } 100{,}000 \text{ h}$$

Hence, the allowable external pressure for this head at 100,000 h is 40.4 psi.

The allowable external pressure for this head using the methodology of ASME VIII-1 and Figure HA-1 of ASME II-D with no creep consideration is 60.0 psi.

9.6 Design of Structural Columns

The elastic buckling of a column of effective length (KL) and applied load F Bleich (1952), Timoshenko and Gere (2009) is given by the Euler equation

$$F_c = \frac{\pi^2 EI}{(KL)^2} \tag{9.29}$$

where

E = elastic modulus
F_c = critical buckling load
I = moment of inertia
K = length factor
= 1.0 column pinned at both ends
= 0.5 column fixed at both ends
= 0.7 column fixed at one end and pinned at the other end
= 2.0 column fixed at one end and free at the other end
= 1.0 column fixed at one end and free with zero rotation at the other end
= 2.0 column pinned at one end and free with zero rotation at the other end
L = length of column.

Equation (9.29) can be written in terms of critical stress as

$$S_c = F_c A_r$$

$$S_c = \frac{\pi^2 E}{(KL/r)^2} \tag{9.30}$$

or, in terms of critical strain as

$$\varepsilon_c = \frac{\pi^2}{(KL/r)^2} \tag{9.31}$$

where

A_r = cross-sectional area

ε_c = critical strain

r = radius of gyration ($A_r r^2 = I$)

S_c = critical stress.

The values of A_r, I, and r for various cross sections are shown in Table 9.1.

The elastic buckling strength given by Eq. (9.30) tends to approach infinity as the slenderness ratio (KL/r) becomes small, as in the case of short columns. However, by using an external pressure chart similar to the one in Figure 9.2, and defining Eq. (9.31) as

$$A = \frac{\pi^2}{(KL/r)^2} \tag{9.32}$$

the stress for short columns (with large A values) is limited by rupture stress rather than elastic buckling stress. The elastic buckling Eq. (9.30) is applicable up to the proportional limit, P_L. In the case of Figure 9.2 for type 304 stainless steel at 100,000 h and 1200°F, the applicability of Eq. (9.30) is limited to $A = 0.00005$ where the proportional limit is located. This A value corresponds to $(KL/r)_{PL} = 444$, obtained from Eq. (9.30). Values of (KL/r) smaller than 444 will result in A values larger than 0.00005 and the compressive stress Sc must be obtained from the External Pressure Chart in the nonlinear zone. A plot of Sc from Eq. (9.30) and Figure 9.2 is shown in Figure 9.8. The value of the proportional limit P_L depends on the type of material, the temperature, and design life of the component. Hence, the value of (KL/r), where Eq. (9.30) is applicable, varies from case to case and is not a fixed quantity, as is the case in the time-independent regime where the (KL/r) value is fixed around 120.

ASME uses a design factor of 1.90 for long columns and gradually decreases it to 1.50 for short columns similar to the methodology given in the AISC manual AISC (1991)

Accordingly, the design procedure for axial compression of structural columns in the creep regime is as follows:

1) Establish length, length factor, operating temperature, modulus of elasticity, and cross section dimensions.
2) Calculate strain factor, A, from the equation

$$A = \frac{\pi^2}{(KL/r)^2} \tag{9.33}$$

Table 9.1 Values of A, I, and r for various cross sections.

shape	A	I	r
Rectangle (B, t)	Bt	$Bt^3/12$	$\frac{t}{(12)^{0.5}}$
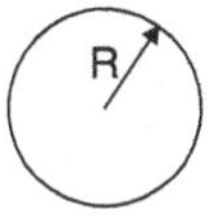 Solid circle (R)	πR^2	$\pi R^4/4$	$R/2$
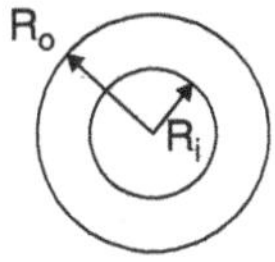 Hollow circle (R_o, R_i)	$\pi(R_o^2 - R_i^2)$	$\pi(R_o^4 - R_i^4)/4$	$(R_o^2 + R_i^2)^{0.5}/2$
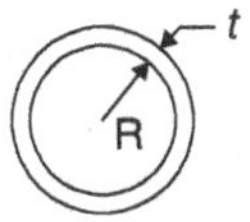 Thin ring (R, t)	$2\pi Rt$	$\pi R^3 t$	$\frac{R}{(2)^{0.5}}$

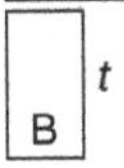

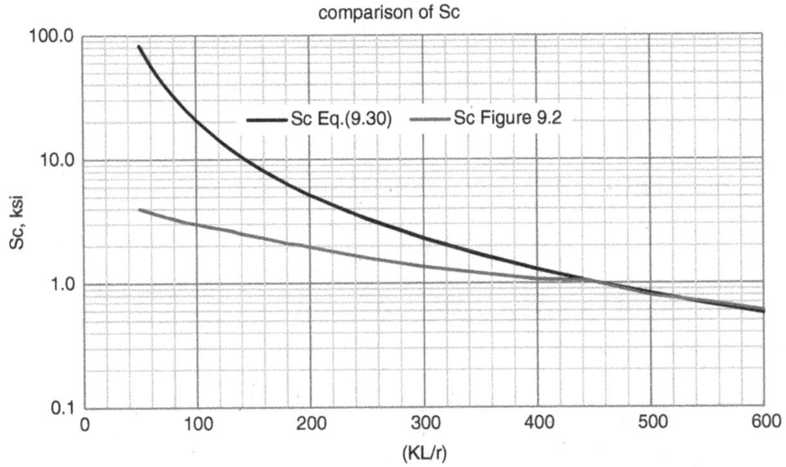

Figure 9.8 Compressive stress for type 304 stainless steel at 100,000 h and 1200°F.

1) Determine the critical compressive stress, S_c, from a time-dependent external pressure chart or external pressure equation for two operating time conditions. The first condition is at less than 1 h (hot tensile) and the second condition at the operating time.
2) ASME uses a factor of safety of 1.90 for long columns and gradually decreases it to 1.50 for short columns. The definition of short and long

columns is a function of (KL/r). The factor of safety, FS, is determined as follows:

$$FS = 1.5 \quad \text{for} \quad 0 < (KL/r) \leq 10 \tag{9.34}$$

$$FS = (2.78 \times 10^{-5})(KL/r)^2 + 1.5 \quad \text{for} \quad 10 < (KL/r) \leq 120 \tag{9.35}$$

$$FS = 1.9 \quad \text{for} \quad 120 < (KL/r) \tag{9.36}$$

Calculate the allowable compressive stress from the smaller of the following two equations for each of the two operating conditions

$$S = AE / FS \tag{9.37}$$

$$S = S_c / FS \tag{9.38}$$

The critical compressive stress, S_c, is obtained from external pressure charts or external pressure equations.

Example 9.4 A pressure vessel operating at 1200°F has a 304 stainless steel internal support bracket as shown in Figure 9.9. Determine the size of member AB due to an applied load of 450 lb and life of 100,000 h. Use Figure 9.2 for External Pressure Chart and let $E = 20{,}930$ ksi,

Solution

The load on member AB from Figure 9.9 is 900 lbs.

Try a 1.0625-in. diameter bar.

From Table 9.1, $A_r = 0.887\text{in.}^2$, $I = 0.0626\text{in.}^4$, $r = 0.266\text{in.}$

$$S = 900 / 0.887 = 1015\,\text{psi}$$

$$KL / r = 2.0(20) / 0.266 = 150$$

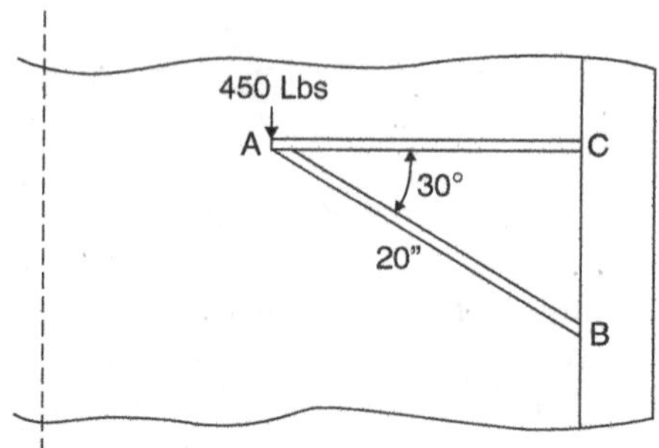

Figure 9.9 Internal bracket.

From Eq. (9.36),

$$FS = 1.9$$

From Eq. (9.33),

$$A = \frac{\pi^2}{150^2} = 0.00044$$

From Figure 9.2, S_c = 2200 psi

The allowable stress is obtained from the smaller of the values given by Eqs. (9.37) and (9.38):

$$S = (0.00044)(20{,}930{,}000)/1.55 = 5940 \text{ psi}$$

$S = 2200/1.9 = 1160$ psi > 1015 psi controls.

Use a 1.0625-in. diameter rod.

It is of interest to note that a rod diameter of 3/4-in. would have been adequate if creep was not a design consideration and External Pressure Chart HA-1 in ASME II-D is used.

9.7 Construction of External Pressure Charts (EPC) Using the Remaining Life Method

The remaining life method, using the omega factor discussed in Chapter 7, can also be used to construct external pressure charts. The equations needed are obtained by taking the derivative of the total strain given by Eq. (7.20) to find the tangent modulus E_t. The derivative of Eq. (7.20) is

$$d\varepsilon_t / d\sigma = d\varepsilon_e / d\sigma + d\varepsilon_p / d\sigma + d\varepsilon_c / d\sigma. \tag{9.39}$$

The tangent modulus, E_t, is defined as

$$\begin{aligned} E_t &= d\sigma / d\varepsilon_t = 1/\left(d\varepsilon_t / d\sigma\right) \\ &= 1/\left(d\varepsilon_e / d\sigma + d\varepsilon_p / d\sigma + d\varepsilon_c / d\sigma\right) \end{aligned} \tag{9.40}$$

In the elastic region, the quantity $d\varepsilon_e/d\sigma$ is calculated from Eq. (7.21) as

$$d\varepsilon_e / d\sigma = 1/E. \tag{9.41}$$

In the plastic region, the quantity $d\varepsilon_p/d\sigma$ is obtained by taking the derivative of Eq. (7.22) using a recursive computer program with symbolic solution. The result is

$$d\varepsilon_p / d\sigma = d_{\gamma_1} / d\sigma + d_{\gamma_2} / d\sigma \tag{9.42}$$

where

$$d\gamma_1/d\sigma = \left(2\alpha_2\alpha_6\sigma_t\right)^{-1}\{(\sigma_t/\alpha_3)^{(1/\alpha_2)}[\tanh((2\alpha_5-2\sigma_t)/\alpha_6)+1] \\ [\alpha_6-2\alpha_2\sigma_t+2\alpha_2\sigma_t\tanh\left(\left(2\alpha_5-2\sigma_t\right)/\alpha_6\right)]\} \tag{9.43}$$

$$d\gamma_2/d\sigma = \left(-2\alpha_6\alpha_7\sigma_t\right)^{-1}\{(\sigma_t/\alpha_8)^{(1/\alpha_7)}[\tanh((2\alpha_5-2\sigma_t)/\alpha_6)-1] \\ [\alpha_6+2\alpha_7\sigma_t+2\alpha_7\sigma_t\tanh\left(\left(2\alpha_5-2\sigma_t\right)/\alpha_6\right)]\} \tag{9.44}$$

and

$$\alpha_1 = R = \sigma_{ys}/\sigma_{ult} \tag{9.45}$$

$$\alpha_2 = m_1 = \left[\ln(R)+\left(\varepsilon_p'-\varepsilon_{ys}\right)\right]/\left\{\ln\left[\left(\ln\left(1+\varepsilon_p'\right)\right)/\left(\ln\left(1+\varepsilon_{ys}\right)\right)\right]\right\} \tag{9.46}$$

$$\alpha_3 = A_1 = \left[\sigma_{ys}\left(1+\varepsilon_{ys}\right)\right]/\left[\ln\left(1+\varepsilon_{ys}\right)\right]^{m_1} \tag{9.47}$$

$$\alpha_4 = K = 1.5R^{1.5}-0.5R^{2.5}-R^{3.5} \tag{9.48}$$

$$\alpha_5 = \sigma_{ys}+K\left(\sigma_{ult}-\sigma_{ys}\right) \tag{9.49}$$

$$\alpha_6 = K\left(\sigma_{ult}-\sigma_{ys}\right) \tag{9.50}$$

$$\alpha_7 = m_2 \tag{9.51}$$

$$\alpha_8 = A_2 = \sigma_{uts}e^{m_2}/\left(m_2^{m_2}\right) \tag{9.52}$$

$$\sigma_t = \left(1+\varepsilon_{es}\right)\sigma_{es}. \tag{9.53}$$

The expressions ε'_p in Eq. (9.46) and m_2 in Eqs. (9.52) and (9.53) have temperature limitations for various materials, as shown in Table E.2. When the temperature limit is exceeded, the plastic value of strain is set to zero. Also, γ_1 and γ_2 are set to zero when the stress results in a strain that is larger than ε'_p.

In the creep regime, the quantity $d\varepsilon_c/d\sigma$ is obtained by taking the derivative of Eq. (7.6) using a recursive computer program with a symbolic solution. The result is

$$d\varepsilon_c/d\sigma = C_{10}/C_7+C_8/K_1 \tag{9.54}$$

where

A_o through A_4 = constants given in Table E.1 of Appendix E
B_o through B_4 = constants given in Table E.1 of Appendix E

$$A_{00} = A_0 + \Delta\Omega^{sr} \quad (9.55)$$

$$B_{00} = B_0 + \Delta\Omega^{cd} \quad (9.56)$$

$$K_1 = 460 + T \quad (9.57)$$

$$S_1 = \log_{10}(\sigma) \text{ in the linear portion of the stress-strain curve} \quad (9.58a)$$

$$S_1 = \log_{10}(0.8\sigma) \text{ in the nonlinear portion of the stress-strain curve} \quad (9.58b)$$

$$C_1 = B_{00} + \left(B_1 + B_2S_1 + B_3S_1^2 + B_4S_1^3\right) / K_1 \quad (9.59)$$

$$C_2 = A_{00} + \left(A_1 + A_2S_1 + A_3S_1^2 + A_4S_1^3\right) / K_1 \quad (9.60)$$

$$C_3 = B_2 + 2B_3S_1 + 3B_4S_1^2 \quad (9.61)$$

$$C_4 = A_2 + 2A_3S_1 + 3A_4S_1^2 \quad (9.62)$$

$$C_5 = \left(10^{C_1}\right)\left(1/10^{C_2}\right)(\tau)[\ln(10)](C_3) \quad (9.63)$$

$$C_6 = [\ln(10)]\left\{\ln\left[1 - \left(10^{C_1}\right)\left(1/10^{C_2}\right)(\tau)\right]\right\} \quad (9.64)$$

$$C_7 = \left(10^{C_1}\right)\left(1/10^{C_2}\right)(\tau) - 1 \quad (9.65)$$

$$C_8 = \left(1/10^{C_1}\right)\left(C_6\right)\left(C_4\right) \quad (9.66)$$

$$C_9 = C_5 / K_1 - \left(10^{C_1}\right)\left(1/10^{C_2}\right)(\tau)[\ln(10)](C_4) / K_1 \quad (9.67)$$

$$C_{10} = \left(1/10^{C_1}\right)\left(C_9\right) \quad (9.68)$$

where C_1 through C_{10} are terms of the strain derivative $d\varepsilon_c/d\sigma$.

Equations (9.41), (9.42), and (9.54) are substituted into Eq. (9.40) to determine the tangent modulus E_t for various σ_t values. Factor A for constructing external pressure line is then obtained from the equation

$$A = \sigma / E_t \tag{9.69}$$

The external pressure curve correlating A to S_t is obtained by calculating A for various values of σ and plotting the result on a log–log graph correlating S_c and Factor A.

Example 9.5 Develop an external pressure chart for Type 304 stainless steel at 1200°F and 100,000 h using the remaining life method.

Solution

The constants needed to develop the chart are shown in Table 9.2. The values of A and B are obtained from Appendix E for Type 304 stainless steel. The value of Δ_Ω^{sr} is set to +0.5 for the upper value of the material property scatter band and the value of Δ_Ω^{sr} is set to 0.0 for ductile material. The value of E is obtained from ASME II-D.

Calculations were performed using a spreadsheet program for obtaining the value of E_t from Eq. (9.40) for various stress values. Equation (9.69) is then used to calculate the relationship between Factor A and stress S_c. A plot, shown in Figure 9.10, of Factor A versus S_c gives the external pressure chart for Type 304 stainless steel at 1200°F and 100,000 h using the Remaining Life Method.

Table 9.2 Parameters for developing an external pressure chart for Type 304 stainless steel at 1200°F and 100,000 h.

Constant	Value	Constant	Value
A_o	−19.17	A_1	53,762.0
A_2	−13,442.40	A_3	3162.6
A_4	−1685.2	B_o	−3.40
B_1	11,250.0	B_2	−5635.8
B_3	3380.4	B_4	−993.6
T	1200	K_1	1660
t	100,000	Δ_Ω^{sr}	+0.5
Δ_Ω^{cd}	0.0	A_{oo}	−19.67
B_{oo}	−3.7	E	20,930

A comparison of the external pressure chart for Type 304 stainless steel at 1200°F and 100,000 h, obtained from the isochronous stress-strain curve (ISSC) and from the Remaining Life Method (RLM), is shown in Figure 9.11. The RLM curve shows higher stresses compared to the ISSC curve for Factor *A* values larger than 0.00005. This difference is due, in part, to the following:

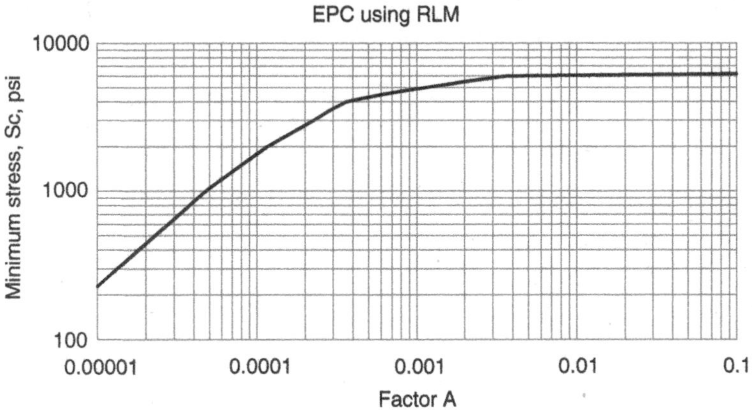

Figure 9.10 External Pressure chart for Type 304 Stainless Steel at 100,000 h and 1200°F obtained from Remaining Life Method.

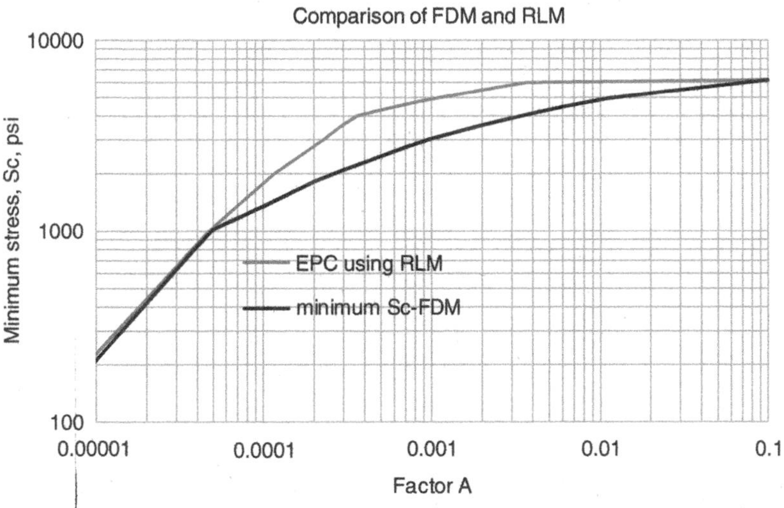

Figure 9.11 Comparison of external pressure chart for Type 304 stainless steel at 1200°F and 100,000 h obtained by the finite difference method (FDM) and Remaining Life Method (RLM).

The isochronous stress-strain curves in API 579/ASME FFS-1 are based on trend curves drawn from factors A and B for various materials and anchored by three points: 0-0, yield stress, and tensile stress. The isochronous stress-strain curves from ASME II-D are based on actual data.

The materials data base in API 579/ASME FFS-1 is slightly different than the data base in ASME II-D.

The Omega equations for constructing the external pressure charts are approximate while the Finite Difference Method is based on using actual stress-strain curves.

Appendix A

ASME VIII-2 Supplemental Rules for Creep Analysis

Case 2843-2

Analysis of Class 2 Components in the Time-Dependent Regime

Section VIII, Division 2

Inquiry: May Section VIII, Division 2, Class 2 components operating in the time-dependent regime and constructed of annealed 2.25Cr-lMo steel, 9Cr-lMo-V steel (Grade 91), Type 304 and 316 stainless steel, and nickel alloy 800H materials be analyzed by the strain deformation method?

Reply: It is the opinion of the Committee that Section VIII, Division 2, Class 2 components operating in the time-dependent regime, and constructed of annealed 2.25Cr-lMo steel, 9Cr-lMo-V steel (Grade 91), Type 304 and 316 stainless steel, and nickel alloy 800H materials may be analyzed by the strain deformation method as follows.

1 Scope

The design rules of this method are only applicable to the materials listed in Table 1 and the temperature range and maximum service life listed in Table 2.

Bolting is not included in the scope of this Case.

Analysis of ASME Boiler, Pressure Vessel, and Nuclear Components in the Creep Range, Second Edition. Maan H. Jawad and Robert I. Jetter.
© 2022 John Wiley & Sons Ltd. Published 2022 by John Wiley & Sons Ltd.

Table 1 Permissible Base Materials for Structures Other Than Bolting.

Base Material	Spec. No.	Product form	Types, Grades, or Classes
Types 304 SS and 316 SS	SA-182	Fittings & Forgings	F 304, F 304H, F 316, F 316H
[Note (1)], [Note (2)], [Note (3)]	SA-213	Smls Tube	TP 304, TP 304H,TP 316, TP 316H
	SA-240	Plate	304, 316, 304H, 316H
	SA-249	Welded Tube	TP 304, TP 304H, TP 316, TP 316H
	SA-312	Welded & Smls, Pipe	TP 304, TP 304H, TP 316, TP 316H
	SA-358	Welded Pipe	304, 316, 304H, 316H
	SA-376	Smls Pipe	TP 304, TP 304H, TP 316, TP 316H
	SA-403	Fittings	WP 304, WP 304H, WP 316, WP 316H, WP 304H, WP 304 HW, WP 316 W, WP 316 HW
	SA-479	Bar	304, 304H, 316, 316H
	SA-965	Forgings	F 304, F 304H, F 316, F 316H
	SA-430	Forged & Bored Pipe	FP 304, FP 304H, FP 316, FP 316H
Nl·Fe·Cr (Alloy 800H) [Note (4)]	SB-163	Smls Tubes	UNS N08810
	SB-407	Smls Pipe & Tube	UNS N08810
	SB-408	Rod & Bar	UNS N08810
	SB-409	Plate, Sheet. & Strip	UNS N08810
	SB-564	Forgings	UNS N08810
2¼Cr-1Mo [Note(5)]	SA-182	Forgings	F 22. Class 1
	SA-213	Smls Tube	T22
	SA-234	Piping Fittings	WP 22, WP 22 W [Note (6)]
	SA-335	Forg. Pipe	P22
	SA-336	Fittings. Forgings	F22a
	SA-369	Forg. Pipe	FP22
	SA-387	Plate	Gr 22, Class 1
	SA-691	Welded Pipe	Pipe 2¼Cr-1Mo (SA-387, Gr. 22, CI. 1)
9Cr·lMo·V	SA-182	Forgings	F91
	SA-213	Smls Tube	T91
	SA-335	Smls Pipe	P91
	SA-387	Plate	91

(*Continued*)

Table 1 (Continued)

Notes
1) These materials shall have a minimum specified room temperature yield strength of 30,000 psi (207 MPa) and a minimum specified carbon content of 0.04%.
2) For use at temperatures above 1000°F (S40°C). These materials may be used only if the material is beat treated by heating to a minimum temperature of 1900°F (1040°C) and quenching in water or rapidly cooling by other means.
3) Section IL, Part D, Table E-100.23-1 provides nonmandatory guidelines on additional specification restrictions to improve performance in certain service applications.
4) These materials shall have a total aluminum-plus-titanium content of at least 0.50% and shall have been heat-treated at a temperature of 2050°F (1120°C) or higher.
5) This material shall have a minimum specified room temperature yield strength of 30,000 psi (207 MPa), a minimum specified room temperature ultimate strength of 60,000 psi (414 MPa), a maximum specified room temperature ultimate strength of 85,000 psi (S86 MPa), and a minimum specified carbon content of 0.07%.
6) The material allowed under SA-234 shall correspond to one of:
   a) SA-33S, Grade P 22
   b) SA-387, Grade 22. Class 1
   c) SA-182, Grade F 22. Class 1 in compliance with [Note (4)].

Table 2 Temperature and Service Life Limitations.

Material	Low Temperature, °F	High Temperature, °F	Life, hr
Type 304SS	800	1500	300,000
Type 316SS	800	1500	300,000
Nl-Fe·Cr (Alloy 800H)	800	1400	300,000
2.25Cr-1Mo	700	1100	300,000
9Cr·1Mo·V	700	1200	300,000

2 Strain Deformation Method

The general outline for this procedure is shown in Figure 1.

3 Materials and other Properties

3.1 Materials

Materials that may be used at elevated temperatures in this Case are presently limited to those in Table 1.

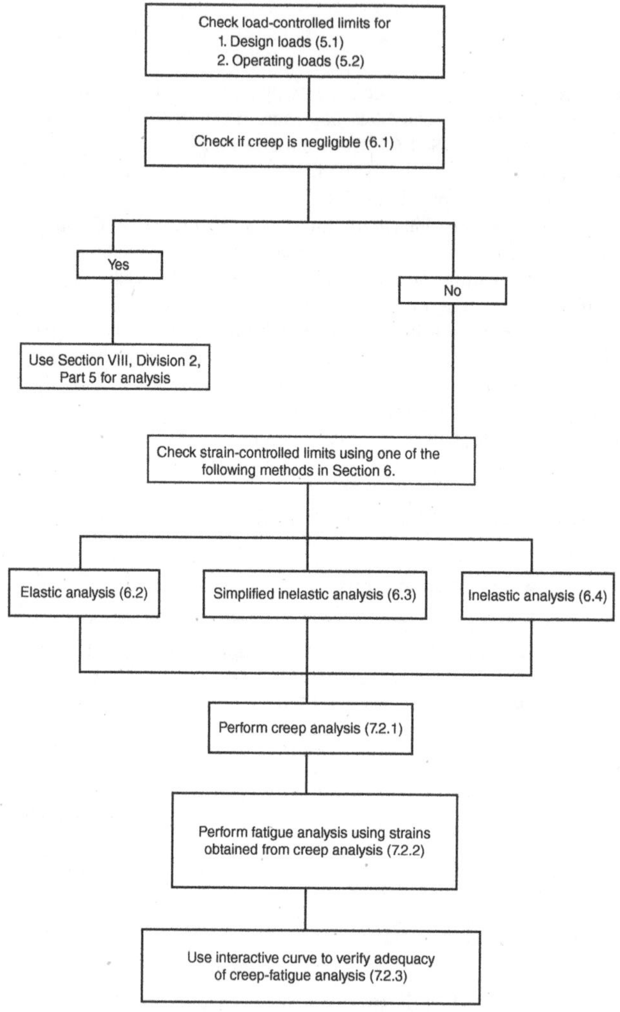

Figure 1 General outline of Strain deformation procedure for Creep Fatigue analysis.

3.2 Weld Materials

Weld materials that may be used at elevated temperatures in this Case are presently limited to those in Table 3.

3.3 Design Fatigue Strain Range

The fatigue data for the materials listed in Table 1 are shown in Section II, Part D, Nonmandatory Appendix E, Figures E-100.16-1 through E-100.16-5.

Table 3 Permissible Weld Materials.

Ease Material	Spec. No.	Class
Types 304 SS and 316 SS	SFA-5.4	E 308, E 308L, E 316, E 316L, E 16-8-2
	SFA-5.9	ER 308, ER 308L, ER 316, ER 316L, ER 16-8-2
	SFA-5.22	E 308, E 308T, E 308LT, E 316T, E316LT-1 EXXXT-G (16-8-2 chemistry)
Ni-Fe-Cr (Alloy 800H)	SFA-5.11	ENiCrFe-2
	SFA-5.14	ERNiCr-3
2¼Cr-lMo	SFA-5.5	E 90XX-B3 (>0.05% Carbon)
	SFA-5.23	EB 3, ECB 3
	SFA-5.28	E 90C-B3 (>0.05% Carbon), ER 90S-B3
	SFA-5.29	E 90T-B3 (>0.05% Carbon)
9Cr·1Mo·V	SFA-5.5	E90XX-B9
	SFA-5.23	EB9
	SFA-5.28	ER90S-B9

Table 4 Stress Values.

Stress	Table/Figure in Section II. Part D. Nonmandatory Appendix E
Allowable stress, S_{mt}	Figures E-100.4-1 through E-100.4-5
Allowable stress, S_t	Figures E-100.5-1 through E-100.5-5
Yield stress, S_y	Table E-100.6-1
Stress-to-rupture, S_r	Figures E-100.7-1 through E-100.7-4 and E-100.7-6
Stress-to-rupture for welds, S_r	Tables E-100.8-1 through E-100.8·3, Tables E-100.9-1 through E-100.9-3. and Tables E-100.10-1, E-100.10-2. E-100.11-1. and E-100.12-1
Isochronous curves	Figures E-100.18-1 through E-100·18·15. Figures E-100.19-1 through E-100.19-15. Figures E-100.20-1 through E-100.20-12. Figures E-100.21-1 through E-100.22-1 and Figures E-100.22-11 through E-100.22-11

3.4 Stress Values

The stress designations in Table 4 are defined in Section 3.5 and their numerical values listed in Section II, Part D, Nonmandatory Appendix E.

The allowable shear stress used in this Case is the same as that used in Section VIII, Division 2, Part 5, with the exception of substituting S_{mt} for S. The allowable bearing stress shall be the smaller of

(*a*) yield stress at operating temperature

(*b*) the stress at 0.2% offset strain as obtained from the isochronous stress-strain curve for the operating temperature and for the time duration equal to the total life.

3.5 Stress Terms

The terms shown in 8, the nomenclature of this Code Case, in addition to the nomenclature in Part 5 of Section VIII, Division 2, arc applicable to this Case.

4 Design Criteria

Evaluation of components in the creep regime consists of satisfying load-controlled limits, strain-controlled limits, and creep-fatigue consideration. Satisfying load-controlled limits assures maintaining stress levels below code allowable values. Satisfying strain controlled limits assures against failure due to ratcheting. Satisfying creep-fatigue criterion establishes the expected life of the components.

4.1 Short-Term Loads

The tensile stress values, S_u, and the yield stress values, S_y, used in this Case when operating in the time-dependent regime may need to be reduced due to material deterioration when the applied loads are short term, such as wind and seismic loads. Table 5 lists reduction factor C_1 for various materials and temperatures.

When the yield and ultimate tensile strengths are reduced by the operating elevated temperature, it is necessary to appropriately reduce the values of S_{mt} and S_m. The reduced values shall be taken as the smallest of

a) the S_{mt} and S_m values taken from Section II, Part D, Figures E-100.4-l through E-100.4-5
b) the product of 1/3 of the tensile strength at temperature (Section II. Part D, Tables E-100.1-l and E-100.1-2) and the tensile strength reduction factor C_1 (Table 5)
c) the product of 2/3 of the yield stress at temperature (Section II. Part D, Table E-100.6-l) and the yield stress reduction factor C_1 (Table 5)

Table 5 Short-Term Reduction Factor C_1.

Material	Operating Temperature, °F	Reduction Factor, S_y	Reduction Factor, S_m
Type 304SS	≥900	1.0	0.8
Type 316SS	≥900	1.0	0.8
Ni·Fe·Cr (Alloy 800H)	≥1350	0.9	0.9
2.25Cr-1Mo	800	1.0	0.94
	850	0.92	0.88
	900	0.86	0.82
	950	0.80	0.77
	1000	0.74	0.72
	1050	0.67	0.67
	1100	0.63	0.62
9Cr·1Mo·V	900	1.0	0.97
	950	1.0	0.93
	1000	1.0	0.90
	1050	1.0	0.84
	1100	1.0	0.84
	1150	1.0	0.81
	1200	1.0	0.78

5 Load-Controlled Limits

Load-controlled limits shall be satisfied for both design and operating conditions. The criterion is outlined in Figure 2.

5.1 Design Load Limits

The stress calculations required for the analysis of design loadings shall be based on a linearly elastic material model. The calculated equivalent stress values shall satisfy the limits of *(a)* and *(b)*:

(a) The general primary membrane equivalent stress derived from P_m shall not exceed S:

$$P_m \leq S. \tag{1}$$

The value of S is determined at the design temperature.

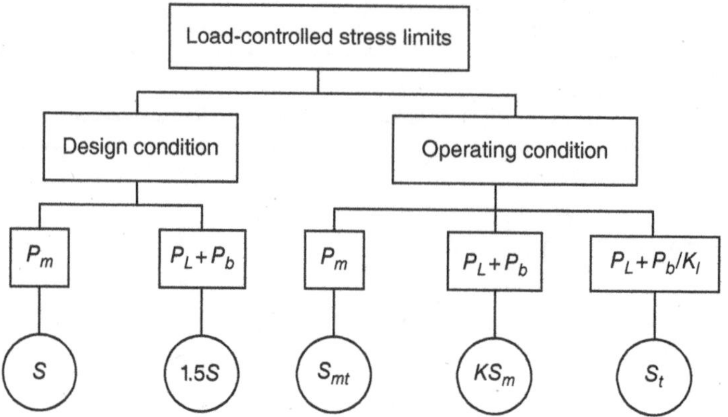

Figure 2 Flow diagram for Load-controlled stress limits.

(b) The combined local primary membrane plus primary bending equivalent stresses derived from P_L and P_b shall not exceed 1.5*S*:

$$P_L + P_b \leq 1.5S. \tag{2}$$

5.2 Operating Load Limits

The stress calculations required for the analysis of operating loadings shall be based on a linearly elastic material model. The calculated equivalent stress values shall satisfy the limits of *(a)* through *(e)*:

(a) The general primary membrane equivalent stress derived from P_m shall not exceed S_{mt}:

$$P_m \leq S_{mt}. \tag{3}$$

The value of S_{mt} is determined at the maximum wall averaged temperature that occurs during the particular loading event

(b) The combined local primary membrane plus primary bending equivalent stresses derived from P_L and P_b shall satisfy the following limits:

$$\left(P_L + P_b\right) \leq KS_m \tag{4}$$

$$\left(P_L + P_b / K_t\right) \leq S_t. \tag{5}$$

The values of S_m and S_t are determined at the maximum wall averaged temperature that occurs during the particular loading event

(c) In Eq. (5), the S_t value is determined for the time, *t*, corresponding to the total duration of the combined equivalent stress derived from P_L and P_b/K_t.

(d) When time, t, in (a) and (b) is less than the total specified service life of the component the cumulative effect of all the loadings shall be evaluated by the following use-fraction sum equation. In addition, it is permissible and often advantageous to subdivide loading history into several load levels and into several temperatures at any given load level

$$\Sigma_1 \left(\frac{t_i}{t_{im}} \right) \leq 1.0. \tag{6}$$

(e) The use of Section II. Part D, Figures E-100.5-l through E-100.5-5 for determining $t_{i\,m}$ for two loading conditions at two different temperatures is shown schematically in Figure 3. In this figure P_{mt} ($i = 1, 2, 3$, etc.) represents the calculated membrane equivalent stress for the loading condition and temperature in question: and T_i represents the maximum local wall averaged temperature during t_i. Note that it may be desirable to consider that a given equivalent stress, P_{mt}, acts during several time periods, t_i, to take credit for the fact that the temperature varies with time.

(f) When time, t, in (b) is less than the total service life of the component, the cumulative effect of all $[P_L + (P_b/K_t)]$ loadings shall be evaluated by the following use-fraction sum equation. The use of Section II, Part D, Figure E-100.15-3 and Figures E-100.16-1 through E-100.16-5 for determining t_{ib} for two loading conditions at two different temperatures is shown schematically in Figure 4. It is permissible and often advantageous to separate a loading history into several load levels and into several temperatures at any given load level.

$$\Sigma_i \left(\frac{t_i}{t_{im}} \right) \leq 1.0. \tag{7}$$

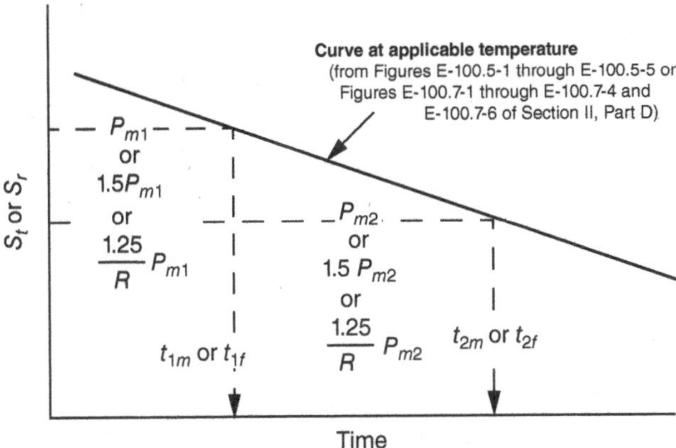

Figure 3 Use-fractions for Membrane stress.

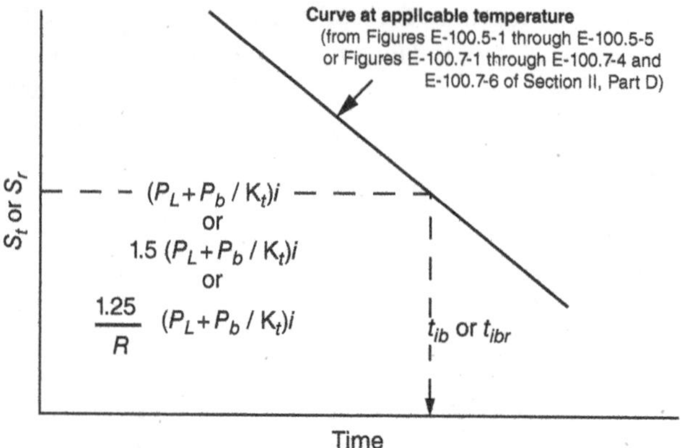

Figure 4 Use-fractions for Membrane plus bending stress.

6 Strain Limits

The requirements of strain limits are summarized in Figure 5. Where creep effects are presumed significant, inelastic analysis is generally required to provide a quantitative assessment of deformations and strains. However, elastic analysis, or simplified inelastic analysis, may be justified and used to establish conservative bounds for deformations, strains, strain ranges, and maximum stress to reduce the number of locations in a structure requiring detailed inelastic analysis.

6.1 Test A-1 Alternative Rules if Creep Effects are Negligible

Test A-1 is a screening criterion for determining whether or not creep is negligible. Compliance with this test indicates that creep is not an issue, and the rules in Section VIII, Division 2, Part 5 may be used directly at the elevated temperature.

To comply with this test, the requirements of *(a)*, *(b)*, and *(c)* shall be met:

(a) Eq. (8) as follows:

$$\Sigma_i\left(\frac{t_i}{rt_{\text{id}}}\right)\leq 0.1. \tag{8}$$

(b) Eq. (9) as follows:

$$\sum \varepsilon_i \leq 0.2\%. \tag{9}$$

If *(a)* and *(b)* are satisfied, then the requirements of Section VIII, Division 2, 5.5.6 may be applied in accordance with *(c)* and *(d)*.

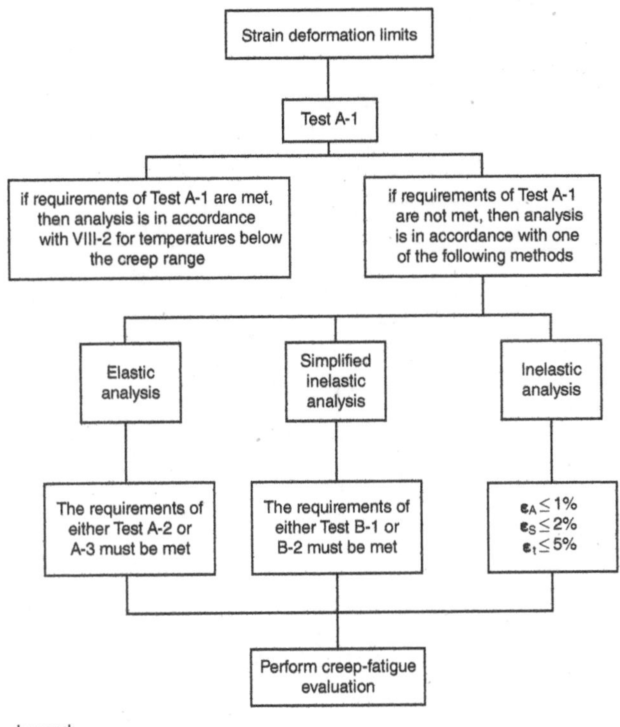

Figure 5 Requirements for Strain limits.

(c) The combined equivalent ($P_L + P_b + Q$) stresses (see Figure 6) shall satisfy the following:

$$(P_L + P_b + Q) \leq \text{lesser of } (3S_m \text{ or } 3\overline{S}_m). \tag{10}$$

The value of S_m is determined at the maximum wall temperature at the high end of the cycle. The value of $3\overline{S}_m$ is determined from the following:

(1) when part of the cycle falls below the creep temperature given in Table 7

$$3\overline{S}_m = 1.5S_{mL} + 0.5S_{tH}.$$

(2) when both temperature extremes of the cycle are above the temperature shown in Table 7

$$3\overline{S}_m = 0.5S_{tL} + 0.5S_{tH}.$$

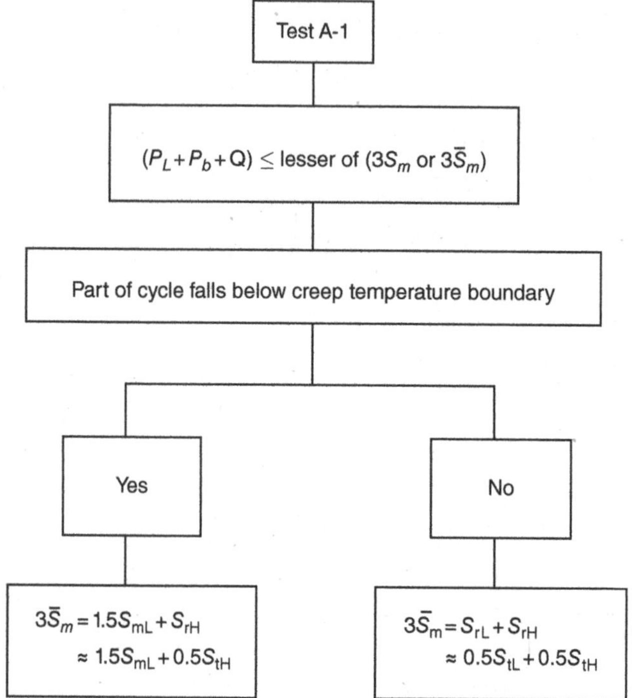

Figure 6 Stress criteria for Test A-1.

where

S_{mL} = the value of S_m determined at the maximum wall temperature at the low end of the cycle

S_{tH} = the value of S_t at the maximum wall temperature at the hot end of the cycle

S_{tL} = the value of S_t at the maximum wall temperature at the low end of the cycle

$0.5S_{tH}$ = an approximate conservative value for the relaxation stress, $S_{tH,}$ at the hot end of the cycle

Table 6 Values of Parameters *r* and *s*.

Material	**Parameter *r***	**Parameter *s***
Type 304SS	1.0	1.5
Type 316SS	1.0	1.5
Ni·Fe·Cr (Alloy 800H)	1.0	1.5
2.25Cr-1Mo	1.0	1.5
9Cr·1Mo·V	0.1	1.0

Table 7 Minimum Temperature for Test A-1.

Material	Temperature, °F (°C)
Type 304 SS	800 (427)
Type 316 SS	800 (427)
Ni-Fe-Cr (Alloy 800H)	800 (427)
2.25 Cr-1Mo	700 (371)
9Cr-1Mo-V	700 (371)

$0.5S_{tL}$ = an approximate conservative value for the relaxation stress, St_L, at the cold end of the cycle

The quantity $3\overline{S}_m$ is a creep shakedown criterion. The $1.5S_m$ value represents the yield strength and the $0.5S_t$ value represents a conservative approximation of the hot relaxation stress S_{rH}.

(d) Thermal stress ratcheting shall be kept below a certain limit defined by the following equations:

(1) for linear thermal distribution

(-a) for $0 < x < 0.5$

$$y' = \frac{1}{x} \tag{11}$$

(-b) for $0.5 < x < 1.0$

$$y' = 4(1 - x) \tag{12}$$

(2) for parabolic thermal distribution

(−a) for $0.3 < x < 0.615$

$$y' = 10.3855e^{-26785x} \tag{13}$$

(−b) for $0.615 < x < 1.0$

$$y' = 52(1 - x). \tag{14}$$

6.2 Strain Limits – Elastic Analysis

6.2.1 General Requirements

Satisfying the strain limits, described herein as Tests A-2 and A-3, is mandatory for creep-fatigue analysis. Figure 7 shows a flow diagram for Tests A-2 and A-3. Only one of these two tests shall be satisfied to comply with the strain limits

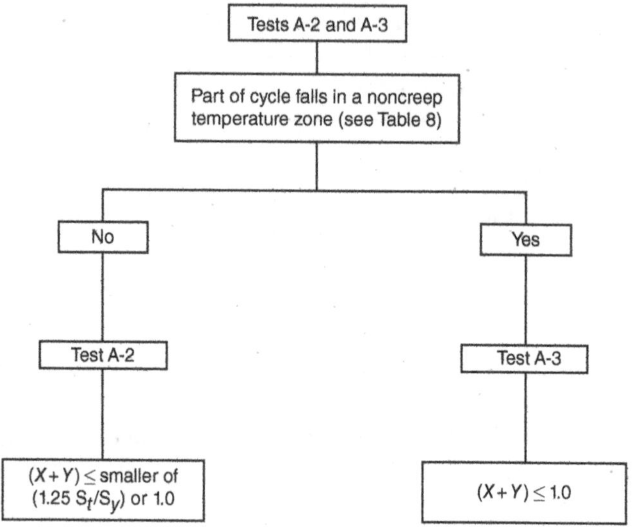

Figure 7 General requirements for Tests A-2 and A-3.

requirement for elastic analysis. Test A-2 is applicable to all operating temperatures while Test A-3 is applicable for those cycles during which the average wall temperature at one of the stress extremes defining the maximum secondary stress range Q_{max} is below the applicable temperature of Table 8.

The following definitions apply to Tests A-2 and A-3.

$$X = \frac{(P_L + P_b / K_t)_{max}}{S_{ya}} \tag{15}$$

$$r = \frac{Q_{range}}{S_{ya}} \tag{16}$$

where

$(P_L + P_b/K_t)_{max}$ = the maximum value of the local primary equivalent stress, adjusted for bending via K_t during the cycle being considered

Table 8 Temperature at which S = S_t at 10^5 h for Tests A-3, B-1, and B-2.

Material	Temperature, °F (°C)
Type 304 SS	948 (509)
Type 316 SS	1011 (544)
Ni-Fe-Cr (Alloy 800H)	1064 (573)
2.25 Cr-1Mo	801 (427)
9Cr-1Mo-V	940 (504)

Q_{range} = the maximum value of the secondary equivalent stress range during the cycle being considered

$S_{ya} \equiv$ the average of the S_y values at the maximum and minimum wall averaged temperatures during the cycle being considered.

6.2.2 Test A-2

This test is applicable to all operating temperatures. For this test, the following requirement shall be met:

$$(X+Y) \leq \frac{S_a}{S_{ya}} \tag{17}$$

where S_a is the lesser of:

(a) $1.25S_t$ using the highest wall averaged temperature during the cycle and a time value of 10^4 h

(b) the average of the two S_y values associated with the maximum and minimum wall averaged temperatures during the cycle.

6.2.3 Test A-3

This test is applicable to those cycles during which the average wall temperature at one of the stress extremes defining the maximum secondary stress range Q_{max} is below the applicable temperature of Table 8. For this test the following requirement shall be met:

$$X+Y \leq 1. \tag{18}$$

6.3 Strain Limits – Simplified Inelastic Analysis

6.3.1 General Requirements

The requirements of Tests B-1 and B-2 are intended to keep the stresses in components below ratcheting levels.

6.3.2 General Requirements for Tests B-1 and B-2

(a) The strain limits of 6.4 are considered to have been satisfied if the limits of Tests B-1 or B-2 in 6.3.3 are satisfied, in addition to the following:

(1) Tests B-1 and B-2 are applicable when the average wall temperature at one of the stress extremes defining each secondary equivalent stress range Q is below the applicable temperature of Table 7.

(2) The quantity X_1 in Figures 8 and 9 is given by

$$X_1 = \frac{\left(P_{1L} + P_{1b} / K_t\right)_{max}}{S_{yL}}. \tag{19}$$

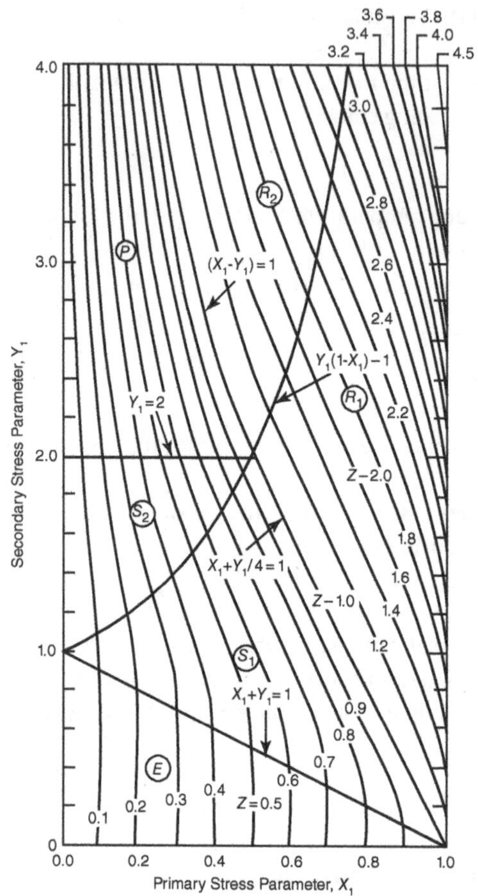

GENERAL NOTE: This figure is divided into the following six regimes: E, S_1, S_2, P, R_1, and R_2, The characteristics of these regimes are as follows:

(1) Regime E
- (a) Stress is elastic below the creep range.
- (b) Ratcheting does not occur below the creep range.
- (c) Stress redistributes to elastic value above the creep range.

(2) Regime P
- (a) Shakedown is not possible below as well as in the creep range.
- (b) Failure occurs due to low cycle fatigue below the creep range.

(3) Regime S_1
- (a) Ratcheting does not occur below the creep range.
- (b) Ratcheting occurs in the creep range.
- (c) Shakedown is not possible in the creep range.

(4) Regime S_2
- (a) Ratcheting does not occur below the creep range.
- (b) Ratcheting occurs in the creep range.
- (c) Shakedown is not possible in the creep range.

(5) Regime R_1
- (a) Ratcheting occurs below as well as in the creep range.
- (b) Shakedown is not possible below as well as in the creep range.

(5) Regime R_2
- (a) Ratcheting occurs below as well as in the creep range.
- (b) Shakedown is not possible below as well as in the creep range.

Figure 8 Effective Creep-stress parameter, Z, for simplified elastic analysis using Tests B-1 and B-2.

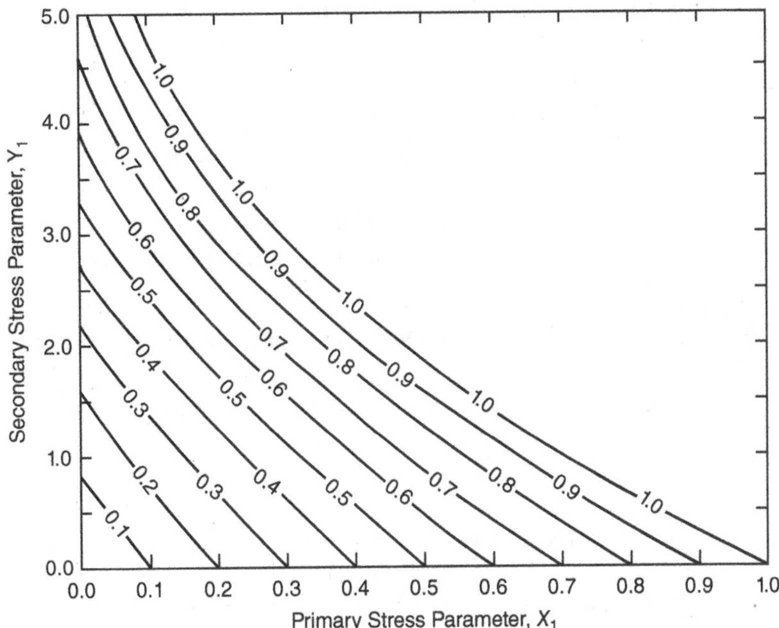

Figure 9 Effective Creep-stress parameter, Z, for simplified inelastic analysis using Test B-2.

(3) The quantity Y_t in Figures 8 and 9 is given by

$$Y_1 = \frac{Q_1}{S_{yL}}. \tag{20}$$

(4) The quantities X_t and Y_t used in Tests B-1 and B-2 are calculated using S_{yL} in lieu of S_y.

(5) The value of effective creep stress parameter, Z, is obtained from Figure 8 or 9 as described in 6.3.3.

(6) The effective creep stress σ_c is determined from the following equation.

$$\sigma_c = Z \times S_{yL}. \tag{21}$$

(7) The creep ratcheting strain, ε_{tr}, is determined by multiplying σ_c by 1.25 and entering the isochronous stress-strain chart at the designated temperature and time with a value of $1.25\sigma_c$ to obtain the creep ratcheting strain. The 1.25 factor is used to increase the tabulated minimum values of S_{yL} in Eq. (21) when entering the isochronous curves, which are based on average values.

(8) The elastic strain, ε_e, is determined from the equation

$$\varepsilon_e = \frac{1.25\sigma_c}{E}. \tag{22}$$

The value of E is determined from an isochronous stress-strain curve at the maximum wall averaged temperature.

(9) The inelastic creep strain, ε_i, is calculated from the equation $\varepsilon_i = \varepsilon_{cr} - \varepsilon_e$,

(10) The inelastic creep strain, ε_i, is limited to 1% for parent metal and ½% for weld metal

(b) The following procedural requirements are used in complying with *(a)(1)* through *(a)(10)*.

(1) The individual cycle as defined in the design specification cannot be split into subcycles.

(2) The time used to enter the isochronous curves for individual cycles or time blocks shall always sum to the entire life, regardless of whether all or only part of the cycles are evaluated under these procedures.

(3) The total service life may be subdivided into temperature–time blocks. The value of σ_c may differ from one block to another, but remains constant throughout each block service time. When σ_c is reduced at the end of a block of loading, the time of the block of loading must be longer or equal to the time needed for σ_c to relax at constant total strain to the σ_c value for the subsequent block. The creep strain increment for each block may be evaluated separately. The times used in selecting the isochronous curves can be entered at the initial strain accumulated throughout the prior load history. The creep strain increments for each time-temperature block shall be added to obtain the total ratcheting creep strain.

6.3.3 Applicability of Tests B-1 and B-2

6.3.3.1 Test B-1

(a) This test is applicable for structures in which the peak stress is negligible.

(b) This test can only be applied when σ_c is less than the yield stress of $S_{y\,H}$.

(c) The wall membrane forces from overall bending of a pipe section or vessel can be conservatively included as axisymmetrical forces.

(d) The dimensionless effective creep stress parameter Z for any combination of loading is given in Figure 8, in which the effective creep stress parameter Z is given by the following:

(1) Regime E

$$Z = X. \tag{23}$$

(2) Regime S_l

$$Z = Y + 1 - 2\sqrt{(1-X)Y}. \tag{24}$$

(3) Regimes S_2 and P

$$Z = XY. \tag{25}$$

6.3.3.2 Test B-2

(a) This test is more conservative than Test B-1 and is applicable to any structure and loading.
(b) This test can only be applied when σ_c is less than the yield stress of $S_{y\,\mathrm{H}}$.
(c) The dimensionless effective creep stress parameter Z for any combination of loading is given in Figure 9.

6.4 Strain Limits – Inelastic Analysis

In regions expecting elevated temperatures, the maximum accumulated positive principal inelastic strain shall meet the following requirements:

(a) strains averaged through the thickness ≤ 1%
(b) strains at the surface due to an equivalent linear distribution of strain through the thickness ≤ 2%
(c) local strains at any point ≤ 5%

7 Creep Fatigue Evaluation

7.1 General Requirements

The rules for creep fatigue evaluation discussed in this section are applicable as follows:

(a) The rules in 6.2 for Tests A-2 and A-3 are met and/or the rules for 6.3 for Tests B-1 and B-2 with $Z < 1.0$ are met. However, the contribution of stress due to radial thermal gradients to the secondary stress range may be excluded of the applicability of elastic creep fatigue rules in Tests A-2 and A-3.
(b) The rule for $(P_L + P_b + Q) \leq 3S$ is met using for $3S$ the lesser of $3S_m$, and $3\bar{S}_\mathrm{m}$ as defined in Test A-1.
(c) Pressure-induced membrane and bending stresses and thermal-induced membrane stresses are classified as primary (load-controlled) stresses.

7.2 Creep Fatigue Procedure

7.2.1 Creep Procedure

The creep procedure is performed as follows:

Step 1. Determine the maximum equivalent strain range, $\Delta\varepsilon_{max}$ from the equation

$$\Delta\varepsilon_{max} = \frac{2S_{alt}}{E}. \tag{26}$$

The value of E is determined from Section II, Part D at the same temperature as the applicable isochronous stress-strain curve.

Step 2. Determine the maximum strain, $\Delta\varepsilon_{mod}$, including the effect of stress concentration factors. The concentration factors take into account plasticity and creep effects. The modified maximum equivalent strain range, $\Delta\varepsilon_{mod}$, is calculated using the procedure specified in any one of *(a)*, *(b)*, or *(c)*.

(a) The modified maximum equivalent strain range, $\Delta\varepsilon_{mod}$, may be calculated from a composite stress-strain diagram as shown in Figure 10, which is constructed by adding the elastic stress-strain curve for the stress range, S_{rH}, to the appropriate time-independent isochronous stress-strain curve from Section II, Part D, Nonmandatory Appendix E, Figures E-100.18-1 through E-100.18-15, Figures E-100.19-1 through E-100.19-15, Figures E-100.20-1 through E-100.20-12, Figures E-100.21-1 through E-100.21-11, and Figures E-100.22-1 through E-100.22-11. A conservative determination of the modified maximum

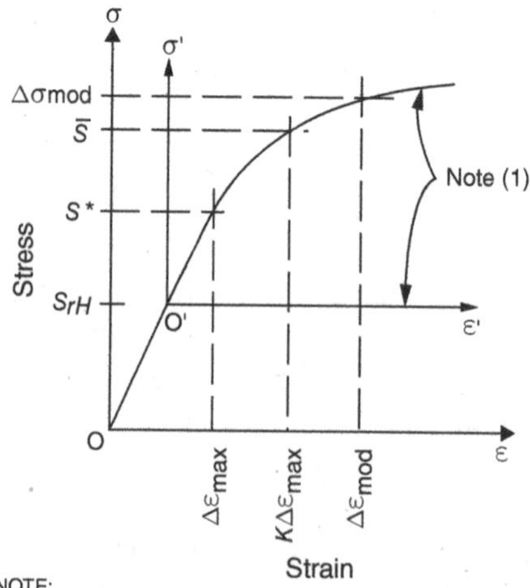

NOTE:

(1) Zero time σ'-e' cure with σ' ,e' coordinates same as Figure E-100.18-1 through E-100.18-15, Figures E-100.19-1 through E-100.19-15, Figures E-100.20-1 through E-100.20-12, Figures E-100.21-1 through E-100.21-11, and Figures E-100.22-1 through E-100.22-11 isochronous curves.

Figure 10 Stress-strain relationship.

equivalent strain range, $\Delta\varepsilon_{mod}$, relative to the maximum equivalent strain range, $\Delta\varepsilon_{max}$, is given by

$$\Delta\varepsilon_{mod} = \frac{S^*}{S} \times K_S^2 \times \Delta\varepsilon_{max}. \tag{27}$$

(b) The modified maximum equivalent strain range, $\Delta\varepsilon_{mod}$, may be calculated more accurately and less conservatively than Eq. (27) using the following equation:

$$\Delta\varepsilon_{mod} = K_S^2 \times S^* \times \frac{\Delta\varepsilon_{max}}{\Delta\sigma_{mod}}. \tag{28}$$

The unknowns $\Delta\varepsilon_{mod}$ and $\Delta\sigma_{mod}$ in Eq. (28) must be solved graphically or analytically by curve fitting the appropriate composite stress-strain curve.

(c) The most conservative estimate of the modified maximum equivalent strain range, $\Delta\varepsilon_{mod}$, may be obtained as

$$\Delta\varepsilon_{mod} = K_e \times K_s \times \Delta\varepsilon_{max} \tag{29}$$

where

$$K_e = 1.0 \text{ if } K_s \times \Delta\varepsilon_{max} \leq \frac{3\overline{S}}{E}$$

$$= K_s \times \Delta\varepsilon_{max} \times \frac{E}{3\overline{S}} \text{ if } K_s \times \Delta\varepsilon_{max} > \frac{3\overline{S}}{E}$$

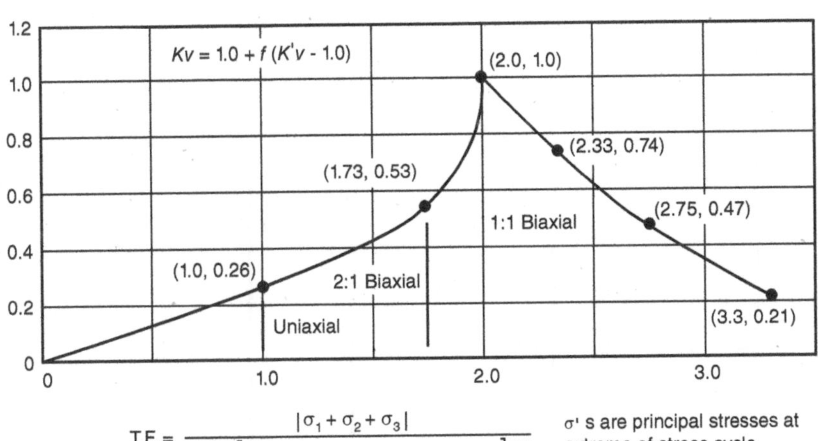

$$\text{T.F.} = \frac{|\sigma_1 + \sigma_2 + \sigma_3|}{\frac{1}{\sqrt{2}}\left[(\sigma_1 - \sigma_2)^2 + (\sigma_2 - \sigma_3)^2 + (\sigma_2 - \sigma_1)^2\right]^{1/2}}$$

σ' s are principal stresses at extreme of stress cycle.

Figure 11 Inelastic multiaxial adjustments.

Step 3. Determine the creep strain increment, $\Delta\varepsilon_c$ for the stress cycle due to load-controlled stresses by using a stress intensity equal to 1.25 times the effective creep stress $\sigma_c = Z\,S_y$. The value of Z is obtained from 6.3.2 with the following exception: The stress cycle time including hold time between transients shall be used instead of the entire service life.

Enter the appropriate isochronous stress-strain curve for the maximum metal temperature during the stress cycle time-temperature block with the $1.25\sigma_c$ stress held constant throughout each temperature-time block of the stress cycle. The value of $\Delta\varepsilon_c$ is equal to:

(a) The sum of the creep strain increment accumulated in one stress cycle time; or alternatively,

(b) The creep strain accumulated during the entire service life divided by the number of stress cycles during the entire service life;

(c) The $\Delta\varepsilon_c$ value used need not exceed 1% divided by the total number of stress cycles.

Step 4. Determine the multiaxial plasticity and Poisson's ratio adjustment factor, K_ν, defined as

$$K_\nu = 1.0 + f(K_V' - 1.0) \tag{30}$$

where

f = the triaxiality factor

K_V' = the plastic Poisson ratio adjustment factor obtained from Figure 12.

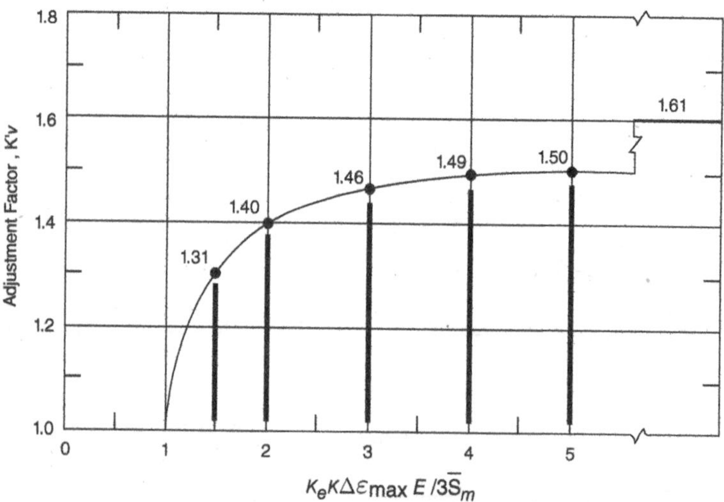

Figure 12 Adjustment for inelastic biaxial Poisson's ratio.

Step 5. Determine the total strain value, ε_t, that includes elastic, plastic, and creep considerations. The total strain is given by:

$$\varepsilon_t = K_v \Delta\varepsilon_{\text{mod}} + K_s \Delta\varepsilon_c. \tag{31}$$

Step 6. Select the time-independent isochronous stress-strain curve from Section II, Part D, Figures E-100.18-1 through E-100.18-15, Figures E-100.19-1 through E-100.19-15, Figures E-100.20-1 through E-100.20-12, Figures E-100.21-1 through E-100.21-1 1, and Figures E-100.22-1 through E-100.22-11, that corresponds to the hold-time temperature, T_{HT}. Enter that stress-strain curve at strain level ε_t and establish the corresponding initial stress level, S_j.

Step 7. Account for stress relaxation, $\bar{s}_r$, during the average cycle time $\bar{t}_j$. This stress relaxation evaluation is to be performed at a constant temperature equal to T_{HT} and initial stress S_j for cycle type j. The stress relaxation history may be determined by entering the appropriate isochronous stress-strain curves of Section II, Part D, Figures E-100.18-1 through E-100.18-15, Figures E-100.19-1 through E-100.19-15, Figures E-100.20-1 through E-100.20-12, Figures E-100.21-1 through E-100.21-11, and Figures E-100.22-1 through E-100.22-11 at a strain level equal to ε_t and determining corresponding stress levels at varying times, as illustrated in Figure 13. This stress relaxation process shall not be permitted to proceed to a stress level less than 1.25 times the core stress σ_c. The stress relaxation procedure results in a stress-time history similar to that illustrated in Figure 14.

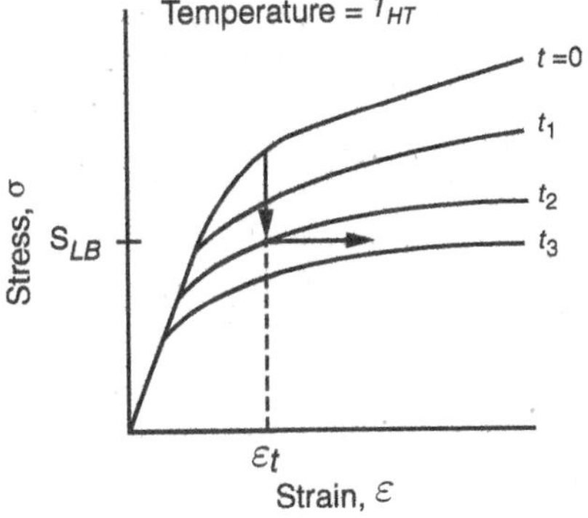

Figure 13 Method for determining stress relaxation from isochronous curves.

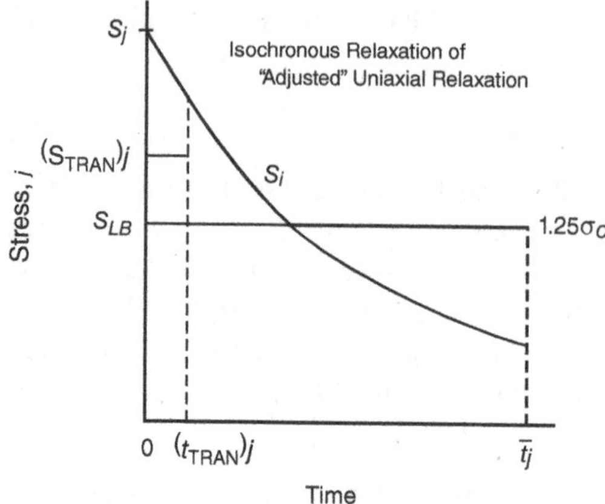

Figure 14 Stress-relaxation limits for Creep damage.

Step 8. Figure 14, which is a composite stress/temperature time history envelope, is divided into q time intervals to facilitate the evaluation of creep damage. These q time intervals are selected to conveniently represent the composite stress/temperature history as a stepwise function of time. During each of these time intervals, $(\Delta t)_k$, the stress, $(S)_k$, and temperature, $(T)_k$, are assumed to be constant and are selected to represent the most damaging stress/temperature combination that could exist during that time interval.

Step 9. Divide the stress $(S)_k$ for each time interval, $(\Delta t)_k$ obtained from Step 8 by the quantity K' obtained from Table 9. K' is an approximate adjustment factor correlating results obtained from this simplified approach to results obtained from more sophisticated analyses. The quantity $(S)_k/K'$ for each time interval $(\Delta t)_k$ is then used in Table 4 to obtain a time-to-rupture value, (T_d).This process is repeated for each time increment $(\Delta t)_k$ and the following equation is solved:

$$\Sigma\left(\frac{\Delta t}{T_d}\right)_k \tag{32}$$

7.2.2 Fatigue Procedure

The fatigue procedure is performed as follows:

Step 1. Determine the number of applied repetitions, $(nc)_j$, of cycle type, j.

Step 2. Determine the number of design allowable cycles, $(N_d)_j$, for cycle type, j, obtained from design fatigue data in Section II, Part D, Figures E-100.16-1 through E-100.16-5, using the maximum strain value during the cycle obtained from Eq. (31).

Table 9 *K′* Values.

Material	*K′* Elastic Analysis	*K′* Inelastic Analysis
2.25Cr·1Mo	0.9	0.67
9Cr·1Mo·V	1.0	0.87
Ni·Fe·Cr (Alloy 800H)	0.9	0.67
Types 304SS and 316SS	0.9	0.67

Step 3. Calculate the ratio of n_c/N_d for all cycle types. This ratio must be less than 1.0.

$$\Sigma\left(\frac{n_C}{N_d}\right)_f \leq 1.0. \tag{33}$$

7.2.3 Creep-Fatigue Interaction

Section VII, Division 2. Eq. (5-G.39) for creep rupture and Eq.(5-G.33) for fatigue are combined as follows:

$$\Sigma\left(\frac{\Delta t}{r_d}\right)_k + \Sigma\left(\frac{n_c}{N_d}\right)_j \leq D. \tag{34}$$

Compliance with Eq. (34) is obtained from Figure 15.

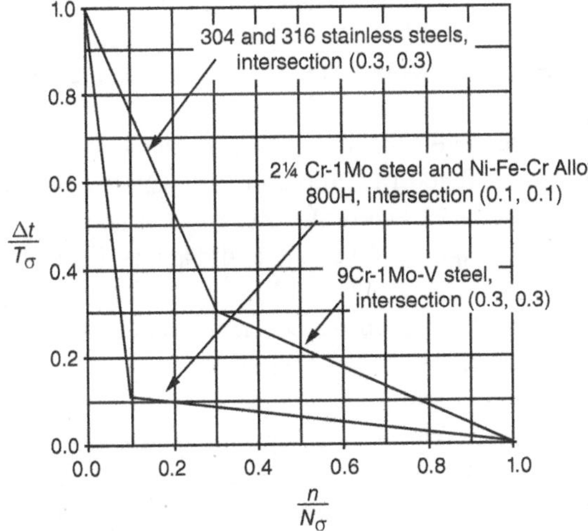

Figure 15 Creep-Fatigue damage envelope.

8 Nomenclature

D = value in Figure 15 encompassed by the horizontal and vertical axes, and the curve associated with a given material

$\Delta\varepsilon_{max}$ = the maximum equivalent strain range as obtained from Eq. (31)

$\Delta\varepsilon_{mod}$ = the modified maximum equivalent strain range that accounts for the effects of local plasticity and creep

ε_i = the creep strain that would be expected from a stress level of $1.25S_y$ at the operating temperature applied for the total time duration during the service lifetime that the metal is at operating temperature

E = the elastic modulus of elasticity

f = the triaxiality factor obtained from Figure 11

K = shape factor based in plastic analysis, for rectangular sections, $K = 1.5$

K_s = either the equivalent stress concentration factor, as determined by test or analysis, or the maximum value of the theoretical elastic stress concentration factor in any direction for the local area under consideration. The equivalent stress concentration factor is defined as the effective (von Mises) primary plus secondary plus peak stress divided by the effective primary plus secondary stress. Note that fatigue strength reduction factors developed from low temperature continuous cycling fatigue tests may not be acceptable for defining K_t When creep effects are not negligible.

K_t = Shape factor based on creep analysis. For rectangular sections, $K_t = (3n)/(1 + 2n)$ Where n is the exponent in Norton's creep equation. K_t is equal to 1.25 in this code case

K_v = multiaxial plasticity and passion ratio adjustment factor

K'_v = plastic poisson ratio adjustment factor determined from figure 12

O = origin of the composite isochronous stress-strain curve, Figure 10, used in the analysis

O' = origin of the time-independent isochronous Stress-strain curve of section II, part D, Figures E-100.18-1 through E-100.18-15, Figures E-100.19-1 through E-100.19-15, Figures E-100.20-1 through E-100.20-12, Figures E-100.21 through E-100.21-11, and Figures E-100.22-1 through E-100.22-11

P_{1L} = Pressure induced and thermally induced local membrane stress

P_{1b} = Pressure induced primary plus secondary bending stress

Q_1 = thermally induced secondary stress

R = the appropriate ratio of the weld metal creep rupture strength to the base metal creep rupture strength. Values are Shown in Section II, Part D, Figures E-100.5-1 through E-100.5-5. The lowest S_t Value of the adjacent base metal Shall be Utilized for the weldment.

r = defined in t_{ld}

S = the maximum allowable value of general primary membrane equivalent stress. Values are shown in Section II, part D, Table 5A and 5B

S^* = the stress indicator determined by entering the stress-strain curve of Section II, Part D, Figures E- 100.18-1 through E-100.18-15, Figures E-100.19-1 through E-100.19-15, Figures E-100.20-1 through E-100.20-12, Figures E-100.21-1 through E-100.21-11, and Figures E-100.22-1 through E-100.22-11 at strain range of $\Delta\varepsilon_{max}$

$\bar{S}$ = the stress indicator determined by entering the stress-strain curve of Section II, Part D, Figures E- 100.18-1 through E-100.18-15, Figures E-100.19-1 through E-100.19-15, Figures E-100.20-1 through E-100.20-12, Figures E-100.21-1 through E-100.21-11, and Figures E-100.22-1 through E-100.22-11 at a strain range of $K\Delta\varepsilon_{max}$

$2S_{alt} = (P_L + Pb + Q)$ at a given point

S_m = the lowest stress value at a given temperature among the time independent strength quanties in Section II, Part D and extended to elevated temperature. Stress values are listed in Section II, Part D, Figures E-100.4-1 through E- 100.4-5

S_{mt} = for base material, the allowable limit of general Primary membrane equivalent stress. Values Are shown in Section II, Part D, Figures E-100.4-1 through E-100.4-5. The S_{mt} values are the lower of two equivalent stress values, S_m (time-independent) and S_t (time-dependent).

= for weldments, it shall be taken as the lower of the S_{mt} values in Section II, Part D, Figures E-100.4-1 through E-100.4-5 or $0.8S_r$ x R.

S_r = the expected minimum stress-to-rupture strength. Values for base materials are shown in Section II, Part D, Figures E-100.7-1 through E-100.7-6. Weld materials strength reduction values are shown in Section II, Part D, Tables E-100.8-1 through E-100.8-3, Tables E-100.9-1 through E-100.9-3, and Tables E-100.10-1, E-100.10-2, E-100.11-1 and E-100.12-1.

S_{rH} = relaxation stress at the hot end of a cycle

S_{rL} = relaxation stress at the cold end of a cycle

S_t = for base material, a temperature and time-dependent equivalent stress limit. Values are shown in Section II, Part D, Figures E-100.5-1 through E-100.5-5. The data considered in establishing these values are obtained from long-term constant-load uniaxial tests. For each specific time, t, the S_t values shall be the least of:

(a) 100% of the average stress required to obtain a total (elastic, plastic, primary, and secondary creep) strain of 1%

(*b*) 80% of the minimum stress to cause initiation of tertiary creep
(*c*) 67% of the minimum stress of cause rupture

= for weldments, it shall be taken as the lower of The St Values in section II, part D, Figures E-100.5-1 through E-100.5.5 or 0.8 S_t x R.

S_u = tensile strength of the material. Values are Shown in Section II, part D, Table E-100.1-1

S_y = yield stress. Values are Shown in section II, part D, table E-100.6-1

S_{ya} = the average of the S_y values at the maximum and minimum wall averaged temperatures during the cycle being considered

S_{yH} = yield stress of the wall averaged temperature at the hot end of the cycle

S_{yL} = yield stress of the wall averaged temperature at The low end of the cycle

S = defined in t_{id}

t_i = the total duration of a specific loading P_{mi}, at elevated temperature, T, during the entire service life of the component

t_{ib} = the time value determined by entering Section II, Part D, Figures E-100.7-1 through E-100.7-6 at a value of stress equal to $(P_L + P_b/K_t)$

t_{id} = the time value determined by entering Section II, Part D. Figures E-100.7-1 through E-100.7-6 (stress to rupture) at a value of stress equal to a factor, *s*, times the yield stress, S_y, associated with T_i. The *s* factor is based on the combined effects of cyclic strain hardening, a factor of 1.2, (or softening, a factor of 0.8) and a factor of 1.25 to convert from minimum to average yield strength. The time to rupture factor, *r*, accounts for the effect of cyclic softening on minimum creep rupture strength. The values of *r* and s are shown in Table 6. If S_y is above the stress values in Figures E-100.7-1 through E-100.7-6, this test cannot be satisfied. When S_y is below the lowest stress value provided in Figures E-100.7-1 through E-100.7-6, the constant temperature line may be extrapolated to larger t_{id} values using the steepest slope on Figure 3 for the material.

t_{im} = maximum allowed time under the load, S_t, as determined from Section II. Part D, Figures E-100.5-1 through E-100.5-5

$X = P_m/S_y$

P_m = general membrane stress

$Y = \Delta Q/S_y$

ΔQ = range of thermal stress calculated elastically

Appendix B

Equations for Average Isochronous Stress-Strain Curves

B.1 Type 304 Stainless Steel Material

B.1.1 304 Customary Units

The total strain is given by

$$\varepsilon = \varepsilon_e + \varepsilon_p + \varepsilon_c \tag{B.1}$$

where ε_e, ε_p, and ε_c are elastic, plastic, and creep strain, respectively.

Define

T = temperature, °F

σ = stress, ksi

E = elastic modulus, ksi

(a) Elastic strain

This is defined by the equation

$$\varepsilon_e = \sigma / E \tag{B.2}$$

with

$$E = 33{,}675 - 13.8229T_k. \tag{B.3}$$

(b) Plastic strain

The plastic strain is defined by the following equations

$$\varepsilon_p = 0 \qquad \sigma \le \sigma_p \tag{B.4}$$

$$\varepsilon_p = [(\sigma - \sigma_p)/K]^{1/m} \quad \sigma > \sigma_p \tag{B.5}$$

Analysis of ASME Boiler, Pressure Vessel, and Nuclear Components in the Creep Range, Second Edition. Maan H. Jawad and Robert I. Jetter.
© 2022 John Wiley & Sons Ltd. Published 2022 by John Wiley & Sons Ltd.

where

$$K = 69.9629 - 0.02545T_k \quad \text{(B.6)}$$

$$m = 0.25824 + 7\,749 \times 10^{-5} T_k \quad \text{(B.7)}$$

$$\sigma_p = \sigma_y + (-12.1979 + 0.00671T_k) \quad \text{(B.8)}$$

$$\sigma_y = 32.3128 - 0.01602T_k \quad \text{(B.9)}$$

$$T_k = (T / 1.8) + 255.3722. \quad \text{(B.10)}$$

(c) Creep strain

The expression for creep strain is given by

$$\varepsilon_c = (0.01)\{\varepsilon_s[1 - \text{Exp}(-st)] + \varepsilon_r[1 - \text{Exp}(-rt)] + \dot{\varepsilon}_m t\} \quad \text{(B.11)}$$

with

$$A = 1.38 \times 10^{13}$$

$$B = 2.518 \times 10^{13}$$

$$\beta_e = -\ 3.652 \times 10^{-4} + 7.518 \times 10^{-7} T_k \quad \text{(B.12)}$$

$$\beta_r = -\ 2.252 \times 10^{-4} + 5.401 \times 10^{-7} T_k \quad \text{(B.13)}$$

C = obtained from Table B.4
D = obtained from Table B.3

$$\varepsilon_r = C\dot{\varepsilon}_m / r \quad \text{(B.14)}$$

$$\varepsilon_s = 0 \qquad \sigma \leq 6.0 \quad \text{(B.15)}$$

$$\sigma_s = G + 1000H\sigma \quad \sigma > 6.0 \quad \text{(B.16)}$$

$$\dot{\varepsilon}_m = A\{[\sinh((1000\beta_e\sigma)/n_e)]^{n_e}\}[\text{Exp}(-Q / RT_k)] \quad \text{(B.17)}$$

G = obtained from Table B.1
H = obtained from Table B.2

$$n_e = 6.0$$

$$n_r = 3.5$$

Table B.1 Values of *G* versus temperature, °F, for Type 304 stainless steel.

Temperature range, °F	*G*
$800 \leq T < 850$	0.0
$850 \leq T < 1000$	$2.24449 - 3.08547 \times 10^{-3}T_k$
$1000 \leq T < 1100$	−0.257143
$1100 \leq T < 1500$	−0.257143

Table B.2 Values of *H* versus temperature, °F, Type 304 stainless steel.

Temperature range, °F	*H*
$800 \leq T < 850$	0.0
$850 \leq T < 1000$	$-3.74081 \times 10^{-4} + 5.14244 \times 10^{-7}T_k$
$1000 \leq T < 1100$	4.28571×10^{-5}
$1100 \leq T < 1500$	4.28571×10^{-5}

Table B.3 Values of *D* versus temperature, °F, Type 304 stainless steel.

Temperature range, °F	*D*
$800 \leq T < 850$	2.266×10^{15}
$850 \leq T < 1000$	2.266×10^{15}
$1000 \leq T < 1100$	$3.19663 \times 10^{16}\ \text{Exp}(-3.66218 \times 10^{-3}T_k)$
$1100 \leq T < 1500$	2.518×10^{14}

Table B.4 Values of *C* versus temperature, °F, Type 304 stainless steel.

Temperature range, °F	*C*
$800 \leq T < 850$	$2.469 \times 10^{-3}\ \text{Exp}(6580.986/T_k)$
$850 \leq T < 1000$	$2.469 \times 10^{-3}\ \text{Exp}(6580.986/T_k)$
$1000 \leq T < 1100$	$56.2405 - 5.91691 \times 10^{-2}T_k$
$1100 \leq T < 1500$	5.0

$Q = 67{,}000$

$R = 1.987$

$$r = B\left\{[\sinh((1000\beta_r\sigma)/n_r)]^{n_r}\right\}\left[\mathrm{Exp}\left(-Q/RT_k\right)\right] \quad \text{(B.18)}$$

$$s = D\left\{[\sinh((1000\beta_r\sigma)/n_r)]^{n_r}\right\}\left\{\mathrm{Exp}\left[-Q/\left(RT_k\right)\right]\right\} \quad \text{(B.19)}$$

$t =$ time, hours.

B.1.2 304 SI Units

The total strain is given by Eq. (B.1). The equations for elastic, plastic, and creep strains are defined below.

Define,
T = temperature, °C
σ = stress, MPa
E = elastic modulus, MPa

(a) Elastic strain

The elastic strain is defined by the equation

$$\varepsilon_e = \sigma / E \quad \text{(B.2)}$$

with

$$E = 232{,}181 - 95.3063T_k. \quad \text{(B.20)}$$

(b) Plastic strain

The plastic strain is defined by the following equations

$$\varepsilon_p = 0 \qquad \sigma \le \sigma_p \quad \text{(B.4)}$$

$$\varepsilon_p = [(\sigma - \sigma_p)/K]^{1/m} \qquad \sigma > \sigma_p \quad \text{(B.5)}$$

where

$$K = 482.38 - 0.17547T_k \quad \text{(B.21)}$$

$$m = 0.25824 + 7\,749 \times 10^{-5}T_k \quad \text{(B.22)}$$

$$\sigma_p = \sigma_y + \left(-84.1023 + 0.046259T_k\right) \quad \text{(B.23)}$$

$$\sigma_y = 222.79 - 0.11047T_k \quad \text{(B.24)}$$

$$T_k = T + 273.15 \quad \text{(B.25)}$$

(c) Creep strain

The expression for creep strain is given by

$$\varepsilon_c = (0.01)\left\{\varepsilon_s\left[1-\mathrm{Exp}(-st)\right]+\varepsilon_r\left[1-\mathrm{Exp}(-rt)\right]+\varepsilon`_m t\right\} \tag{B.11}$$

with

$$A = 1.38\times10^{13}$$

$$B = 2.518\times10^{13}$$

$$\beta_e = -\ 3.652\times10^{-4} + 7.518\times10^{-7}T_k \tag{B.12}$$

$$\beta_r = -\ 2.252\times10^{-4} + 5.401\times10^{-7}T_k \tag{B.13}$$

$C =$ Table B.8

$D =$ Table B.7

$$\varepsilon_r = C\dot{\varepsilon}_m / r \tag{B.14}$$

$$\varepsilon_s = 0 \qquad \sigma \le 41.3686 \tag{B.26}$$

$$\varepsilon_s = G + 145.037681H\sigma \quad \sigma > 41.3686 \tag{B.27}$$

$$\dot{\varepsilon}_m = A\left\{[\sinh((145.037681\beta_e\sigma)/n_e)]^{n_e}\right\}\left[\mathrm{Exp}(-Q/RT_k)\right. \tag{B.28}$$

G = Table B.5
H = Table B.6

$$n_e = 6.0$$

$$n_r = 3.5$$

$$Q = 67\,000$$

$$R = 1.987$$

$$r = B\left\{[\sinh\ ((145.037681\beta_r\sigma)/n_r)]^{n_r}\right\}\left[\mathrm{Exp}(-Q/RT_k)\right] \tag{B.29}$$

$$s = D\left\{[\sinh\ ((145.0376811\beta_r\sigma)/n_r)]^{n_r}\right\}\left\{\mathrm{Exp}\left[-Q/(RT_k)\right]\right\} \tag{B.30}$$

$t =$ time, hours.

Table B.5 Values of *G* versus temperature, °C, Type 304 stainless steel.

Temperature range, °C	*G*
$427 \le T < 454$	0.0
$454 \le T < 538$	$2.24449 - 3.08547 \times 10^{-3} T_k$
$538 \le T < 593$	−0.257143
$593 \le T < 816$	−0.257143

Table B.6 Values of *H* versus temperature, °C, Type 304 stainless steel.

Temperature range, °C	*H*
$427 \le T < 454$	0.0
$454 \le T < 538$	$-3.74081 \times 10^{-4} + 5.14244 \times 10^{-7} T_k$
$538 \le T < 593$	4.28571×10^{-5}
$593 \le T < 816$	4.28571×10^{-5}

Table B.7 Values of *D* versus temperature, °C, Type 304 stainless steel.

Temperature range, °C	*D*
$427 \le T < 454$	2.266×10^{15}
$454 \le T < 538$	2.266×10^{15}
$538 \le T < 593$	$3.19663 \times 10^{16}\ \mathrm{Exp}(-3.66218 \times 10^{-3} T_k)$
$593 \le T < 816$	2.518×10^{14}

Table B.8 Values of *C* versus temperature, °C, Type 304 stainless steel.

Temperature range, °C	*C*
$427 \le T < 454$	$2.469 \times 10^{-3}\ \mathrm{Exp}(6580.986/T_k)$
$454 \le T < 538$	$2.469 \times 10^{-3}\ \mathrm{Exp}(6580.986/T_k)$
$538 \le T < 593$	$56.2405 - 5.91691 \times 10^{-2} T_k$
$593 \le T < 816$	5.0

B.2 Type 316 Stainless Steel Material

B.2.1 316 Customary Units

The total strain is given by Eq. (B.1). The equations for elastic, plastic, and creep strains are defined below.

Define,
T = temperature, °F
σ = stress, ksi
E = elastic modulus, ksi

(a) Elastic strain
The elastic strain is defined by the equation

$$\varepsilon_e = \sigma / E \tag{B.2}$$

with

$$E = 33.675 - 13.8229T_k \tag{B.3}$$

(b) Plastic strain
The plastic strain is defined by the following equations

$$\varepsilon_p = 0 \qquad \sigma \leq \sigma_p \tag{B.4}$$

$$\varepsilon_p = [(\sigma - \sigma_p)/K]^{1/m} \qquad \sigma > \sigma_p \tag{B.5}$$

where

$$K = 60.78463 - 0.01380T_k \tag{B.31}$$

$$m = 0.30950 + 6.1328 \times 10^{-5} T_k \tag{B.32}$$

$$\sigma_p = \sigma_y + \left(-8.18851 + 0.00351T_k\right) \tag{B.33}$$

$$\sigma_y = 29.25163 - 0.01006T_k \tag{B.34}$$

$$T_k = T/1.8 + 255.37 \tag{B.10}$$

(c) Creep strain
The expression for creep strain is given by

$$\varepsilon_c = (0.01)\left\{\varepsilon_s\left[1 - \text{Exp}(-\text{st})\right] + \varepsilon_r\left[1 - \text{Exp}(-\text{rt})\right] + \dot{\varepsilon}_m t\right\} \tag{B.11}$$

with

A = Table B.16
B = Table B.14

$$\beta = -\ 4.257 \times 10^{-4} + 7.733 \times 10^{-7} T_k \tag{B.35}$$

$C =$ Table B.13
$D =$ Table B.11

$$\varepsilon_r = C\,\dot{\varepsilon}_m / r \tag{B.14}$$

$$\varepsilon_s = 0 \qquad \sigma \leq 4.0 \tag{B.36}$$

$$\varepsilon_s = G + 1000H\sigma \quad \sigma > 4.0 \tag{B.37}$$

$$\dot{\varepsilon}_m = A\left\{\left[\sinh\left(\left(1000\beta\sigma\right)/n\right)\right]^n\right\}\left[\mathrm{Exp}\left(-Q/RT_k\right)\right] \tag{B.38}$$

$G =$ Table B.9
$H =$ Table B.10
$L =$ Table B.15
$n =$ Table B.12

$Q = 67\,000$

$R = 1.987$

$$r = \max\left(r_1, r_2\right) \tag{B.39}$$

$$r_1 = L(1000\sigma)^{n-3.6} \tag{B.40}$$

$$r_2 = B\{[\sinh\,((1000\beta\sigma)/n)]^n\}[\mathrm{Exp}\left(-Q/RT_k\right)] \tag{B.41}$$

$$s = \max\left(s_1, s_2\right) \tag{B.42}$$

$$s_1 = 2.5 \times 10^{-2} \tag{B.43}$$

$$s_2 = D\{[\sinh\,((1000\beta\sigma)/n)]^n\}\{\mathrm{Exp}[-Q/\left(RT_k\right)]\} \tag{B.44}$$

Table B.9 Values of *G* versus temperature, °F, for Type 316 stainless steel.

Temperature range, °F	*G*
$800 \leq T < 1000$	0.0
$1000 \leq T < 1075$	$1.28221 - 1.58103 \times 10^{-3} T_k$
$1075 \leq T < 1100$	$1.28221 - 1.58103 \times 10^{-3} T_k$
$1100 \leq T < 1200$	$-0.271855 + 2.13509 \times 10^{-4} T_k$
$1200 \leq T < 1300$	$-0.692411 + 6.69643 \times 10^{-4} T_k$
$1300 \leq T \leq 1500$	$-0.704318 + 6.61818 \times 10^{-4} T_k$

Table B.10 Values of *H* versus temperature, °F, for Type 316 stainless steel.

Temperature range, °F	*H*
$800 \le T < 1000$	0.0
$1000 \le T < 1075$	$-3.2055 \times 10^{-4} + 3.9525 \times 10^{-7} T_k$
$1075 \le T < 1100$	$-3.2055 \times 10^{-4} + 3.9525 \times 10^{-7} T_k$
$1100 \le T < 1200$	$6.796 \times 10^{-5} - 5.338 \times 10^{-8} T_k$
$1200 \le T < 1300$	$1.731 \times 10^{-4} - 1.674 \times 10^{-7} T_k$
$1300 \le T \le 1500$	$1.7608 \times 10^{-4} - 1.70455 \times 10^{-7} \mathrm{T_k}$

Table B.11 Values of *D* versus temperature, °F, for Type 316 stainless steel.

Temperature range, °F	*D*
$800 \le T < 1000$	5.7078×10^{13}
$1000 \le T < 1075$	$-4.499 \times 10^{17} + 5.548 \times 10^{14} T_k$
$1075 \le T < 1100$	$2.869 \times 10^{17} - 3.093 \times 10^{14} T_k$
$1100 \le T < 1200$	$2.869 \times 10^{17} - 3.093 \times 10^{14} T_k$
$1200 \le T < 1300$	$1.3369 \times 10^{10} \mathrm{Exp}(10878.5/T_k)$
$1300 \le T \le 1500$	$1.3369 \times 10^{10} \mathrm{Exp}(10878.5/T_k)$

Table B.12 Values of *n* versus temperature, °F, for Type 316 stainless steel.

Temperature range, °F	*n*
$800 \le T < 1000$	4.6
$1000 \le T < 1075$	$-80.9236 + 0.105455 T_k$
$1075 \le T < 1100$	$50.1136 - 0.0482143 T_k$
$1100 \le T < 1200$	$50.1136 - 0.0482143 T_k$
$1200 \le T < 1300$	$14.4647 - 9.54954 \times 10^{-3} T_k$
$1300 \le T \le 1500$	$14.4647 - 9.54954 \times 10^{-3} T_k$

Table B.13 Values of *C* versus temperature, °F, for Type 316 stainless steel.

Temperature range, °F	*C*
$800 \le T < 1000$	7.1
$1000 \le T < 1075$	$25.5318 - 0.0227273 T_k$
$1075 \le T < 1100$	$25.5318 - 0.0227273 T_k$
$1100 \le T < 1200$	$54.5625 - 0.05625 T_k$
$1200 \le T < 1300$	$7.68378 - 5.4054 \times 10^{-3} T_k$
$1300 \le T \le 1500$	$7.68378 - 5.4054 \times 10^{-3} T_k$

Table B.14 Values of *B* versus temperature, °F, for Type 316 stainless steel.

Temperature range, °F	*B*
$800 \le T < 1000$	5.7078×10^{13}
$1000 \le T < 1075$	$-3.922 \times 10^{16} + 4.84416 \times 10^{13} T_k$
$1075 \le T < 1100$	$1.44225 \times 10^{-8} \mathrm{Exp}(45475.8/T_k)$
$1100 \le T < 1200$	$1.44225 \times 10^{-8} \mathrm{Exp}(45475.8/T_k)$
$1200 \le T < 1300$	$2.85517 \times 10^{8} \mathrm{Exp}(10878.5/T_k)$
$1300 \le T \le 1500$	$2.85517 \times 10^{8} \mathrm{Exp}(10878.5/T_k)$

Table B.15 Values of *L* versus temperature, °F, for Type 316 stainless steel.

Temperature range, °F	*L*
$800 \le T < 1000$	$[\mathrm{Exp}(43.1255)][\mathrm{Exp}(-49995.0/T_k)]$
$1000 \le T < 1075$	$\mathrm{Exp}(-1153.4 + 16.446 T_z - 0.075433 T_z^2 + 0.00010780 T_z^3)$
$1075 \le T < 1100$	$\mathrm{Exp}(-274.235 + 1.1560 T_z - 0.001160 T_z^2)$
$1100 \le T < 1200$	$\mathrm{Exp}(-274.235 + 1.1560 T_z - 0.001160 T_z^2)$
$1200 \le T < 1300$	$\mathrm{Exp}(-54.603 + 0.1185 T_W - 8.6357 \times 10^{-6} T_W^2)$
$1300 \le T \le 1500$	$\mathrm{Exp}(-54.603 + 0.1185 T_W - 8.6357 \times 10^{-6} T_W^2)$

$T_W = T_k - 680$
$T_z = T_k - 610$

Table B.16 Values of *A* versus temperature, °F, for Type 316 stainless steel.

Temperature range, °F	*A*
$800 \le T < 1000$	5.6229×10^{12}
$1000 \le T < 1075$	$-7.8535 \times 10^{15} + 9.6933 \times 10^{12} T_k$
$1075 \le T < 1100$	$5.28787 \times 10^{-6} \mathrm{Exp}(39057.1/T_k)$
$1100 \le T < 1200$	$5.28787 \times 10^{-6} \mathrm{Exp}(39057.1/T_k)$
$1200 \le T < 1300$	$6.03371 \times 10^{10} \mathrm{Exp}(4967.76/T_k)$
$1300 \le T \le 1500$	$6.03371 \times 10^{10} \mathrm{Exp}(4967.76/T_k)$

B.2.2 316 SI Units

The total strain is given by Eq. (B.1). The equations for elastic, plastic, and creep strains are defined below.

Define,
T = temperature, °C
σ = stress, MPa
E = elastic modulus, MPa

(a) Elastic strain
The elastic strain is defined by the equation

$$\varepsilon_e = \sigma / E \tag{B.2}$$

with

$$E = 232{,}181 - 95.3063T_k \tag{B.20}$$

(b) Plastic strain
The plastic strain is defined by the following equations

$$\varepsilon_p = 0 \qquad \sigma \leq \sigma_p \tag{B.4}$$

$$\varepsilon_p = [(\sigma - \sigma_p)/K]^{1/m} \qquad \sigma > \sigma_p \tag{B.5}$$

where

$$K = 419.11 - 0.095119T_k \tag{B.45}$$

$$m = 0.30950 + 6.1328 \times 10^{-5}T_k \tag{B.46}$$

$$\sigma_p = \sigma_y + \left(-56.4598 + 0.024225T_k\right) \tag{B.47}$$

$$\sigma_y = 201.69 - 0.069361T_k \tag{B.48}$$

$$T_k = T + 273.15 \tag{B.25}$$

(c) Creep strain
The expression for creep strain is given by

$$\varepsilon_c = (0.01)\left\{\varepsilon_s\left[1 - \text{Exp}(-\text{st})\right] + \varepsilon_r\left[1 - \text{Exp}(-\text{rt})\right] + \dot{\varepsilon}_m t\right\} \tag{B.11}$$

with

A = Table B.24
B = Table B.22

$$\beta = -\ 4.257\text{x}10^{-4} + 7.733 \times 10^{-7}T_k \tag{B.49}$$

C = Table B.21
D = Table B.19

$$\varepsilon_r = C\,\dot{\varepsilon}_m / r \tag{B.14}$$

Table B.17 Values of *G* versus temperature, °C, for Type 316 stainless steel.

Temperature range, °C	*G*
$427 \leq T < 538$	0.0
$538 \leq T < 579$	$1.28221 - 1.58103 \times 10^{-3} T_k$
$579 \leq T < 593$	$1.28221 - 1.58103 \times 10^{-3} T_k$
$593 \leq T < 649$	$-0.271855 + 2.13509 \times 10^{-4} T_k$
$649 \leq T < 704$	$-0.692411 + 6.69643 \times 10^{-4} T_k$
$704 \leq T < 816$	$-0.704318 + 6.61818 \times 10^{-4} T_k$

Table B.18 Values of *H* versus temperature, °C, for Type 316 stainless steel.

Temperature range, °C	*H*
$427 \leq T < 538$	0.0
$538 \leq T < 579$	$-3.2055 \times 10^{-4} + 3.9525 \times 10^{-7} T_k$
$579 \leq T < 593$	$-3.2055 \times 10^{-4} + 3.9525 \times 10^{-7} T_k$
$593 \leq T < 649$	$6.796 \times 10^{-5} - 5.338 \times 10^{-8} T_k$
$649 \leq T < 704$	$1.731 \times 10^{-4} - 1.674 \times 10^{-7} T_k$
$704 \leq T < 816$	$1.7608 \times 10^{-4} - 1.70455 \times 10^{-7} T_k$

$$\varepsilon_s = 0 \qquad \sigma \leq 27.5790 \tag{B.50}$$

$$\varepsilon_s = G + 145.037681 H\sigma \qquad \sigma > 27.5790 \tag{B.51}$$

$$\dot{\varepsilon}_m = A\{[\sinh\left((145.037681\beta\sigma)/n\right)]^n\}\left[\text{Exp}\left(-Q/RT_k\right)\right] \tag{B.52}$$

G = Table B.17
H = Table B.18
L = Table B.23
n = Table B.20

$Q = 67{,}000$

$R = 1.987$

$$r = \max(r_1, r_2) \tag{B.39}$$

$$r_1 = L(145.037681\sigma)^{n-3.6} \tag{B.53}$$

Table B.19 Values of *D* versus temperature, °C, for Type 316 stainless steel.

Temperature range, °C	*D*
$427 \le T < 538$	5.7078×10^{13}
$538 \le T < 579$	$-4.499 \times 10^{17} + 5.548 \times 10^{14} T_k$
$579 \le T < 593$	$2.869 \times 10^{17} - 3.093 \times 10^{14} T_k$
$593 \le T < 649$	$2.869 \times 10^{17} - 3.093 \times 10^{14} T_k$
$649 \le T < 704$	$1.3369 \times 10^{10} \text{Exp}(10878.5/T_k)$
$704 \le T < 816$	$1.3369 \times 10^{10} \text{Exp}(10878.5/T_k)$

Table B.20 Values of *n* versus temperature, °C, for Type 316 stainless steel.

Temperature range, °C	*n*
$427 \le T < 538$	4.6
$538 \le T < 579$	$-80.9236 + 0.105455 T_k$
$579 \le T < 593$	$50.1136 - 0.0482143 T_k$
$593 \le T < 649$	$50.1136 - 0.0482143 T_k$
$649 \le T < 704$	$14.4647 - 9.54954 \times 10^{-3} T_k$
$704 \le T < 816$	$14.4647 - 9.54954 \times 10^{-3} T_k$

Table B.21 Values of *C* versus temperature, °C, for Type 316 stainless steel.

Temperature range, °C	*C*
$427 \le T < 538$	7.1
$538 \le T < 579$	$25.5318 - 0.0227273 T_k$
$579 \le T < 593$	$25.5318 - 0.0227273 T_k$
$593 \le T < 649$	$54.5625 - 0.05625 T_k$
$649 \le T < 704$	$7.68378 - 5.4054 \times 10^{-3} T_k$
$704 \le T < 816$	$7.68378 - 5.4054 \times 10^{-3} T_k$

$$r_2 = B\{[\sinh((145.037681\beta\sigma)/n)]^n\}\left[\text{Exp}\left(-Q/RT_k\right)\right] \quad \text{(B.54)}$$

$$s = \max\left(s_1, s_2\right) \quad \text{(B.42)}$$

$$s_1 = 2.5 \times 10^{-2} \quad \text{(B.43)}$$

$$s_2 = D\left\{\left[[\sinh((145.0376811\beta\sigma)/n)\right]^n\}\{\text{Exp}\left[-Q/\left(RT_k\right)\right]\right\} \quad \text{(B.55)}$$

Table B.22 Values of *B* versus temperature, °C, for Type 316 stainless steel.

Temperature range, °C	*B*
$427 \le T < 538$	5.7078×10^{13}
$538 \le T < 579$	$-3.922 \times 10^{16} + 4.84416 \times 10^{13} T_k$
$579 \le T < 593$	$1.44225 \times 10^{-8} \mathrm{Exp}(45475.8/T_k)$
$593 \le T < 649$	$1.44225 \times 10^{-8} \mathrm{Exp}(45475.8/T_k)$
$649 \le T < 704$	$2.85517 \times 10^{8} \mathrm{Exp}(10878.5/T_k)$
$704 \le T < 816$	$2.85517 \times 10^{8} \mathrm{Exp}(10878.5/T_k)$

Table B.23 Values of *L* versus temperature, °C, for Type 316 stainless steel.

Temperature range, °C	*L*
$427 \le T < 538$	$[\mathrm{Exp}(43.1255)][\mathrm{Exp}(-49995.0/Tk)]$
$538 \le T < 579$	$\mathrm{Exp}(-1153.4 + 16.446 T_z - 0.075433 T_z^2 + 0.00010780 T_z^3)$
$579 \le T < 593$	$\mathrm{Exp}(-274.235 + 1.1560 T_z - 0.001160 T_z^2)$
$593 \le T < 649$	$\mathrm{Exp}(-274.235 + 1.1560 T_z - 0.001160 T_z^2)$
$649 \le T < 704$	$\mathrm{Exp}(-54.603 + 0.1185 T_W - 8.6357 \times 10^{-6} T_W^2)$
$704 \le T < 816$	$\mathrm{Exp}(-54.603 + 0.1185 T_W - 8.6357 \times 10^{-6} T_W^2)$

$T_W = T_k - 680$ $T_z = T_k - 610$

Table B.24 Values of *A* versus temperature, °C, for Type 316 stainless steel.

Temperature range, °C	*A*
$427 \le T < 538$	5.6229×10^{12}
$538 \le T < 579$	$-7.8535 \times 10^{15} + 9.6933 \times 10^{12} T_k$
$579 \le T < 593$	$5.28787 \times 10^{-6} \mathrm{Exp}(39057.1/T_k)$
$593 \le T < 649$	$5.28787 \times 10^{-6} \mathrm{Exp}(39057.1/T_k)$
$649 \le T < 704$	$6.03371 \times 10^{10} \mathrm{Exp}(4967.76/T_k)$
$704 \le T < 816$	$6.03371 \times 10^{10} \mathrm{Exp}(4967.76/T_k)$

B.3 Low Alloy 2.25Cr–1Mo Annealed Steel

B.3.1 2.25Cr–1Mo Customary Units

The total strain is given by Eq. (B.1). The equations for elastic, plastic, and creep strains are defined below.

Define

T = temperature, °F

σ = stress, ksi

E = elastic modulus, ksi

(a) Elastic strain

The elastic strain is defined by the equation

$$\varepsilon_e = \sigma / E \tag{B.2}$$

with

E = obtained from Section II-D

(b) Plastic strain

The plastic strain is defined by the following equations

$$\varepsilon_p = 0 \qquad \sigma \le \sigma_y \tag{B.56}$$

$$\varepsilon_p = (1/C)\{\ln[(\sigma - B)/(\sigma_y - B)]\} \qquad \sigma > \sigma_y \tag{B.57}$$

where B, C, and σ_y are given in Table B.25

Table B.25 Parameter values versus temperature, °F, for 2.25Cr-1Mo steel.

T_F, °F	σ_y	B	C
700	30.50	83.66	−25.59
750	28.05	78.47	−32.09
800	28.94	69.33	−39.90
850	28.16	70.76	−35.38
900	26.66	87.22	−26.45
950	26.22	60.70	−45.35
1000	26.58	56.45	−40.45
1050	25.20	46.00	−60.25
1100	24.57	43.63	−52.16
1150	22.23	40.26	−58.78
1200	19.42	34.79	−65.17

(c) Creep strain

The expression for creep strain is given by

ε_c = creep strain obtained from Figure B.1

where

$$a = -\ 13.528 + (6.5196)(U)/T_k + 23349/T_k - (2472.78/T_k)(\ln(6.895\sigma)) \tag{B.58}$$

$$b = -\ 11.098 - (28.2357\sigma)/U + 11965/T_k \tag{B.59}$$

$$C_1 = 10^{d_1} \tag{B.60}$$

$$C_2 = 10^{d_2} \tag{B.61}$$

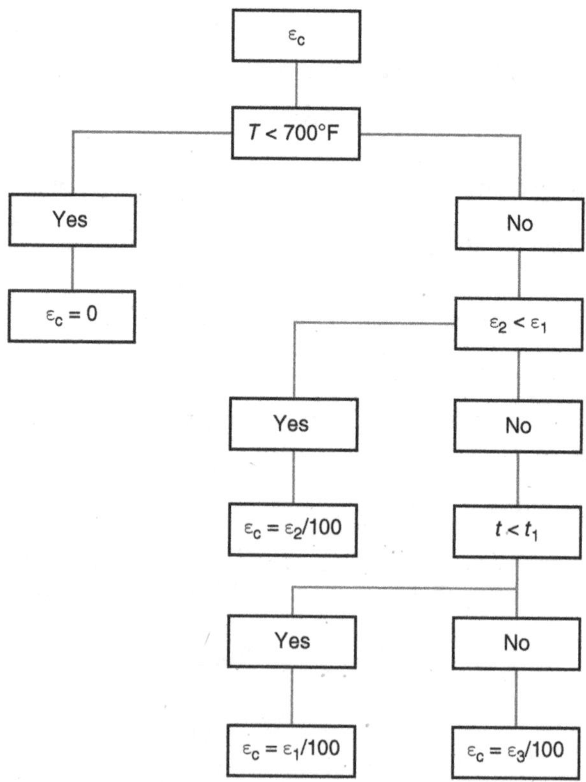

Figure B.1 Evaluation of creep strain – Customary units.

$$d_1 = 1.0328 + 168680/(U)(T_k) - 0.023772U + 0.003437(U)(\ln(6.895\sigma)) \tag{B.62}$$

$$d_2 = -0.051086 + 140730/(U)(T_k) - 0.01U + 0.0016219(U)(\ln(6.895\sigma)) \tag{B.63}$$

$$e_1 = 6.7475 + 0.07878\sigma + (428.9604/U)(\ln(6.895\sigma)) - 13494/T_k \tag{B.64}$$

$$e_2 = 11.498 - (8.2226)(U)/T_k - 20448/T_k + (2546.0/T_k)(\ln(6.895\sigma)) \tag{B.65}$$

$$\varepsilon_1 = [C_1 p_1 t/(1 + p_1 t)] + \dot{\varepsilon}_1 t \tag{B.66}$$

$$\varepsilon_2 = [C_2 p_2 t/(1 + p_2 t)] + \dot{\varepsilon}_2 t \tag{B.67}$$

$$\varepsilon_3 = [C_2 p_2 t'/(1 + p_2 t')] + \dot{\varepsilon}_2 t \tag{B.68}$$

$$\dot{\varepsilon}_1 = 10^{e_1} \tag{B.69}$$

$$\dot{\varepsilon}_2 = 10^{e_2} \tag{B.70}$$

$$\varepsilon_1' = (C_1 p_1 t_1)/(1 + p_1 t_1) + \dot{\varepsilon}_1 t_1 \tag{B.71}$$

$$p_1 = 10^{q_1} \tag{B.72}$$

$$p_2 = 10^{q_2} \tag{B.73}$$

$$q_1 = 7.6026 + 1.45037(\ln(6.895\sigma)) - 12323/T_k \tag{B.74}$$

$$q_2 = 8.1242 + 0.0179678\sigma + (175.728/U)(\ln(6.895\sigma)) - 11659/T_k \tag{B.75}$$

$\sigma =$ stress

$t =$ time, hours

$$t_1 = t_1^{(a)} \text{ for } T \leq 850^\circ\text{F} \tag{B.76}$$

$$= t_1^{(a)} + \left(t_1^{(b)} - t_1^{(a)}\right)(T_F/1.8 - 471.78)/56 \text{ for } 850^\circ\text{F} < T \leq 950^\circ\text{F} \tag{B.77}$$

Table B.26 Values of U versus temperature, °F, for 2.25Cr-1Mo steel.

T_F, °F	U
700	471
750	468
840	452
930	418
1020	364
1110	284
1150	300
1200	270

$$= t_1^{(b)} \text{ for } T > 950°\text{F} \tag{B.78}$$

$$t_1^{(a)} = 10^{a} \tag{B.79}$$

$$t_1^{(b)} = 10^{b} \tag{B.80}$$

$$t_c = \{-\dot{\varepsilon}_2 - C_2 p_2 + \varepsilon_1' p_2 + [4\varepsilon_1' \dot{\varepsilon}_2 p_2 + (\dot{\varepsilon}_2 + (C_2 - \varepsilon_1')p_2)^2]^{0.5} / (2\dot{\varepsilon}_2 p_2) \tag{B.81}$$

$$t' = t - (t_1 - t_c) \tag{B.82}$$

$$T_k = T_F / 1.8 + 255.37 \tag{B.10}$$

U = given in Table B.26

B.3.2 2.25 Cr–1Mo Steel SI Units

The total strain is given by Eq. (B.1). The equations for elastic, plastic, and creep strains are defined below.

Define,

T = temperature, °C

σ = stress, MPa

E = elastic modulus, MPa

(a) Elastic strain

The elastic strain is defined by the equation

$$\varepsilon_e = \sigma / E \tag{B.2}$$

with, E obtained from Section II-D

(b) Plastic strain

The plastic strain is defined by the following equations

$$\varepsilon_p = 0 \qquad \sigma \le \sigma_y \tag{B.56}$$

$$\varepsilon_p = (1/C)\{\ln[(\sigma - B)/(\sigma_y - B)]\} \quad \sigma > \sigma_y \tag{B.57}$$

where B, C, and σ_y are given in Table B.27

(c) Creep strain

The expression for creep strain is given by

ε_c = creep strain obtained from Figure B.2.

where

$$a = -\ 13.528 + (6.5196)(U)/T_k + 23349/T_k - (2472.8/T_k)(\ln\sigma) \tag{B.83}$$

$$b = -\ 11.098 - (4.0951)(\sigma)/U + 11965/T_k \tag{B.84}$$

$$C_1 = 10^{d_1} \tag{B.60}$$

$$C_2 = 10^{d_2} \tag{B.61}$$

$$d_1 = 1.0328 + 168680/(U)(T_k) - 0.023772U + 0.003437(U)(\ln\sigma) \tag{B.85}$$

$$d_2 = -0.051086 + 140730/(U)(T_k) - 0.01U + 0.001622(U)(\ln\sigma) \tag{B.86}$$

Table B.27 Parameter values versus temperature, °C, for 2.25Cr-1Mo steel.

T, °C	σ_y	*B*	*C*
371	210.26	576.81	−25.59
399	193.42	541.04	−32.09
427	199.53	477.98	−39.90
454	194.18	487.86	−35.38
482	183.83	601.37	−26.45
510	180.78	418.51	−45.35
538	183.29	389.23	−40.45
566	173.76	317.16	−60.25
593	169.40	300.79	−52.16
621	153.30	277.60	−58.78
649	133.88	239.85	−65.17

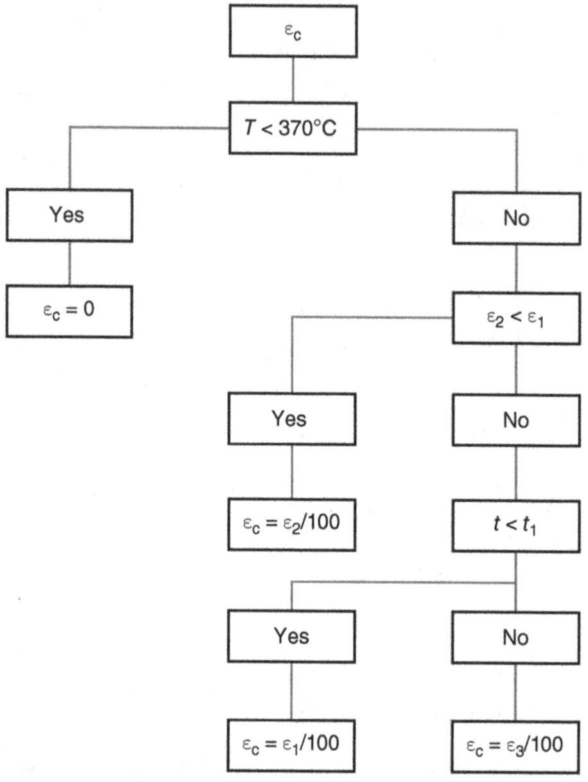

Figure B.2 Evaluation of creep strain – SI units.

$$e_1 = 6.7475 + 0.011426\sigma + (428.961/U)(\ln\sigma) - 13494/T_k \quad \text{(B.87)}$$

$$e_2 = 11.498 - (8.2226)(U)/T_k - 20448/T_k + (2546.0/T_k)(\ln\sigma) \quad \text{(B.88)}$$

$$\varepsilon_1 = [C_1 p_1 t/(1 + p_1 t)] + \dot{\varepsilon}_1 t \quad \text{(B.66)}$$

$$\varepsilon_2 = [C_2 p_2 t/(1 + p_2 t)] + \dot{\varepsilon}_2 t \quad \text{(B.67)}$$

$$\varepsilon_3 = [C_2 p_2 t'/(1 + p_2 t')] + \dot{\varepsilon}_2 t' \quad \text{(B.68)}$$

$$\dot{\varepsilon}_1 = 10^{e_1} \quad \text{(B.69)}$$

$$\dot{\varepsilon}_2 = 10^{e_2} \quad \text{(B.70)}$$

$$\varepsilon_1' = (C_1 p_1 t_1)/(1 + p_1 t_1) + \dot{\varepsilon}_1 t_1 \quad \text{(B.71)}$$

$$p_1 = 10^{q_1} \quad \text{(B.72)}$$

$$p_2 = 10^{q_2} \tag{B.73}$$

$$q_1 = 7.6026 + 1.45037(\ln \sigma) - 12323 / T_k \tag{B.89}$$

$$q_2 = 8.1242 + 0.0179678\sigma + (175.73 / U)(\ln \sigma) - 11659 / T_k \tag{B.90}$$

$\sigma =$ stress

$t =$ time, hrs

$$t_1 = t_1^{(a)} \quad \text{for} \qquad T \le 45 \tag{B.91}$$

$$= t_1^{(b)} \quad \text{for} \qquad 454°\text{C} < T \le 510°\text{C} \tag{B.92}$$

$$= t_1^{(a)} + \left(t_1^{(b)} - t_1^{(a)}\right)(T - 454)/56 \quad \text{for} \quad T > 510°\text{C} \tag{B.93}$$

$$t_1^{(a)} = 10^{a} \tag{B.79}$$

$$t_1^{(b)} = 10^{b} \tag{B.80}$$

$$t_c = \{-\dot{\varepsilon}_2 - C_2 p_2 + \varepsilon_1' p_2 + [4\varepsilon_1' \dot{\varepsilon}_2 p_2 + (\dot{\varepsilon}_2 + (C_2 - \varepsilon_1')p_2)^2]^{0.5} / (2\dot{\varepsilon}_2 p_2) \tag{B.81}$$

$$t' = t - (t_1 - t_c) \tag{B.82}$$

$$T_k = T + 273.15 \tag{B.25}$$

$U =$ given in Table B.28

Table B.28 Values of U versus temperature, °C, for 2.25Cr-1Mo steel.

T, °C	U
371	471
400	468
450	452
500	418
550	364
600	284
621	300
649	270

B.4 Low Alloy 9Cr–1Mo-V Steel

B.4.1 9Cr–1Mo-V Customary Units

The total strain is given by Eq.(B.1). The equations for elastic, plastic, and creep strains are defined below.

Define
T = temperature, °F
σ = stress, ksi
E = elastic modulus, ksi

(a) Elastic strain

The elastic strain is defined by the equation

$$\varepsilon_e = \sigma / E \tag{B.2}$$

with
E = obtained from Table B.29

Table B.29 Values of *E* versus temperature, °F, for 9Cr-1Mo-V steel.

Temperature *T*, °F	Modulus of elasticity *E*, ksi
700	27,266
750	26,686
801	26,106
849	25,381
900	24,656
932	24,075
950	23,640
1000	22,770
1022	21,755
1051	21,610
1099	20,450
1112	20,160
1150	19,289
1200	18,129

(b) Plastic strain

The plastic strain is defined by the following equations

$$\varepsilon_p = 0 \qquad \sigma \le RP \tag{B.94}$$

$$\varepsilon_p = (1/100b)\{\ln[(RP - RU)/(\sigma - RU)]\}^2 \quad RP < \sigma < RU \tag{B.95}$$

$$\varepsilon_p = \infty \qquad \sigma \ge RU \tag{B.96}$$

where

$$R = 1.25(Y_1 / Y) \tag{B.97}$$

Other factors are shown in Table B.30

Table B.30 Parameter values versus temperature, °F, for 9Cr-1Mo-V steel.

T, °F	*P*	*U*	*b*	*Y*	Y_1
700	45.98	87.45	3.73	70.05	53.66
750	44.96	84.99	4.49	69.47	51.92
801	43.94	82.52	4.77	68.02	50.76
849	42.93	78.46	5.88	66.57	48.73
900	41.04	73.53	8.24	64.54	46.27
932	39.01	70.05	9.30	62.51	44.53
950	38.00	68.02	10.59	61.06	43.22
1000	33.07	62.51	14.11	57.00	39.59
1022	30.02	59.46	15.70	54.39	38.00
1051	26.98	55.98	17.36	51.49	35.53
1099	21.03	49.02	14.84	43.94	31.04
1112	20.01	46.99	14.80	42.93	29.88
1150	15.95	41.04	14.70	36.55	26.54
1200	12.04	35.24	10.97	29.01	22.04

(c) Creep strain

The expression for creep strain is given by

$$\varepsilon_c = (0.01)[D\sigma(\mathrm{Exp}(V_o\sigma))(\mathrm{Exp}(-Q_o / T_k))(t^{1/3}) + (C)(6.895\sigma)^n(\mathrm{Exp}(V\sigma))(\mathrm{Exp}(-Q / T_k))(t)] \tag{B.98}$$

where

$$C = 2.25 \times 10^{22}$$

$$D = 5{,}840{,}065 \qquad T < 900°\mathrm{F} \tag{B.99}$$

$$= 566{,}741\,(T/1.8 - 500) + 5{,}840{,}065 \quad 900^\circ\text{F} < T < 1000^\circ\text{F} \tag{B.100}$$

$$= 37{,}577{,}750 \quad T \geq 1000^\circ\text{F} \tag{B.101}$$

$$n = 5.0$$

$$Q = 77{,}280$$

$$Q_o = 25{,}330 \quad T < 1000^\circ\text{F} \tag{B.102}$$

$$= 23{,}260 \quad T \geq 1000^\circ\text{F} \tag{B.103}$$

$$T_k = T_F/1.8 + 255.37 \tag{B.10}$$

$$V = 0.26201$$

$$V_o = 0.15859.$$

B.4.2 9Cr–1Mo-V SI Units

The total strain is given by Eq.(B.1). The equations for elastic, plastic, and creep strains are defined below.

Define,
T = temperature, °C
σ = stress, MPa
E = elastic modulus, MPa

(a) Elastic strain
The elastic strain is defined by the equation

$$\varepsilon_e = \sigma/E \tag{B.2}$$

with
E = obtained from Table B.31

(b) Plastic strain
The plastic strain is defined by the following equations

$$\varepsilon_p = 0 \quad \sigma \leq RP \tag{B.94}$$

$$\varepsilon_p = (1/100b)\{\ln[(RP - RU)/(\sigma - RU)]\}^2 \quad RP < \sigma < RU \tag{B.95}$$

$$\varepsilon_p = \infty \quad \sigma \geq RU \tag{B.96}$$

Table B.31 Values of E versus temperature, °C, for 9Cr-1Mo-V steel.

Temperature T, °C	Modulus of elasticity E, MPa
371	188,000
399	184,000
427	180,000
454	175,000
482	170,000
500	166,000
510	163,000
538	157,000
550	150,000
566	149,000
593	141,000
600	139,000
621	133,000
649	125,000

where

$$R = 1.25\left(Y_1 / Y\right) \quad \text{(B.97)}$$

Other factors are shown in Table B.32

(c) Creep strain

The expression for creep strain is defined by

$$\varepsilon_c = \left(0.01\right)[D\sigma(\text{Exp}(V_o\sigma))\left(\text{Exp}\left(-Q / T_k\right)\right)\left(t^{1/3}\right) + C\sigma^n(\text{Exp}(V\sigma))\left(\text{Exp}\left(-Q / T_k\right)\right)\left(t\right)] \quad \text{(B.104)}$$

where

$$C = 2.25 \times 10^{22}$$

$$D = 847{,}000 \qquad T < 482^\circ\text{C} \quad \text{(B.105)}$$

$$= 82{,}196\left(T - 482\right) + 487{,}000 \qquad 482^\circ\text{C} < T < 537^\circ\text{C} \quad \text{(B.106)}$$

$$= 5{,}450{,}000 \qquad T \geq 537^\circ\text{C} \quad \text{(B.107)}$$

$$n = 5.0$$

$$Q = 77{,}280$$

Table B.32 Parameter values versus temperature, °C, for 9Cr-1Mo-V steel.

T, °C	P	U	b	Y	Y_1
371	317	603	3.73	483	370
399	310	586	4.49	479	358
427	303	569	4.77	469	350
454	296	541	5.88	459	336
482	283	507	8.24	445	319
500	269	483	9.30	431	307
510	262	469	10.59	421	298
538	228	431	14.11	393	273
550	207	410	15.70	375	262
566	186	386	17.36	355	245
593	145	338	14.84	303	214
600	138	324	14.80	296	206
621	110	283	14.70	252	183
649	83	243	10.97	200	152

$$Q_o = 25{,}330 \quad T < 537^{\circ}\text{C} \tag{B.108}$$

$$= 23{,}260 \quad T \geq 537^{\circ}\text{C} \tag{B.109}$$

$$T_k = T + 273.15 \tag{B.25}$$

$$V = 0.038$$

$$V_o = 0.023.$$

B.5 Nickel Alloy 800H

B.5.1 Alloy 800H Customary Units

The total strain is given by Eq. (B.1). The equations for elastic, plastic, and creep strains are defined below.

Define,

T = temperature, °F

σ = stress, ksi

E = elastic modulus, ksi

(a) Elastic strain

The elastic strain is defined by the equation

$$\varepsilon_e = \sigma / E \tag{B.2}$$

with E obtained from ASME II-D conventional

(b) Plastic strain

The plastic strain is defined by the following equations

$$\varepsilon_p = 0 \qquad \sigma \leq \sigma_p(\varepsilon_e) \tag{B.110}$$

$$\varepsilon_p = \varepsilon_i - \varepsilon_e \quad \sigma > \sigma_p(\varepsilon_e) \tag{B.111}$$

with

$$\sigma_p(\varepsilon_i) = \sigma = \text{Exp}\left\{B_1 + B_2 \ln(100\varepsilon_i) + B_3\left[\ln(100\varepsilon_i)\right]^2 + B_4\left[\ln(100\varepsilon_i)\right]^3\right\} \tag{B.112}$$

where

$$B_1 = B_{11} + B_{12}\gamma + B_{13}\gamma^2 + B_{14}\gamma^3 + B_{15}\gamma^4$$

$$B_2 = B_{21} + B_{22}\gamma + B_{23}\gamma^2 + B_{24}\gamma^3 + B_{25}\gamma^4$$

$$B_3 = B_{31} + B_{32}\gamma + B_{33}\gamma^2 + B_{34}\gamma^3 + B_{35}\gamma^4$$

$$B_4 = B_{41} + B_{42}\gamma + B_{43}\gamma^2 + B_{44}\gamma^3 + B_{45}\gamma^4$$

$$\gamma = (T - 800)/600. \tag{B.113}$$

The coefficients of B_{ij} are shown in Table B.33

Table B.33 Values of B_{ij}.

B_{11}	3.18312201E + 00	B_{21}	1.7229733E-01
B_{12}	-1.94465649E-01	B_{22}	-6.54008012E-04
B_{13}	-2.53179862E-02	B_{23}	2.66280150E-02
B_{14}	7.99351461E-02	B_{24}	-9.36658919E-02
B_{15}	-4.54091288E-02	B_{25}	5.86915202E-02
B_{31}	3.75671539E-02	B_{41}	4.50390345E-03
B_{32}	1.32897986E-02	B_{42}	4.98519009E-03
B_{33}	-2.96225929E-02	B_{43}	4.82573850E-02
B_{34}	6.14678657E-03	B_{44}	-1.01054504E-01
B_{35}	5.12025443E-03	B_{45}	5.47052447E-02

(c) Creep strain

The expression for creep strain is given by

$$\varepsilon_c = (0.01)[\text{Exp}[(\ln t - C_1 \ln(\sigma) - C_3)/C_2] \tag{B.114}$$

with

$$C_1 = \left(-1.84503305 \times 10^4\right)/\left(1.8T_k\right)$$

$$C_2 = \left(1.09662615 \times 10^4\right)/\left(1.8T_k\right) - 4.62117596$$

$$C_3 = \left(1.77459417 \times 10^5\right)/\left(1.8T_k\right) - 67.5590904$$

$$T_k = T/1.8 + 255.37. \tag{B.10}$$

B.5.2 Alloy 800H SI Units

The total strain is given by Eq. (B.1). The equations for elastic, plastic, and creep strains are defined below.

Define,

T = temperature, °C

σ = stress, MPa

E = elastic modulus, MPa

(a) Elastic strain

The elastic strain is defined by the equation

$$\varepsilon_e = \sigma/E \tag{B.2}$$

with E obtained from ASME II-D metric

(b) Plastic strain

The plastic strain is defined by the following equations

$$\varepsilon_p = 0 \qquad \sigma \leq \sigma_p(\varepsilon_e) \tag{B.110}$$

$$\varepsilon_p = \varepsilon_i - \varepsilon_e \quad \sigma > \sigma_p(\varepsilon_e) \tag{B.111}$$

with

$$\sigma_p(\varepsilon_i) = \sigma = 6.895\,\text{Exp}\left\{B_1 + B_2 \ln(100\varepsilon_i) + B_3[\ln(100\varepsilon_i)]^2 + B_4[\ln(100\varepsilon_i)]^3\right\} \tag{B.115}$$

where B_{ij} are as defined in Eq. (B.112).

$$\gamma = \left(1.8T - 768\right)/600 \tag{B.116}$$

The coefficients of B_{ij} are shown in Table B.33.

(c) Creep strain

The expression for creep strain is given by

$$\varepsilon_c = (0.01)\left[\mathrm{Exp}\left[(\ln t - C_1 \ln(\sigma/6.895) - C_3)/C_2\right]\right. \quad \text{(B.114)}$$

with

$$C_1 = \left(-1.84503305 \times 10^4\right)/\left(1.8T_k\right)$$

$$C_2 = \left(1.09662615 \times 10^4\right)/\left(1.8T_k\right) - 4.62117596$$

$$C_3 = \left(1.77459417 \times 10^5\right)/\left(1.8T_k\right) - 67.5590904$$

$$T_k = T + 273.15. \quad \text{(B.25)}$$

Appendix C

Equations for Tangent Modulus, E_t

C.1 Tangent Modulus, E_t

Equation (B.1) in Appendix B is used to determine the tangent modulus, E_t. The derivative of Eq. (B.1) is given by

$$d\varepsilon / d\sigma = d\varepsilon_e / d\sigma + d\varepsilon_p / d\sigma + d\varepsilon_c / d\sigma \tag{C.1}$$

The tangent modulus, E_t, is defined as

$$E_t = d\sigma / d\varepsilon = \frac{1}{d\varepsilon / d\sigma} \tag{C.2}$$

substituting Eq. (C.1) into Eq. (C.2) gives

$$E_t = \frac{1}{d\varepsilon_e / d\sigma + d\varepsilon_p / d\sigma + d\varepsilon_c / d\sigma} \tag{C.3}$$

C.2 Type 304 Stainless Steel Material

In order to simplify taking the derivatives of the equations for Type 304 stainless steel in Section B.1.1, they are rewritten as follows:

Elastic strain

Let Eq.(B.10) be represented by

$$C_1 = T_k = (T / 1.8) + 255.3722 \tag{C.4}$$

Analysis of ASME Boiler, Pressure Vessel, and Nuclear Components in the Creep Range, Second Edition. Maan H. Jawad and Robert I. Jetter.
© 2022 John Wiley & Sons Ltd. Published 2022 by John Wiley & Sons Ltd.

Eq. (B.3) becomes

$$E = 33{,}679 - 13.823C_1 \tag{C.5}$$

and Eq. (B.2) is

$$\varepsilon_e = \sigma / E \tag{C.6}$$

Plastic strain

The plastic strain Eqs. (B.9), (B.8), (B.6), and (B.7) in Appendix B are written as follows:

$$C_2 = \sigma_y = 32.3128 - 0.01602C_1 \tag{C.7}$$

$$C_3 = \sigma_p = C_2 + \left(-12.1979 + 0.00671C_1\right) \tag{C.8}$$

$$C_4 = K = 69.9629 - 0.02545C_1 \tag{C.9}$$

$$C_5 = m = 0.25824 + 7.749 \times 10^{-5} C_1 \tag{C.10}$$

Let

$$C_6 = 1 / C_5 \tag{C.11}$$

Then the plastic strain can be expressed as

$$\begin{aligned} \varepsilon_p &= \left[(\sigma - C_3)/C_4\right]^{c6} & \sigma > \sigma_p \\ &= 0 & \sigma \le \sigma_p \end{aligned} \tag{C.12}$$

Creep strain

The terms for creep strain in Section B.1.1 of Appendix B are expressed as follows:

$$C_7 = G = \left(\text{Table B.1}\right) \tag{C.13}$$

$$C_8 = 1000H = 1000\left(\text{Table B.2}\right) \tag{C.14}$$

$$\varepsilon_s = C_7 + C_8\sigma \tag{C.15}$$

$$C_9 = 0.01$$

$$C_{10} = \text{t} = \text{time}$$

$$C_{11} = D = \left(\text{Table B.3}\right) \tag{C.16}$$

$$C_{12} = \beta_r = -2.252 \times 10^{-4} + 5.401 \times 10^{-7} C_1 \tag{C.17}$$

$$C_{13} = \text{Exp}\left(-33{,}719.17/C_1\right) \tag{C.18}$$

$$C_{14} = C_{11}C_{13} \tag{C.19}$$

$$C_{15} = 285.71\, C_{12} \tag{C.20}$$

$$s = C_{14}[\sinh(C_{15}\sigma)]^{3.5} \tag{C.21}$$

$$C_{16} = B = 2.518 \times 10^{13} \tag{C.22}$$

$$C_{17} = C_{16}C_{13} \tag{C.23}$$

$$r = C_{17}[\sinh(C_{15}\sigma)]^{3.5} \tag{C.24}$$

$$C_{18} = \beta_e = -3.652 \times 10^{-4} + 7.518 \times 10^{-7} C_1 \tag{C.25}$$

$$C_{19} = A = 1.38 \times 10^{13} \tag{C.26}$$

$$C_{20} = C_{13}C_{19} \tag{C.27}$$

$$C_{21} = 166.67C_{18} \tag{C.28}$$

$$\dot{\mathcal{E}}_m = C_{20}[\sinh(C_{21}\sigma)]^{6.0} \tag{C.29}$$

$$C = \text{Table B.4} \tag{C.30}$$

$$C_{22} = C C_{20} / C_{17} \tag{C.31}$$

$$\varepsilon_r = \left(C_{22}\right)\frac{[\sinh(C_{21}\sigma)]^{6.0}}{[\sinh(C_{15}\sigma)]^{3.5}} \tag{C.32}$$

$$\varepsilon_c = \left(C_9\right)\begin{Bmatrix}(\sigma_1)[1 - \text{Exp}(-C_{10}\sigma_2)] + \\ (\sigma_3)[1 - \text{Exp}(-C_{10}\sigma_4)] + C_{10}\sigma_5\end{Bmatrix} \tag{C.33}$$

where,

$$\sigma_1 = C_7 + C_8\sigma \tag{C.34}$$

$$\sigma_2 = C_{14}[\sinh(C_{15}\sigma)]^{3.5} \tag{C.35}$$

$$\sigma_3 = \left(C_{22}\right)\frac{[\sinh(C_{21}\sigma)]^{6.0}}{[\sinh(C_{15}\sigma)]^{3.5}} \tag{C.36}$$

$$\sigma_4 = C_{17}[\sinh(C_{15}\sigma)]^{3.5} \tag{C.37}$$

$$\sigma_5 = C_{20}[\sinh(C_{21}\sigma)]^{6.0} \tag{C.38}$$

The derivative of the elastic strain Eq. (B.2) is

$$d\varepsilon_e / d\sigma = 1/E \tag{C.39}$$

The derivative of the plastic strain Eq. (B.5) is

$$d\varepsilon_p / d\sigma = C_6 \frac{[(\sigma - C_3)/C_4]^{C6}}{\sigma - C_3} \tag{C.40}$$

The derivative of the creep strain Eq. (B.11) is complicated. It is obtained from a commercial Symbolic Math program as

$$d\varepsilon_c / d\sigma = A - B + \left(CD\right)/E - \left(FG\right)/H + I/J + K/L \tag{C.41}$$

where

$$A = 6C_{10}C_{20}C_{21}C_9 \cosh(C_{21}\,\sigma)\sinh(C_{21}\sigma)^5 \tag{C.42}$$

$$B = C_8C_9 \frac{1}{\text{Exp}[C_{10}C_{14}\sinh(C_{15}\,\sigma)^{7/2}]} - 1 \tag{C.43}$$

$$C = 7\ C_{15}C_{22}C_9 \cosh(C_{15}\,\sigma)\sinh(C_{21}\,\sigma)^6 \tag{C.44}$$

$$D = \frac{1}{\text{Exp}[C_{10}C_{17}\sinh(C_{15}\,\sigma)^{7/2}]} - 1 \tag{C.45}$$

$$E = 2\sinh(C_{15}\,\sigma)^{9/2} \tag{C.46}$$

$$F = 6C_{21}C_{22}C_9 \cosh(C_{21}\,\sigma)\sinh(C_{21}\,\sigma)^5 \tag{C.47}$$

$$G = \frac{1}{\text{Exp}[C_{10}C_{17}\sinh(C_{15}\,\sigma)^{7/2}]} - 1 \tag{C.48}$$

$$H = \sinh(C_{15}\,\sigma)^{7/2} \tag{C.49}$$

$$I = 7C_{10}C_{14}C_{15}C_9 \cosh(C_{15}\sigma)\sinh(C_{15}\sigma)^{5/2}(C_7 + C_8\sigma) \tag{C.50}$$

$$J = 2\ \mathrm{Exp}[C_{10}C_{14}\sinh(C_{15}\sigma)^{7/2}] \tag{C.51}$$

$$K = 7C_{10}C_{15}C_{17}C_{22}C_9 \cosh(C_{15}\sigma)\sinh(C_{21}\sigma)^6 \tag{C.52}$$

$$L = 2\left\{\mathrm{Exp}[C_{10}C_{17}\sinh(C_{15}\sigma)^{7/2}]\right\}\sinh(C_{15}\sigma) \tag{C.53}$$

Appendix D

Background of the Bree Diagram

D.1 Basic Bree Diagram Derivation

The Bree diagram, Figure D.1, is constructed for the purpose of combining mechanical and thermal stresses in a cylindrical shell in order to establish allowable stress criteria. The diagram is plotted with the mechanical stress as an abscissa and the thermal stress as an ordinate. Zones E, S_1, S_2, P, R_1, and R_2, shown in the figure, correspond to specific mechanical and thermal load groupings. The criteria used to establish each of these zones are described below and are based on Bree (1967, 1968), Burgreen (1975), Kraus (1980), and Wilshire and Owen (1983).

Zone E

It is assumed that all stresses in Zone E remain elastic in the first half (thermal loading) as well as the second half (thermal down loading) of the thermal cycle. Bree assumed a thin cylindrical shell subjected to internal pressure. The average stresses in the cylinder are

$$\sigma_\theta = PR_i / t \tag{D.1}$$

$$\sigma_\ell = PR_i / 2t \tag{D.2}$$

$$\sigma_r = -P, \text{ maximum at the inner surface} \tag{D.3}$$

where
P = internal pressure
R_i = inside radius

Analysis of ASME Boiler, Pressure Vessel, and Nuclear Components in the Creep Range, Second Edition. Maan H. Jawad and Robert I. Jetter.
© 2022 John Wiley & Sons Ltd. Published 2022 by John Wiley & Sons Ltd.

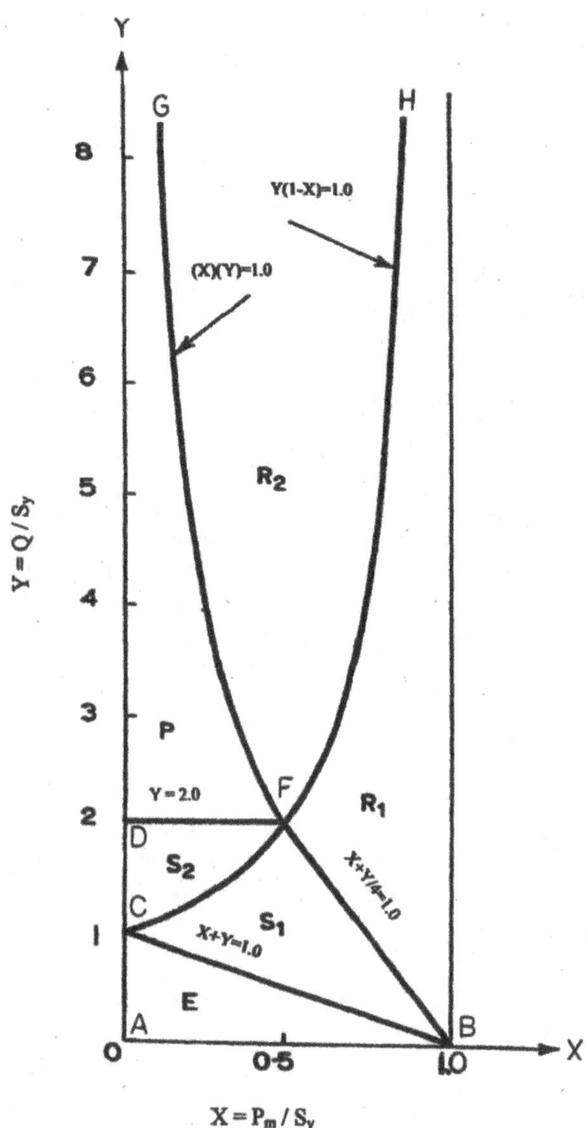

Figure D.1 Bree Diagram Bree (1967). *Source:* Based on Bree, J., 1967. Elastic-Plastic Behaviour of Thin Tubes Subjected to Internal Pressure and Inter-mittent High-Heat Fluxes with Application to Fast-Nuclear-Reactor Fuel Elements. Journal of Strain Analysis, 2(3), pp. 226–238.

t = thickness of cylinder
σ_ℓ = longitudinal stress
σ_r = radial stress
σ_θ = circumferential stress.

Radial thermal gradients in thin-walled cylinder are normally linear in distribution. For a linear radial thermal distribution through the thickness, Figure D.2, the thermal stress equations Burgreen (1975) are expressed as

$$\sigma_\theta = \sigma_\ell = (2x/t)\{E\alpha\Delta T/[2(1-\mu)]\} \quad \text{(D.4)}$$

$$\sigma_r \approx 0 \quad \text{(D.5)}$$

where

E = modulus of elasticity
x = distance from midwall of the cylinder, Figure D.2
α = coefficient of thermal expansion
ΔT= difference in temperature between inside and outside surfaces of the cylinder
μ = Poisson's ratio.

Bree made the following assumptions in order to evaluate the mechanical and thermal stress equations in a practical manner:

- Radial mechanical and thermal stresses, σ_r, are small compared to the circumferential and longitudinal stresses and can thus be ignored.
- Because mechanical stress is considered primary stress, it cannot exceed the yield stress value of the material. Radial thermal stress, on the other hand, is considered secondary stress and can thus exceed the yield stress of the material.

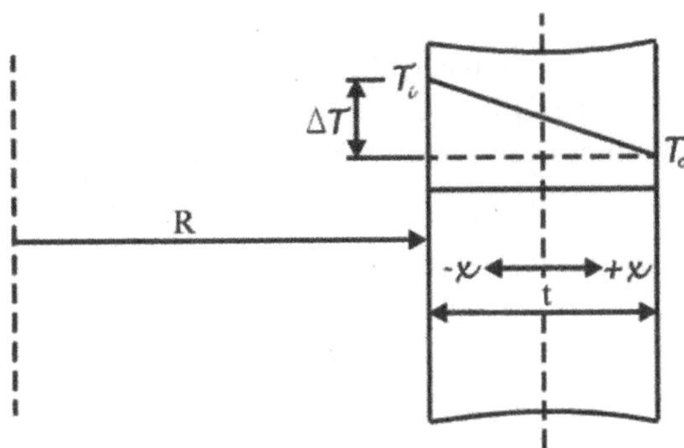

Figure D.2 Thermal gradient in shell.

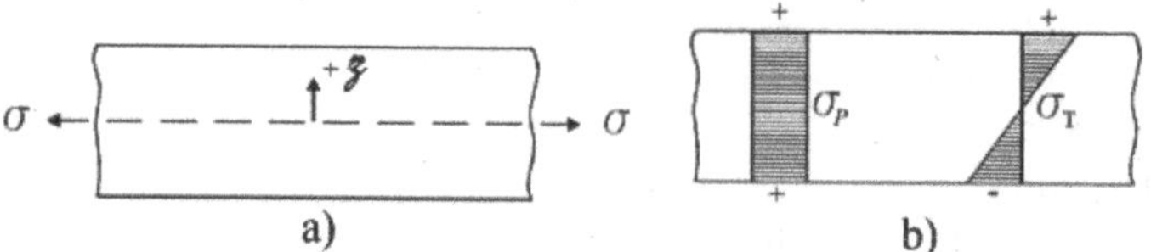

Figure D.3 Stress in the circumferential direction.

- The combination of mechanical and thermal stresses in the θ and ℓ directions may result in stresses past the yield stress in one direction and not the other. Such a condition prevents the formulation of a closed-form solution in a cylindrical shell. Accordingly, Bree made a conservative assumption by setting the stress in the ℓ direction, σ_ℓ, equal to zero.
- With the ℓ and r stresses set to zero, Bree assumed for simplicity that the remaining stress in the θ direction could be applied on a flat plate as shown in Figure D.3 with σ values obtained from Eqs. (D.1) and (D.4) for cylindrical shells.
- An additional assumption was made where the flat plate is not allowed to rotate due to variable thermal stress across the thickness in order to simulate the actual condition in a cylindrical shell.
- It is assumed that the material has an elastic perfectly plastic stress-strain diagram.
- The initial evaluation of the mechanical and thermal stresses in the elastic and plastic region was made without any consideration to relaxation or creep.
- The final results were subsequently evaluated for relaxation and creep effect.
- It is assumed that stress due to pressure is held constant, whereas thermal stress is cycled. Hence, pressure and temperature stress exist at the end of the first half of the cycle, and only pressure exists at the end of the second half of the cycle.

Based on these assumptions, the mechanical and thermal stress in Zone E can now be derived. It is assumed that the stresses remain elastic during the stress cycle. The stress distribution in the first half-cycle, shown in Figure D.4, is

$$\sigma = PR_i/t + (2z/t)\{E\alpha\Delta T/[2(1-\mu)]\} \tag{D.6}$$

or

$$\sigma = \sigma_p + (2z/t)\sigma_T \tag{D.7}$$

where

z = distance from midwall of the flat plate, Figure D.2

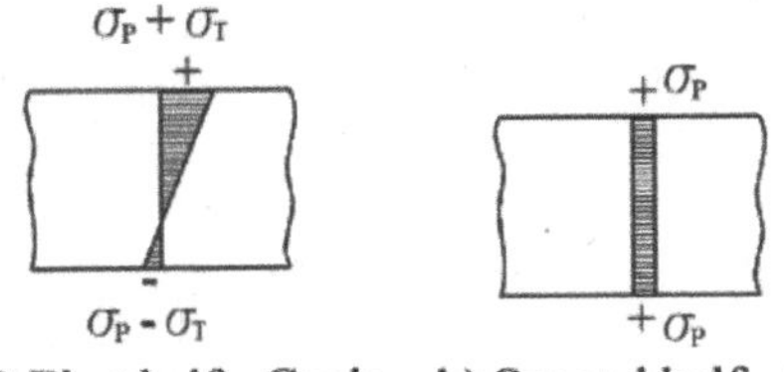

Figure D.4 Stress cycle in Zone E.

σ_p = stress due to pressure, PR_i/t
σ_T = stress due to temperature, $E\alpha\Delta T / 2(1-\mu)$.

It should be noted that, for a flat plate Eq. (D.6) uses the quantity $\Delta T / [2(1-\mu)]$ rather than ΔT in order to simulate thermal stress in a cylinder.

The maximum value of Eq. (D.7) is reached when $z = t/2$. In Zone E, the stress combination of σ_p plus σ_T is kept below σ_y, as shown in Figure D.4. Thus, Eq. (D.7) becomes

$$\sigma_p + \sigma_T < \sigma_y. \tag{D.8a}$$

Define $X = \sigma_p / \sigma_y$ and $Y = \sigma_T / \sigma_y$.
Then, Eq. (D.8a) becomes

$$X + Y = 1.0. \tag{D.8b}$$

In the second half-cycle, the equation becomes

$$\sigma_p < \sigma_y. \tag{D.9}$$

Thus, Zone E, which defines an elastic stress condition, is bound by Eq. (D.8) as well as the x- and y-axes as shown in Figure D.1.

Zone S_1

In Zone S_1, it is assumed that shakedown will occur after the first cycle. It is also assumed that the elastic pressure plus temperature stress combination in the first half-cycle exceeds the yield stress of the material at only one side of the shell, as shown in Figure D.5. Stress in the second half-cycle after the temperature is removed remains elastic. This assures shakedown after the first cycle. The strain distribution in the elastic region is

$$E\varepsilon_1 = \sigma_1 - (2z/t)\sigma_T \quad \text{for} -t/2 < z < v \tag{D.10}$$

where the subscript "1" refers to the first half-cycle and v is to be determined. The strain distribution in the plastic region is

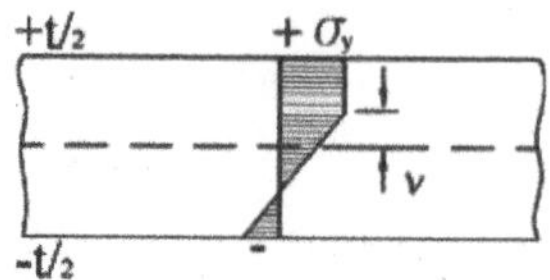

Figure D.5 Stress cycle in Zone S_1 with yielding on one side only.

$$E\varepsilon_1 = \sigma_y - (2z/t)\sigma_T + E\eta \quad \text{for } \nu < z < +t/2 \tag{D.11}$$

where η is the plastic component of the stress.

At point ν, $\sigma_1 = \sigma_y$. Equating Eqs. (D.10) and (D.11) gives

$$E\eta = 2[(z - \nu)/t]\sigma_T. \tag{D.12}$$

And Eqs. (D.10) and (D.11) become

$$\sigma_1 = \sigma_y + 2[(z - \nu)/t]\sigma_T \quad \text{in the elastic region} \tag{D.13}$$

$$\sigma_1 = \sigma_y \quad \text{in the plastic region.} \tag{D.14}$$

The expression for ν is obtained by summing the stress Eqs. (D.13) and (D.14) across the thickness in accordance with the equation

$$\int_{-t/2}^{t/2} \sigma dz = t\sigma_p. \tag{D.15}$$

This gives

$$\nu/t = \left[\left(\sigma_y - \sigma_p\right)/\sigma_T\right]^{1/2} - 1/2. \tag{D.16}$$

A limit of Eq. (D.16) is that ν cannot be less than zero. This gives

$$\sigma_p + \sigma_T/4 < \sigma_y. \tag{D.17a}$$

Equation (D.17a) can also be written as

$$X + Y/4 = 1.0. \tag{D.17b}$$

Another limit of Eq. (D.16) is that ν must be less than $t/2$. This gives

$$\sigma_p + \sigma_T < \sigma_y \tag{D.18a}$$

or

$$X + Y = 1.0. \tag{D.18b}$$

In order for shakedown to take place, the stresses remaining after removal of the temperature in the second half-cycle must remain elastic. The change of strain from the first to second half-cycle is given by

$$E\Delta\varepsilon_2 = \Delta\sigma_2 + (2z/t)\sigma_T \quad \text{for } -t/2 < z < v \tag{D.19}$$

$$= \sigma_2 - \sigma_1 + (2z/t)\sigma_T \tag{D.20}$$

and

$$E\Delta\varepsilon_2 = \Delta\sigma_2 + (2z/t)\sigma_T \quad \text{for } v < z < t/2 \tag{D.21}$$

$$= \sigma_2 - \sigma_y + (2z/t)\sigma_T \tag{D.22}$$

where the subscript "2" refers to the second half-cycle. Multiplying both sides of Eqs. (D.19) and (D.21) by dz and integrating their total sum results in $\Delta\varepsilon 2 = 0$. This is because the integral of the total sum of $\Delta\sigma_2 = 0$ since there is no change in external forces during the second half-cycle. Also, the total sum of the integral $(2z/t)\sigma_T$ from the limits $-t/2$ to $+t/2$ is equal to zero. Hence,

$$\Delta\varepsilon_2 = 0. \tag{D.23}$$

Substituting this quantity and Eq. (D.13) into Eqs. (D.20) and (D.22) results in

$$\sigma_2 = \sigma_y - (2v/t)\sigma_T \quad \text{for } -t/2 < z < v \tag{D.24}$$

$$\sigma_2 = \sigma_y - (2z/t)\sigma_T \quad \text{for } v < z < t/2. \tag{D.25}$$

One of the conditions for shakedown is that $\sigma_2 < \pm\,\sigma_y$. Hence, the above two equations reduce to

$$v > 0 \tag{D.26}$$

and

$$\sigma_T < 2\sigma_y \tag{D.27a}$$

or

$$Y < 2.0. \tag{D.27b}$$

Thus, Zone S_1 is bounded by Eqs. (D.17), (D.18), and (D.27). This is shown by area *BCDF* in Figure D.1.

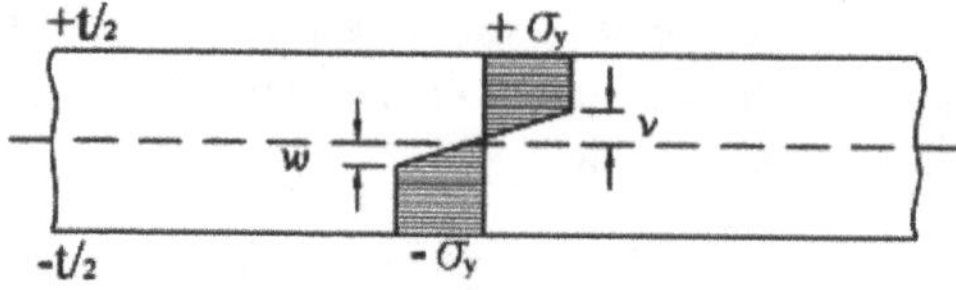

Figure D.6 Stress cycles in Zone S_2 with yielding on both sides.

Zone S_2

In Zone S_2, it is assumed that shakedown will occur after the first cycle. It is also assumed that the elastic pressure plus temperature stress combination in the first half-cycle exceeds the yield stress of the material at both sides of the shell. Stress in the second half-cycle after the temperature is removed remains elastic. This assures shakedown after the first cycle. The derivation of the limiting equations is similar to that in Zone S_1 with the exception that both sides of the shell reach the yield stress, as shown in Figure D.6. Two unknown quantities v and w must be determined. The solution Burgreen (1975) yields

$$v/t = (1/2)\left[\left(\sigma_y/\sigma_T\right) - \left(\sigma_p/\sigma_y\right)\right] \tag{D.28}$$

$$w/t = -(1/2)\left[\left(\sigma_y/\sigma_T\right) + \left(\sigma_p/\sigma_y\right)\right]. \tag{D.29}$$

A limiting value of these two equations is $v < t/2$ and $w < -t/2$. The two equations reduce to

$$\sigma_T\left(\sigma_y + \sigma_p\right) = \sigma_y^2 \tag{D.30}$$

and

$$\sigma_T\left(\sigma_y - \sigma_p\right) = \sigma_y^2 \tag{D.31a}$$

or

$$Y(1 - X) = 1.0. \tag{D.31b}$$

The second of these equations is the prevalent one, because satisfying it will automatically satisfy the first one. Equation (A.27) is also a controlling equation in the S_2 domain. Thus, Eqs. (D.27) and (D.31) form the boundary of Zone S_2. This is shown by area *CDF* in Figure D.1. Notice that this zone falls within Zone S_1. Thus, Zone S_2, where yielding occurs on both sides of the shell, supersedes the condition in Zone S_1 in the range shown.

Zone P

Plasticity is assumed to occur in Zone P. The main characteristics of this zone is that the core section of the shell remains elastic (otherwise, ratcheting may occur) whereas the outer surfaces alternate between tensile and compression yield stress as the temperature is applied then removed as shown in Figure D.7. Figure D.7a shows the stress distribution at the end of the first half-cycle. It resembles the same figure in the second half-cycle of Zone S_2. Figure D.7b shows the stress distribution during the second half-cycle, whereas Figure D.7c shows the stress distribution at the end of the second half-cycle.

Burgreen (1975). Referring to Figure D.7 with the center portion of the shell remaining elastic, $-u \leq z \leq u$, the relationship between the stress and strain Kraus (1980) is given by

$$E\Delta\varepsilon_2 = \Delta\sigma_2 + (2z/t)\sigma_T. \tag{D.32}$$

At the elastic-plastic boundary, the stress increment is $2\sigma_y$ at $z = -u$ and $-2\sigma_y$ at $z = u$. Substituting these two values in Eq. (D.32) and subtracting the resultant two equations, since the strain change is linear, gives

$$u = \left(\sigma_y / \sigma_T\right)t. \tag{D.33}$$

Because u cannot exceed $t/2$, Eq. (D.33) gives

$$\sigma_T \geq 2\sigma_y \tag{D.34a}$$

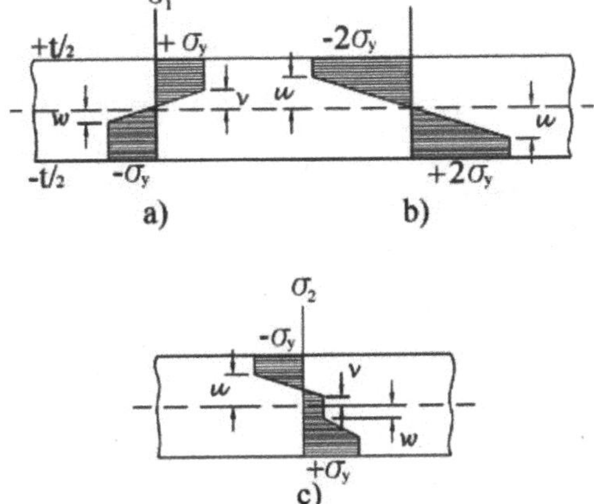

Figure D.7 Stress cycle in Zone P.

or

$$Y \geq 2. \tag{D.34b}$$

The net stress distribution in the core section of the cylinder, Figure D.7c, is

$$\sigma = \sigma_y - 2\sigma_T(v/t) \quad -w \leq z \leq v. \tag{D.35}$$

This stress cannot exceed the yield stress. Thus, Eq. (D.35) gives $v > 0$. Substituting this value in Eq. (D.28) gives

$$\sigma_p \sigma_T \leq \sigma_y^2 \tag{D.36a}$$

or

$$(X)(Y) = 1. \tag{D.36b}$$

Equations (D.34) and (D.36) form the boundary of Zone P. It is defined by area *DFG* of the Bree diagram in Figure D.1.

Zone R_1

Ratcheting is assumed to take place in Zone R_1. The main characteristic of this zone is that yielding extends past the midwall of the shell due to yielding at one of the surfaces of the shell. Thus, shakedown will not take place. Figure D.8 shows the stress distribution in the first and second half-cycles, whereas Figure D.9 shows the stress distribution in the second and third half-cycles.

The first half-cycle, Figure D.8a, is the same as the stress distribution in Zone S_1. The stress-strain relationship for the second half-cycle, Figure D.8c, is given by

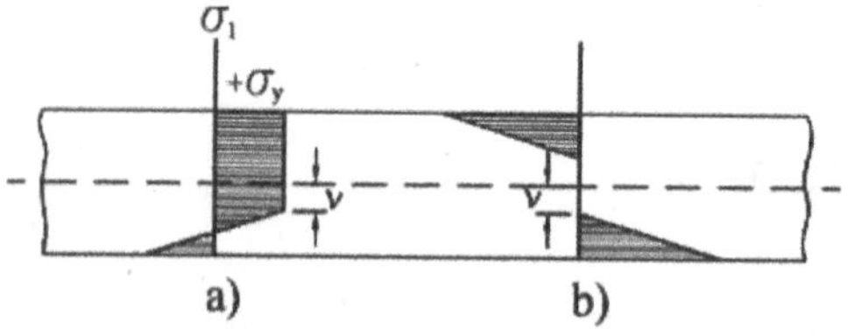

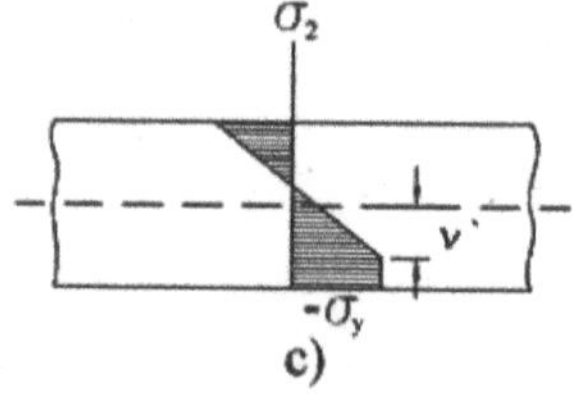

Figure D.8 First and second half-cycles in Zone R_1.

$$E\Delta\varepsilon_2 = \sigma_y - \sigma_1 + (2z/t)\sigma_T + E\Delta\eta_2 \quad -t/2 \leq z \leq v \tag{D.37}$$

$$E\Delta\varepsilon_2 = (2z/t)\sigma_T + E\Delta\eta_2 \quad v \leq z \leq v' \tag{D.38}$$

$$E\Delta\varepsilon_2 = \sigma_2 - \sigma_y + (2z/t)\sigma_T \quad v' \leq z \leq t/2. \tag{D.39}$$

From Eqs. (D.37), (D.38), and (D.39) plus the boundary conditions shown in Figure D.8c, the following equations are obtained after various substitutions:

$$\sigma_2 = \sigma_y + 2\sigma_T\left(v' - z\right)/t \quad v' \leq z \leq t/2 \tag{D.40}$$

$$\sigma_2 = \sigma_y \quad -t/2 \leq z \leq v \tag{D.41}$$

$$v'/t = -v/2 = 0.5 - \left[\left(\sigma_y - \sigma_p\right)/\sigma_T\right)\right]^{1/2} \tag{D.42}$$

$$E\Delta\eta_2 = 4\sigma_T v'/t \quad -t/2 \leq z \leq v \tag{D.43}$$

$$E\Delta\eta_2 = 2\sigma_T\left(v' - z\right)/t \quad v \leq z \leq v'. \tag{D.44}$$

In the third half-cycle, Figure D.9, the temperature stress is re-applied. The stress-strain relationship for the third half-cycle, Figure D.9c, is given by

$$E\Delta\varepsilon_3 = \sigma_3 - \sigma_y - (2z/t)\sigma_T \quad -t/2 \leq z \leq v \tag{D.45}$$

$$E\Delta\varepsilon_3 = -(2z/t)\sigma_T + E\Delta\eta_3 \quad v \leq z \leq v' \tag{D.46}$$

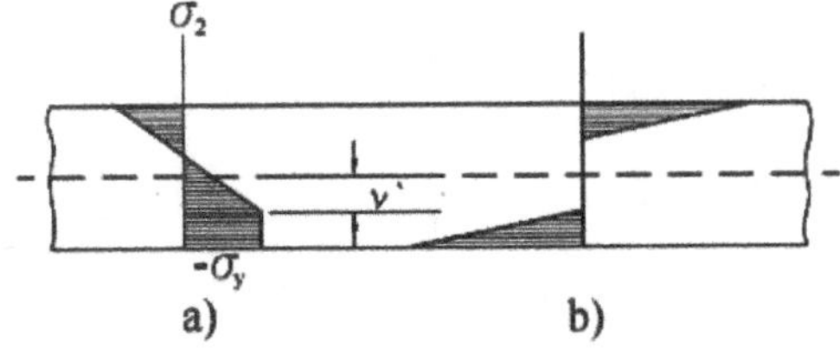

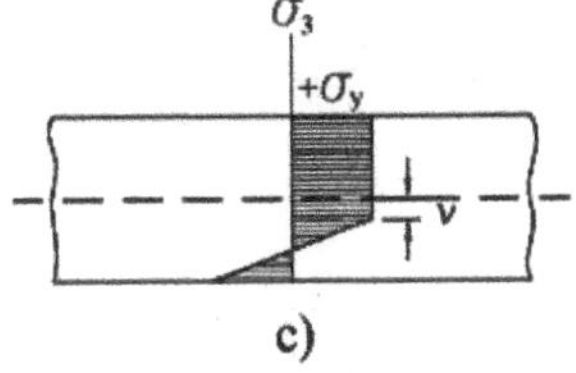

Figure D.9 Second and third half-cycles in Zone R_1.

$$E\Delta\varepsilon_3 = \sigma_y - \sigma_2 - (2z/t)\sigma_T + E\Delta\eta_3 \quad v' \le z \le t/2. \tag{D.47}$$

From Eqs. (D.45), (D.46), and (D.47) plus the boundary conditions shown in Figure D.9c, the following equations are obtained after various substitutions:

$$\sigma_3 = \sigma_y + 2\sigma_T(z - v)/t \quad -t/2 \le z \le v \tag{D.48}$$

$$\sigma_3 = \sigma_y \quad v' \le z \le t/2 \tag{D.49}$$

$$v'/t = -v/2 = 0.5 - \left[\left(\sigma_y - \sigma_p\right)/\sigma_T\right]^{1/2} \tag{D.50}$$

$$E\Delta\eta_3 = 2\sigma_T(z - v)/t \quad v \le z \le v' \tag{D.51}$$

$$E\Delta\eta_3 = 2\sigma_T\left(v' - v\right)/t \quad v' \le z \le t/2. \tag{D.52}$$

The plastic strain through the full cycle is obtained by adding Eqs. (D.44) and (D.52) to obtain

$$E\Delta\eta = 4\sigma_T v'/t. \tag{D.53}$$

Substituting Eq. (D.50) into Eq. (D.53) results in

$$E\Delta\eta = 4\sigma_T\{0.5 - \left[\left(\sigma_y - \sigma_p\right)/\sigma_T\right]^{1/2}\}. \tag{D.54}$$

Ratcheting occurs when the quantity $E\Delta\eta \ge 0$. Hence, Eq. (D.54) becomes

$$\left(\sigma_p + \sigma_T/4\right) \ge \sigma_y \tag{D.55a}$$

or

$$X + Y/4 \ge 1.0. \tag{D.55b}$$

The second requirement is that σ_2 in Eq. (D.40) must be greater than $-\sigma_y$ at $z = t/2$. Similarly, σ_3 in Eq. (D.48) must be greater than $-\sigma_y$ at $z = -t/2$. Using one of these two equations yields

$$\sigma_T\left(\sigma_y - \sigma_p\right) \le \sigma_y^2 \tag{D.56a}$$

or

$$Y(1 - X) \le 1.0. \tag{D.56b}$$

The final requirement is that the stress due to pressure, σ_p, must be less than the yield stress

$$\sigma_p \le \sigma_y \tag{D.57a}$$

or

$$X \leq 1.0. \tag{D.57b}$$

Equations (D.55), (D.56), and (D.57) form the boundary of the Zone R_1. It is defined by area *BFH* of the Bree diagram in Figure D.1.

Zone R_2

Ratcheting is assumed to take place in Zone R_2. The main characteristic of this zone is that yielding extends past the midwall of the shell due to yielding at both surfaces of the shell. Thus, shakedown will not take place. The derivation of the equations is very similar to that for Zone R_1. The area for Zone R_2 is defined by area *FGH* of the Bree diagram in Figure D.1

Appendix E

Factors for the Remaining Life Method

Table E.1 Factors Ai and Bi.

Material	Notes	Strain Rate Parameter – $\dot{\varepsilon}_{co}$		Omega Parameter – Ω	
Carbon Steel $\sigma_{uts}^{min} \leq 55$	• See Notes 1, 2, and 3	A_0	−1.600000E+01	B_0	−1.000000E+00
		A_1	3.519286E+04	B_1	3.124493E+03
		A_2	−3.310228E+03	B_2	−4.807777E+01
		A_3	−2.132352E+03	B_3	−5.389615E+02
		A_4	−4.578665E+02	B_4	1.749835E+02
Carbon Steel $\sigma_{uts}^{min} = 60$	• See Notes 1, 2, and 3	A_0	−1.600000E+01	B_0	−1.000000E+ 00
		A_1	3.524034E+04	B_1	3.171972E+03
		A_2	−3.275595E+03	B_2	−1.267951E+02
		A_3	−2.037106E+03	B_3	−4.437148E+02
		A_4	−4.222948E+02	B_4	1.439052E+02
Carbon Steel $\sigma_{uts}^{min} = 65$	• See Notes 1, 2, and 3	A_0	−1.600000E+01	B_0	−1.000000E +00
		A_1	3.528787E+04	B_1	3.219497E+03
		A_2	−3.241766E+03	B_2	−2.063161E+02
		A_3	−1.940557E+03	B_3	−3.471664E+02
		A_4	−3.872131E+02	B_4	1.123369E+02

(Continued)

Analysis of ASME Boiler, Pressure Vessel, and Nuclear Components in the Creep Range, Second Edition. Maan H. Jawad and Robert I. Jetter.
© 2022 John Wiley & Sons Ltd. Published 2022 by John Wiley & Sons Ltd.

Table E.1 (Continued)

Material	Notes	Strain Rate Parameter – $\dot{\varepsilon}_{co}$		Omega Parameter – Ω	
Carbon Steel $\sigma_{uts}^{min} = 70$	• See Notes 1, 2, and 3	A_0	−1.600000E+01	B_0	−1.000000E+00
		A_1	3.533558E+04	B_1	3.267211E+03
		A_2	−3.209096E+03	B_2	−2.869956E+02
		A_3	−1.842527E+03	B_3	−2.491363E+02
		A_4	−3.526338E+02	B_4	8.026620E+01
Carbon Steel $\sigma_{uts}^{min} \geq 75$	• See Notes 1, 2, and 3	A_0	−1.600000E+01	B_0	−1.000000E+00
		A_1	3.538350E+04	B_1	3.315135E+03
		A_2	−3.177710E+03	B_2	−3.689600E+02
		A_3	−1.742836E+03	B_3	−1.494446E+02
		A_4	−3.186231E+02	B_4	4.762685E+01
Cartoon Steel - Graphitized	• See Notes 1 and 2	A_0	−1.680000E+01	B_0	−1.000000E+00
		A_1	3.806000E+04	B_1	3.060000E+03
		A_2	−9.165000E+03	B_2	1.350000E+02
		A_3	1.200000E+03	B_3	−7.600000E+02
		A_4	−6.000000E+02	B_4	2.470000E+02
C-0.5Mo	• See Notes 1 and 2	A_0	−1.950000E+01	B_0	−1.300000E+00
		A_1	6.100000E+04	B_1	4.500000E+03
		A_2	−4.900000E+04	B_2	2.000000E+03
		A_3	3.300000E+04	B_3	−4.500000E+03
		A_4	−8.000000E+03	B_4	2.000000E+03
1.25Cr-0.5Mo–N&T	• See Notes 1 and 2 • Use only below 538°C (1000°F)	A_0	−2.335000E+01	B_0	−4.400000E+00
		A_1	6.207000E+04	B_1	1.451000E+04
		A_2	−4.752000 E+04	B_2	−2.467100E+04
		A_3	4.380000E+04	B_3	2.938400E+04
		A_4	−1.479000E+04	B_4	−1.063000E+04
1.25Cr-0.5Mo - Annealed	• See Notes 1 and 2	A_0	−2.350000E+01	B_0	−2.650000E+00
		A_1	5.278700E+04	B_1	6.287000E+03
		A_2	−5.965500E+03	B_2	1.534500E+03
		A_3	−2.621000E+03	B_3	−1.871000E+03
		A_4	−3.840000E+02	B_4	1.660000E+02

(Continued)

Table E.1 (Continued)

Material	Notes	Strain Rate Parameter – $\dot{\varepsilon}_{co}$		Omega Parameter – Ω	
2.25Cr-1Mo–N&T	• See Notes 1 and 2 • Use only below 510°C (950°F)	A_0	−2.156000E+01	B_0	−1.120000E+00
		A_1	5.552600E+04	B_1	5.040000E+03
		A_2	−1.098590E+04	B_2	−4.359000E+02
		A_3	−1.560170E+03	B_3	−2.175170E+03
		A_4	−7.386000E+01	B_4	1.136140E+03
2.25Cr-1Mo-Annealed	• See Notes 1 and 2	A_0	−2.186000E+01	B_0	−1.850000E+00
		A_1	5.166900E+04	B_1	7.195000E+03
		A_2	−7.597400E+03	B_2	−2.358000E+03
		A_3	−2.131000E+03	B_3	−1.274000E+02
		A_4	−1.993000E+02	B_4	5.620000E+01
2.25Cr-1Mo-Q&T	• See Notes 1 and 2 • Use only below 510°C (950°F)	A_0	−2.156000E+01	B_0	−1.120000E+00
		A_1	5.553000E+04	B_1	5.044000E+03
		A_2	−1.100580E+04	B_2	−4.558000E+02
		A_3	−1.544000E+03	B_3	−2.159000E+03
		A_4	−7.221000E+01	B_4	1.137790E+03
2.25Cr-1Mo-V	• See Notes 1 and 2 • Use only below 510°C (950°F)	A_0	−2.500000E+01	B_0	−2.520000E+00
		A_1	5.221050E+04	B_1	5.797600E+03
		A_2	−9.367000E+02	B_2	−3.648000E+02
		A_3	−1.814000E+03	B_3	−2.282000E+02
		A_4	−1.171000E+03	B_4	2.949000E+02
5Cr-0.5Mo	• See Notes 1 and 2	A_0	−2.240000E+01	B_0	−1.400000E+00
		A_1	5.166870E+04	B_1	5.019240E+03
		A_2	−7.597370E+03	B_2	−1.207620E+03
		A_3	−2.131000E+03	B_3	2.213200E+02
		A_4	−1.992700E+02	B_4	8.943000E+01
9Cr-1Mo	• See Notes 1 and 2	A_0	−2.255000E+01	B_0	−2.050000E+00
		A_1	5.147300E+04	B_1	6.410800E+03
		A_2	−7.027300E+03	B_2	−1.426600E+03
		A_3	−1.718000E+03	B_3	−6.900000E+01
		A_4	−1.406000E+02	B_4	8.380000E+01

(Continued)

Table E.1 (Continued)

Material	Notes	Strain Rate Parameter – $\dot{\varepsilon}_{co}$		Omega Parameter – Ω	
9Cr-1Mo-V	• See Notes 1 and 2	A_0	−3.400000E+01	B_0	−2.000000E+00
		A_1	7.328500E+04	B_1	7.185000E+03
		A_2	−3.428000E+03	B_2	−1.231720E+03
		A_3	−3.433000E+03	B_3	−4.492800E+02
		A_4	−1.146170E+03	B_4	2.052200E+02
10.5Cr-V (Alloy 115) UNS K91060	• See Notes 1 and 2	A_0	−3.732000E+01	B_0	−1.963000E+00
		A_1	8.496750E+04	B_1	9.403300E+03
		A_2	−2.140980E+04	B_2	−8.867700E+03
		A_3	1.623600E+04	B_3	1.003100E+04
		A_4	−7.606000E+03	B_4	−3.876900E+03
12 Cr	• See Notes 1 and 2	A_0	−3.029000E+01	B_0	−3.298000E+00
		A_1	6.711000E+04	B_1	6.508000E+03
		A_2	−2.109300E+04	B_2	3.016000E+03
		A_3	1.455600E+04	B_3	−2.784000E+03
		A_4	−5.884000E+03	B_4	4.800000E+02
Type 304 & 304H	• See Notes 1 and 2	A_0	−1.917000E+01	B_0	−3.400000E+00
		A_1	5.376240E+04	B_1	1.125000E+04
		A_2	−1.344240E+04	B_2	−5.635800E+03
		A_3	3.162600E+03	B_3	3.380400E+03
		A_4	−1.685200E+03	B_4	−9.936000E+02
Type 316 & 316H	• See Notes 1 and 2	A_0	−1.890000E+01	B_0	−4.163000E+00
		A_1	5.726150E+04	B_1	1.717626E+04
		A_2	−1.865600E+04	B_2	−1.321600E+04
		A_3	3.866860E+03	B_3	4.973798E+03
		A_4	−2.671640E+02	B_4	−6.736400E+01
Type 321	• See Notes 1 and 2	A_0	−1.900000E+01	B_0	−3.400000E+00
		A_1	4.942500E+04	B_1	1.062500E+04
		A_2	−7.417000E+03	B_2	−3.217000E+03
		A_3	1.240000E+03	B_3	1.640000E+03
		A_4	−1.290000E+03	B_4	−4.900000E+02

(Continued)

Table E.1 (Continued)

Material	Notes	Strain Rate Parameter – $\dot{\varepsilon}_{co}$		Omega Parameter – Ω	
Type 321H	• See Notes 1 and 2	A_0	−1.840000E+01	B_0	−3.400000E+00
		A_1	4.942500E+04	B_1	1.062500E+04
		A_2	−7.417000E+03	B_2	−3.217000E+03
		A_3	1.240000E+03	B_3	1.640000E+03
		A_4	−1.290000E+03	B_4	−4.900000E+02
Type 347	• See Notes 1 and 2	A_0	−1.830000E+01	B_0	−3.500000E+00
		A_1	4.714000E+04	B_1	1.000000E+04
		A_2	−5.434000E+03	B_2	−8.000000E+02
		A_3	5.000000E+02	B_3	−1.000000E+02
		A_4	−1.128000E+03	B_4	1.000000E+02
Type 347H	• See Notes 1 and 2	A_0	−1.770000E+01	B_0	−3.650000E+00
		A_1	4.714000E+04	B_1	1.000000E+04
		A_2	−5.434000E+03	B_2	−8.000000E+02
		A_3	5.000000E+02	B_3	−1.000000E+02
		A_4	−1.128000E+03	B_4	1.000000E+02
Type 347LN UNS S34751	• See Notes 1 and 2	A_0	−1.680664E+01	B_0	−4.000000E−01
		A_1	1.067725E+05	B_1	6.488220E+04
		A_2	−1.448239E+05	B_2	−1.391989E+05
		A_3	1.004320E+05	B_3	1.010732E+05
		A_4	−2.401329E+04	B_4	−2.384369E+04
Type 347 AP Advanced UNS S34752	• See Notes 1 and 2	A_0	−1.432000E+01	B_0	−5.500000E−01
		A_1	4.356980E+04	B_1	5.584300E+03
		A_2	−6.471400E+03	B_2	−1.163200E+03
		A_3	−2.949000E+02	B_3	4.087300E+02
		A_4	−4.480000E+00	B_4	1.113900E+02
Alloy 800	• See Notes 1 and 2	A_0	−1.940000E+01	B_0	−3.600000E+00
		A_1	5.554800E+04	B_1	1.125000E+04
		A_2	−1.587700E+04	B_2	−5.635000E+03
		A_3	3.380000E+03	B_3	3.380000E+03
		A_4	−9.930000E+02	B_4	−9.930000E+02

(Continued)

Table E.1 (Continued)

Material	Notes	Strain Rate Parameter – $\dot{\varepsilon}_{co}$		Omega Parameter – Ω	
Alloy 800H	• See Notes 1 and 2	A_0	−1.880000E+01	B_0	−3.600000E+00
		A_1	5.554800E+04	B_1	1.125000E+04
		A_2	−1.587700E+04	B_2	−5.635000E+03
		A_3	3.380000E+03	B_3	3.380000E+03
		A_4	−9.930000E+02	B_4	−9.930000E+02
Alloy 800HT	• See Notes 1 and 2	A_0	−2.025000E+01	B_0	−3.400000E+00
		A_1	5.944500E+04	B_1	1.119500E+04
		A_2	−1.412300E+04	B_2	−4.848500E+03
		A_3	4.142000E+01	B_3	1.556000E+03
		A_4	−4.050000E-02	B_4	8.242000E+01
HK-40	• See Notes 1 and 2	A_0	−1.480000E+01	B_0	−4.400000E+00
		A_1	4.725200E+04	B_1	1.300000E+04
		A_2	−8.684000E+03	B_2	−4.000000E+02
		A_3	−4.060000E+02	B_3	−1.800000E−01
		A_4	−6.330000E+00	B_4	3.004000E−01

Notes
1 Coefficients in this table are estimates of the typical material behavior (center of scatter band) based on the MPC Project Omega Materials data from service-aged materials at design stress levels.
2 The coefficients in this table are intended to describe material behavior in the range of the ASME Code design allowable stress for a given material at a specified temperature. These coefficients may be used to estimate the stress relaxation resulting from creep over a similar stress range.
3 σ^{min}_{UTS} is the minimum specified ultimate tensile strength at room temperature.

Table E.2 Factors m_2 and ε_p.

Material	Temperature Limit	m_2 (2)	ε_p
Ferritic Steel (1)	480°C (900°F)	0.60(1.00–R)	2.0E–5
Stainless Steel and Nickel Base Alloys	480°C (900°F)	0.75(1.00–R)	2.0E–5
Duplex Stainless Steel	480°C (900°F)	0.70(0.95–R)	2.0E–5
Super Alloys (3)	540°C (1000°F)	1.09(0.93–R)	2.0E–5
Aluminum	120°C (250°F)	0.52(0.98–R)	5.0E–6
Copper	65°C (150°F)	0.50(1.00–R)	5.0E–6
Titanium and Zirconium	260°C (500°F)	0.50(0.98–R)	2.0E–5

Notes
1 Ferritic steels include carbon, low alloy, alloy steels, and ferritic, martensitic, and iron-based age-hardening stainless steels.
2 R is the ratio of the engineering yield stress to engineering tensile stress evaluated at the assessment temperature (see Equation (7.23)).
3 Precipitation hardening nickel based austenitic alloys.

Appendix F

Conversion Factors

Multiply	By factor	To get units
bar	14.504	psi
cm	0.3937	inches
inches	25.4	mm
kg	2.205	lb
kg/mm^2	14.22	psi
lb	0.454	kg
MPa	145.03	psi
mm	0.03937	inches
psi	0.06895	bar
psi	0.0704	kg/mm^2
psi	0.006895	MPa
^oC	1.8C + 32	^oF
^oF	(F − 32)/1.8	^oC

$MPa = MN/m^2 = N/mm^2 = 10$ bars.
see: http://www.asme.org/about-asme/terms-of-use.

Analysis of ASME Boiler, Pressure Vessel, and Nuclear Components in the Creep Range, Second Edition. Maan H. Jawad and Robert I. Jetter.
© 2022 John Wiley & Sons Ltd. Published 2022 by John Wiley & Sons Ltd.

References

American Institute of Steel Construction. (1991). *Manual of Steel Construction-Allowable Stress Design*. New York: AISC.

American Society of Mechanical Engineers. (December 1976). *ASME-MPC Symposium on Creep-Fatigue Interaction*, Figure 5, page 55. New York: ASME.

American Society of Mechanical Engineers. (May 1976). *Criteria for Design of Elevated Temperature Class 1 Components in Section III, Division 1*, Figure 34, page 59, New York: ASME.

American Society of Mechanical Engineers. (2016). *Power Piping ASME B31.1*. New York: ASME.

American Society of Mechanical Engineers. (2016). *Process Piping ASME B31.3*. New York: ASME.

American Society of Mechanical Engineers. (2021). *Boiler and Pressure Vessel Code, Section I, Rules for Construction of Power Boilers*. New York: ASME.

American Society of Mechanical Engineers. (2021). *Boiler and Pressure Vessel Code, Section II-D, Properties*. New York: ASME.

American Society of Mechanical Engineers. (2021). *Boiler and Pressure Vessel Code, Section III-5, Rules for Construction of Nuclear Facility Components- High Temperature Reactors*. New York: ASME.

American Society of Mechanical Engineers. (2021). *Boiler and Pressure Vessel Code, Section VIII-1, Rules for Construction of Pressure Vessels- Division 1*. New York: ASME.

American Society of Mechanical Engineers. (2021). *Boiler and Pressure Vessel Code, Section VIII-2, Rules for Construction of Pressure Vessels- Division 2 Alternative Rules*. New York: ASME.

Analysis of ASME Boiler, Pressure Vessel, and Nuclear Components in the Creep Range, Second Edition. Maan H. Jawad and Robert I. Jetter.
© 2022 John Wiley & Sons Ltd. Published 2022 by John Wiley & Sons Ltd.

American Society of Mechanical Engineers. (2021). *Fitness-for-Service. API 579/ ASME FFS-1*. New York: ASME.

ASM. (2000). *Mechanical Testing and Evaluation -volume 8*. Materials Park, Ohio: ASM International.

Baker, J. and Heyman, J. (1969). *Plastic Design of Frames*. Cambridge: University of Cambridge Press.

Bannantine, J.A., Comer, J.J., and Handrock, J.L. (1990). *Fundamentals of Metal Fatigue Analysis*. New Jersey: Prentice-Hall.

Becht, C. et al. (1989). *Structural Design for Elevated Temperature Environments — Creep, Ratchet, Fatigue, and Fracture*. PVP-163. NY: American Society of Mechanical Engineers.

Beedle, L. (1966). *Plastic Design of Steel Frames*. New York: John Wiley Publishing.

Bleich, F. (1952). *Buckling Strength of Metal Structures*. New York: McGraw Hill.

Boyle, J.T. and Spence, J. (1983). *Stress Analysis for Creep*. London: Camelot Press.

Bree, J. (1967). Elastic-plastic behaviour of thin tubes subjected to internal pressure and inter- mittent high-heat fluxes with application to fast-nuclear-reactor fuel elements. *Journal of Strain Analysis* 2 (3): 226–238.

Bree, J. (1968). Incremental growth due to creep and plastic yielding of thin tubes subjected to internal pressure and cyclic thermal stresses. *Journal of Strain Analysis* 3 (2).

Burgreen, D. (1975). *Design Methods for Power Plant Structures*. New York: C. P. Press.

Byke, A. (2018). *ASME Nuclear Codes and Standards Ecosystem*. New York: ASME.

Chen, L., Jawad, M., and Griffin, D. (2021). *Automated External Pressure Chart Generation*. ASME STP-PT-091. New York: ASME.

Chen, W.F. and Zhang, H. (1991). *Structural Plasticity*. New York: Springer-Verlag.

Conway, J.B. (1969). *Stress-Rupture Parameters, Origin, Calculations and Use*. New York: Gordon and Breach Science Publishers.

Curran, R.M. (1976). *Symposium on Creep Fatigue Interaction*. MPC-3, American Society of Mechanical Engineers, New York.

Den Hartog, J.P. (1987). *Advance Strength of Materials*. Massachusetts: Dover Publications.

Dillon, C.P. (2000). *Unusual Corrosion Problems in the Chemical Industry*. Missouri: Materials Technology Institute of the Chemical Process Industries.

Faupel, J.H. and Fisher, F.E. (1981). *Engineering Design*. New York: John Wiley.

Finnie, I. and Heller, W. (1959). *Creep of Engineering Materials*. New York: McGraw Hill.

Frost, N.E., Marsh, K.J., and Pook, L.P. (1974). *Metal Fatigue*. Oxford: Clarendon Press.

Gerard, G. (1962). *Introduction to Structural Stability Theory*. New York: McGraw Hill.

Goodall, I.W. (2003). *Reference Stress Methods - Analysing Safety and Design*. London: Professional Engineering Publishing.

Grant, N.J. and Mullendore, A.W. (1965). *Deformation and Fracture at Elevated Temperatures*. Massachusetts: MIT Press.

Griffin, D. (1981). *Design limits for Creep Buckling in Structural Components.* Published in "Creep in Structures" ed. Ponter and Hayhurst. IUTAM Symposium Leicester UK/1980. Springer-Verlag, New York.

Hill, R. (1950). *The Mathematical Theory of Plasticity*. London: Oxford Press.

Hoff, N.J. (1958). *High Temperature Effects in Aircraft Structures*. New York: Pergamon.

Hult, J. (1966). *Creep in Engineering Materials*. Massachusetts: Blaisdell Publishing Company.

Jawad, M. and Farr, J. (2019). *Structural Analysis and Design of Process Equipment*, 3rd e. New Jersey: John Wiley.

Jawad, M. and Griffin, D. (2011). *External Pressure Design in Creep Range*. ASME STP-PT-029. New York: ASME.

Jawad, M. and Griffin, D. (2012). *Design Limits for Buckling in the Creep Regime*. ASME Journal of Pressure Vessel Technology.

Jawad, M., Swindeman, R., Swindeman, M., and Griffin, D. (2016). *Development of Average Isochronous Stress-Strain Curves and Equations and External Pressure Charts and Equations for 9Cr-1Mo-V Steel*. ASME STP-PT- 080. New York: ASME.

Jawad, M.H. (2018). *Stress in ASME Pressure Vessels, Boilers, and Nuclear Components*. Wiley/ASME Press, New York.

Jetter, R.I. (2018). *Companion Guide to the ASME Boiler & Pressure Vessel Code*, 1 (ed. K.R. Rao), Chapter 12. New York: ASME Press.

Jones, R. M. (2009). *Deformation Theory of Plasticity*. Blacksburg, Virginia: Bull Ridge Publishing.

Kraus, H. (1980). *Creep Analysis*. New Jersey: John Wiley Publishing.

Larsson, L.H. (1992). *High Temperature Structural Design*. ESIS 12. London: Mechanical Engineering Publications.

Markl, A. (1960). *Fatigue Tests of Piping Components, Pressure Vessel and Piping Design, Collected Papers, 1927–1959*. New York: ASME.

Neuber, H. (1961). Theory of stress concentration for Shear-Strained Prismatical bodies with Arbitrary Nonlinear Stress-Strain Law. *ASME Journal of Applied Mechanics* 28: 544–550.

O'Donnell, W.J. and Porowski, J. (1974). Upper bounds for accumulated strains due to creep ratcheting. *Transactions ASME Journal of Pressure Vessel Technology* 96: 150–154.

Odqvist, F.K.G. (1966). *Mathematical Theory of Creep and Creep Rupture*. Oxford: Oxford Mathematical Monographs.

Penny, R.K. and Marriott, D.L. (1995). *Design for Creep*. New York: Chapman and Hall.

Pilkey, W.D. and Pilkey, D.F. (2008). *Peterson's Stress Concentration Factors*. John Wiley.

Prager, M. (1995). Development of the MPC Omega method for life assessment in the creep range. *Journal of Pressure Vessel Technology* 117: 95–103.

Prager, M. (2000). The Omega method - an engineering approach to life assessment. In: *ASME Journal of Pressure Vessel Technology*, 122. New York: American Society of Mechanical Engineers.

Rao, K.R. (2018). *Companion Guide to the ASME Boiler and Pressure Vessel Code*, 5th e. New York: ASME.

Severud, L.K. (1975). *Simplified Methods and Application to Preliminary Design of Piping for Elevated Temperature Service*. ASME 2nd National congress on pressure vessels and piping, San Francisco, ASME.

Severud, L.K. (1980). *Experience with Simplified Inelastic Analysis of Piping Designed for Elevated Temperature Services*. New York: ASME Journal of Nuclear Engineering, ASME.

Severud, L.K. (1991). Creep-Fatigue assessment methods using elastic analysis results and adjustments. *ASME Journal of Pressure Vessel Technology* 114: 34–40.

Severud, L.K. and Winkel, B.V. (1987). *Elastic Creep-Fatigue Evaluation for ASME Code*. Transactions of the 9th International conference on structural mechanics in reactor technology, Lausanne, Switzerland.

Sim, R.G. (1968). *Creep of structures*. Ph.D. dissertation, University of Cambridge, Cambridge, U.K.

Smith, A.I. and Nicolson, A.M. (1971). *Advances in Creep Design*. New York: John Wiley.

Spence, J., Cotton, J., Underwood, B., and Duncan, C., 1991. *Elementary Statistics*. Prentice Hall, Englewood, NJ.

Sturm, R.G. (1941). *A Study of the Collapsing Pressure of Thin-Walled Cylinders*. Experiment Station Bulletin 329. Urbana, Illinois: University of Illinois Engineering.

Thielsch, H. (1977). *Defect and Failures in Pressure Vessels and Piping*. New York: Robert E. Krieger Publishing.

Timoshenko, S.P. and Gere, J.M. (2009). *Theory of Elastic Stability*. Dover Publishing.

Von Karman, T. and Tsien, H. (1960). *The Buckling of Spherical Shells by External Pressure*. Pressure Vessel and Piping Design — Collected Papers 1927–1959. ASME, New York.

Wang, C.K. (1970). *Matrix Methods of Structural Analysis*. Pennsylvania: International Textbook Company.

Weaver, W. and Gere, J. (1990). *Matrix Structural Analysis*. New York: Van Nostrand.

Weeks, R.W. (October 1973). Mechanical properties test data for structural materials quarterly progress reports. ORNL-4936, Oak Ridge National Lab, Oak Ridge, TN.

Wilshire, B. and Owen, D.R.J. (1983). *Engineering Approaches to High Temperature Design*. Swansea, U.K: Pineridge Press.

Winston, B., Burrows, R., Michel, R., and Rankin, A.W. (1954). *A Wall-Thickness Formula for High-Pressure Piping*. New York: Transactions of ASME.

Bibliography of Some Publications Related to Creep in Addition to Those Cited in the References

Arutyunyan, N.K. and Zevin, A.A. (1997). *Design of Structures considering Creep.* Vermont: AA. Balkema Publishers.

Au-Yang, M.K. (1993). *Technology for the 90's. A Decade of Progress.* The ASME Pressure and Piping Division. New York: ASME.

Bernasconi, G. and Piatti, G. (1979). *Creep of Engineering Materials and Structures.* London: Applied Science Publishers Ltd.

Bensussan, P. and Mascarell, J.P. (1990). *Hight Temperature Fracture Mechanisms and mechanics.* London: Mechanical Engineering Publishers Limited.

Chopra, O.K. and Shack, W.J. (2002). *Review of the Margins for ASME Code Fatigue Design Curves – Effect of Surface Roughness and Material Variability.* NUREG/CR 6815, ANL 02/39.

Conway, J.B. (1967). *Numerical Methods for Creep and Rupture Analysis.* Gordon and Breach, Science Publishers, New York.

Cooper, W.E. (1992). *The initial Scope and Intent of the Section III Fatigue Design Procedure* Welding Research Council, Inc. Workshop on Cyclic Life and Environmental Adjustment Effects, Clearwater, Florida, January 20–21.

Findley, W.N., Lai, J.S., and Onaran, K. (1976). *Creep and Relaxation of Nonlinear Viscoelastic Materials.* New York: Dover Publications.

Fistedis, S.H. (1987). *Nuclear Engineering and Design*, Vol. 98, No.3. Amsterdam: North-Holland Publishing.

Flugge, W. (1967). *Viscoelasticity.* Massachusetts: Blaisdell Publishing Company.

Ginsztler, J. and Skelton, R.P. (1998). *Component Reliability Under Creep-Fatigue Conditions.* New York: Springer Wien.

Hoff, N.J. (1960). *Creep in Structures IUTAM Colloquium, Stanford, California.* New York: Springer-Verlag.

Analysis of ASME Boiler, Pressure Vessel, and Nuclear Components in the Creep Range, Second Edition. Maan H. Jawad and Robert I. Jetter.
© 2022 John Wiley & Sons Ltd. Published 2022 by John Wiley & Sons Ltd.

Hyde, T.H. (1994). *Creep of Materials and Structures.* London: Mechanical Engineering Publications.

Jo, B. and Okamoto, K. (2017). *Experimental Investigation Into Creep Buckling of a stainless steel plate Column under axial compression at Extremely Hight Temperatures.* ASME Journal of Pressure Vessel Technology, 011406. ASME, New York.

Koves, W.J., Shimakawa, T., and Hayashi, M. (2004). *Elevated Temperature Design and Analysis, Nonlinear Analysis, and Plastic Components.* PVP-472. New York: ASME.

Larsson, L.H. (1992). *High Temperature Structural Design*, ESIS 12, Mechanical Engineering Publications, London.

Leckie, F.A. (1985). *Damage Studies in Pressure Vessel Components*, WRC Bulletin 310, WRC New York, N.Y.

Liu, F., Gong, J., Gao, F., and Xuan, F. (2019). *A Creep Buckling Design Method of Elliptical Heads Based on the External Pressure Chart.* Journal of Pressure Vessel Technology, 031203. New York: ASME Press.

Ponter, A.R.S. and Hayhurst D.R. (1980). *Creep in Structures.* IUTAM Colloquium, Leicester, UK. Springer-Verlag.

Pressure Vessel Technology. (1992). 7th International Conference on Pressure Vessel Technology. Verband der Technischen Uberwachungs- Vereine e.V.

Rabotnov, Y.N. (1969). *Creep Problems in Structural Members.* New York: Elsevier.

Schaefer, A.O. (1975). *Structural Materials for Service at Elevated Temperatures in Nuclear Power Generation.* MPC-1. New York: ASME.

Yoon, K.K. (1998). *Fatigue, Fracture, and High Temperature Design Methods in Pressure Vessels and Piping.* PVP-365. New York: ASME.

Zamrik, S.Y. (1971). *Symposium on Design for Elevated Temperature Environment.* ASME, New York.

Index

Analysis of ASME Boiler, Pressure Vessel, and Nuclear Components in the Creep Range, Second Edition. Maan H. Jawad and Robert I. Jetter.
© 2022 John Wiley & Sons Ltd. Published 2022 by John Wiley & Sons Ltd.